AF368538

Geological and Mineralogical Sequestration of CO$_2$

Geological and Mineralogical Sequestration of CO$_2$

Editors

Giovanni Ruggieri
Fabrizio Gherardi

MDPI • Basel • Beijing • Wuhan • Barcelona • Belgrade • Manchester • Tokyo • Cluj • Tianjin

Editors
Giovanni Ruggieri
Istituto di Geoscienze e
Georisorse (IGG)—Consiglio
Nazionale delle Ricerche (CNR)
Italy

Fabrizio Gherardi
Istituto di Geoscienze e
Georisorse (IGG)—Consiglio
Nazionale delle Ricerche (CNR)
Italy

Editorial Office
MDPI
St. Alban-Anlage 66
4052 Basel, Switzerland

This is a reprint of articles from the Special Issue published online in the open access journal *Minerals* (ISSN 2075-163X) (available at: https://www.mdpi.com/journal/minerals/special_issues/ CO2_Sequestration).

For citation purposes, cite each article independently as indicated on the article page online and as indicated below:

LastName, A.A.; LastName, B.B.; LastName, C.C. Article Title. *Journal Name* **Year**, *Article Number*, Page Range.

ISBN 978-3-03936-876-1 (Hbk)
ISBN 978-3-03936-877-8 (PDF)

Cover image courtesy of Chiara Boschi.

Contents

About the Editors

Giovanni Ruggieri holds a Ph.D. in Mineralogy and Petrology and was awarded the Johndino Nogara prize of the Italian Society of Mineralogy and Petrology. He is a researcher at IGG-CNR and a member of the Commission on Ore Mineralogy of the IMA. He coordinates the participation of CNR in the EU H2020 project GEMex. Specialized in fluid inclusions, stable isotopes, hydrothermal alteration and minero-petrographic studies, his fields of interests include geothermal exploration, fluid-rock interactions and fluid flow in hydrothermal systems, ore genesis, travertine formation, CO_2 sequestration by serpentine carbonation, and experimental mineralogy.

Fabrizio Gherardi received his PhD from the University of Pisa, Italy, defending a dissertation on gas-water-rock interactions and fluid geochemistry of high-enthalpy geothermal systems. A geochemist with a background in isotope hydrology, fluid geochemistry and gas-water-rock interactions, he uses numerical models to investigate geochemical reactivity in natural and anthropogenically disturbed systems. His recent research focuses on noble gas geochemistry, geochemical and reactive transport modelling of gas-water-rock interactions in hydrothermal/geothermal systems, and sites for geological confinement of greenhouse gases.

Preface to "Geological and Mineralogical Sequestration of CO$_2$"

The rapid increasing of concentrations of anthropologically generated greenhouse gases (primarily CO$_2$) in the atmosphere is responsible for global warming and ocean acidification. Carbon capture and storage (CCS) techniques have been proposed and developed to mitigate the rise of CO$_2$ in the atmosphere. One of the technological solutions is the long-term storage of CO$_2$ in appropriate geological formations, such as deep saline formations and depleted oil and gas reservoirs. A potential alternative to geological CO$_2$ storage is CO$_2$ mineral sequestration through carbonation (ex situ and in situ), leading to the permanent and safe storage of CO$_2$. This Special Issue collects articles covering various aspects of recent scientific advances in the geological and mineralogical sequestration of CO$_2$. In particular, it includes the assessment of the storage potential of candidate injection sites, numerical modelling of geochemical–mineralogical reactions aimed at predicting CO$_2$ leakage, studies of natural analogues, and experimental investigations of carbonation processes.

Giovanni Ruggieri, Fabrizio Gherardi
Editors

Editorial

Editorial for Special Issue "Geological and Mineralogical Sequestration of CO_2"

Giovanni Ruggieri [1],* and Fabrizio Gherardi [2]

[1]   Institute of Geoscience and Earth Resources (IGG), National Research Council of Italy (CNR), 50121 Florence, Italy

[2]   Institute of Geoscience and Earth Resources (IGG), National Research Council of Italy (CNR), 56124 Pisa, Italy; f.gherardi@igg.cnr.it

*   Correspondence: ruggieri@igg.cnr.it

Received: 16 June 2020; Accepted: 29 June 2020; Published: 2 July 2020

Carbon Capture Utilization and Storage (CCUS) has been substantiated by the International Panel on Climate Change (IPCC) [1] as a necessary measure to reduce greenhouse gas emissions in the short-to-medium term. Considered as a "climate change technology", CCUS encompasses an integrated number of different technologies aimed at preventing large amounts of CO_2 from being further released into the atmosphere through the use of fossil fuels. Along with fuel switch, energy efficiency, and use of renewables, CCUS is thus currently considered a key option within the portfolio of approaches required to reduce greenhouse gas emissions.

CCUS basically involves (i) capturing CO_2 from stationary sources of C-gases, (ii) compressing and transporting it at the injection point, and (iii) injecting it in deep geological repositories (Geological Carbon Storage, GCS). Storage options include geological storage, ocean storage, and mineral carbonation [2]. Deep saline formations and depleted oil and gas reservoirs are currently envisaged as the most appropriate targets of geological storage.

Promising, alternative options to GCS, that guarantee a permanent, although on a smaller scale, capture of CO_2, are the in situ and ex-situ fixation of CO_2 in the form of inorganic carbonates by carbonation of mafic and ultramafic rocks and of Mg/Ca-rich fly ash, iron and steel slags, cement waste, and mine tailings [3–5]. Moreover, the industrial utilization of CO_2 as technical fluid for diverse applications in the field of material and chemical engineering may contribute to reduce CO_2 emission.

Since the late 1990s (e.g., Weyburn-Midale, 1996 [6], Sleipner, 2000 [7], projects in Canada and Norway, respectively), a number of large-scale CCUS facilities exist, mostly in North America, Europe, China, Australia and the Arabian Peninsula. Pilot projects are now growing in number around the world that are expected to evolve to an operational stage by the end of the 2020s, and further development is expected over the next years to help in accomplishing the ambitious task of keeping the increase in global average temperature below 1.5 °C [8]. From this perspective, key success factors will be the availability of financial incentives for deployment of CCUS technology, and the demonstrated technical and operational capability to effectively manage the risks of storage.

According to this general framework, this Special Issue assimilates contributions covering various aspects of recent scientific advances in CCUS, GCS in particular, that include the assessment of the storage potential of candidate injection sites, numerical modelling of geochemical–mineralogical reactions and CO_2 flow, studies of natural analogues, and experimental investigations of carbonation processes.

Wide-scale deployment of large storage projects over time requires a preliminary screening and ranking of the geological reservoirs for their suitability for storage. Following the recommendations of the international Carbon Sequestration Leadership Forum [9], an integrated assessment of the geological storage capacity of a prospect area requires integrating geological considerations with engineering, legal, regulatory, infrastructure, and general economic constraints.

The studies by Saftić et al. [10], Koukouzas et al. [11] and Sundal and Hellevang [12] tackle this issue by providing geological information on the storage potential of selected areas in Croatia, Greece and Norway. In particular, Saftić et al. [10] performed a potential assessment of three prospect areas in the southern part of the Pannonian basin, in the Northern and Central Adriatic Sea, by taking advantage of a detailed stratigraphic knowledge of the Croatian territory derived from previous oil prospection activity. A ranking of the three prospect sites was finally proposed based on the integration of geological and infrastructure constraints.

Koukouzas et al. [11] present a preliminary assessment of the trapping potential of two promising geological formations in the Volos area, Central Greece. The investigated lithologies are volcanic rocks (basalts and trachyandesitic lava flows) formed during the Pleistocene back-arc extension of the Aegean Sea. Based on their high porosity, low alteration grade, silica under-saturated alkaline composition, and the presence of Ca-bearing minerals, the authors estimated that the basalts of the Volos area have the appropriate physical-chemical characteristics to act as a storage reservoir, with a maximum capacity of about 110,000 tonnes of CO_2.

By combining reservoir geology, petrographic observation and geochemical modelling techniques, Sundal and Hellevang [12] assessed the storage capacity of a specific reservoir in the Northern Sea, Norway. In this study, the specific reactive areas of minerals used in the numerical simulations were proposed as an additional parameter to be considered for the geological characterization of the reservoir. The target of the study was the Johansen formation, a new CCUS prospect in Norway, licenced for the storage of CO_2 as of 2019.

A complex interplay of multiphase flow, diffusion, and chemical reactions is expected in the storage sites after injection of CO_2 deep underground. A wide range of homogeneous and heterogeneous reactions have the potential to significantly impact on both injection performance and storage security. In this framework, numerical modelling techniques emerge as an efficient tool to integrate fundamental research into the study of real-world complex processes. By applying reactive transport modelling techniques, Wasch and Koenen [13] set up a field-scale wellbore model aimed at predicting CO_2 leakage along possible fractures at the cement–rock interface. Contrasting evolutionary scenarios were predicted, primarily based on variable initial leakage rates considered in the model. Hypotheses were advanced about the most relevant parameters controlling the process of potential leakage, and to design leakage mitigation measures.

Understanding carbonation in natural systems provides constraints to develop efficient engineering strategies for CO_2 sequestration with both in situ and ex situ methodologies [2,4,14]. Following this approach, the studies of Boschi et al. [15] and Picazo et al. [16] furnish information on the processes and the physical-chemical conditions characterizing serpentinite replacement by carbonates in two different environments.

The research of Boschi et al. [15] is focussed on the spontaneous CO_2 mineral sequestration on serpentinite walls of the Montecastelli copper mine located in Southern Tuscany, Italy. On the basis of the analytical data of solid and liquid phases present in the mine and on geochemical modelling, the authors explain the process which triggered the formation of hydromagnesite and kerolite from the interaction of condensed mine waters and a layer of serpentinite powder accumulated on mine walls during the excavation of the mine adits.

Picazo et al. [16] studied the process of serpentinite replacement by carbonates in brecciated serpentinized peridotites, recovered in the frame of the International Ocean Discovery Program (IODP) (site 1277), from the Newfoundland margin. The authors presented micro-textural, micro-chemical and O and C isotopic data. The analytical results coupled with a thermodynamic model of fluid/rock interaction during seawater transport in serpentine were utilized to constrain the most probable temperature condition of carbonation process and to discuss the role of temperature and seawater flows (i.e., influx vs. discharge) for the efficiency of CO_2 mineral sequestration.

Experimental studies are also fundamental to develop reliable methods for the mineralogical sequestration of CO_2 and to better understand the effectiveness and mechanisms of CO_2 geological storage. The papers of De Brueil et al. [17], Martin et al. [18], Kim and Cho [19], and Park et al. [20] deal with this topic.

De Brueil et al. [17] examined the efficiency of the thermal activation of serpentine for mineral carbonation in the presence of the aqueous-phase at ambient temperature and moderate pressure in flue gas conditions. In particular, the study emphasizes the importance of amorphous phases, quantified by means a new original approach based on XRD analyses and Rietveld refinements, which formed during the dihydroxylation processes, and their role on the magnesium leaching during carbonation reaction.

The paper of Martin et al. [18] investigated the possibility to use ceramic construction waste (brick, concrete, tiles) for carbonation reactions. The proposed methodology includes two steps: a sample pre-selection based on in situ carbonation and the mineralogical and chemical characterization of the samples, and laboratory carbonation tests at room temperature and at relatively low-pressure on a brick selected according to the previous analysis. The study highlights the potential use of Ca-silicate-rich bricks as raw material for direct mineral carbonation under surface condition.

The study of Kim and Cho [19] deals with the possibility of storing CO_2 in marine unconsolidated sediments. In this case, CO_2 hydrates-bearing sediments, formed during the CO_2 liquid injection process, would act as cap rocks preventing leakage from the CO_2-stored layer. The feasibility of such a CO_2-storage method was experimentally examined and temperature, pressure, P-wave velocity, and electrical resistance were measured during the experiments. Minimum breakthrough pressure and maximum absolute permeability of CO_2 hydrate-bearing sediment were also estimated.

Park et al. [20] investigated the potential supercritical CO_2 storage capacity of conglomerate and sandstone and the sealing performance of the cap rocks (i.e., dacitic tuff and mudstone) in the Janggi Basin (Korea). To these aims, the authors presented the results of laboratory measurements of the amount of supercritical CO_2 replacing the pore water in each reservoir rock core and of the initial supercritical capillary entry pressure for the cap rocks. Moreover, they also examined the mineralogical changes of the cap rocks related to supercritical CO_2–water–rock reaction.

Acknowledgments: The authors thank the Editorial Board for their suggestions which improved the quality of this editorial.

Conflicts of Interest: The authors declare no conflict of interest.

References

1. IPCC. *Carbon Dioxide Capture and Storage*; Metz, B., Davidson, O., de Coninck, H., Loos, M., Meyer, L., Eds.; Cambridge University Press: Cambridge, UK, 2005; p. 431.
2. Aminu, M.D.; Nabavi, S.A.; Rochelle, C.A.; Manovic, V. A review of developments in carbon dioxide storage. *Appl. Energy* **2017**, *208*, 1389–1419. [CrossRef]
3. Li, J.; Hitch, M.; Power, I.M.; Pan, Y. Integrated Mineral Carbonation of Ultramafic Mine Deposits—A Review. *Minerals* **2018**, *8*, 147. [CrossRef]
4. Power, I.; Harrison, A.L.; Dipple, G.M.; Wilson, S.A.; Kelemen, P.B.; Hitch, M.; Southam, G. Carbon mineralization: From natural analogues to engineered systems. *Rev. Mineral. Geochem.* **2013**, *77*, 305–360. [CrossRef]
5. Kelemen, P.B.; Matter, J.; Streit, E.E.; Rudge, J.F.; Curry, W.B.; Blusztajn, J. Rates and Mechanisms of mineral Carbonation in peridotite: Natural Processes and Recipes for Enhanced, Insitu CO_2 Capture and Storage. *Annu. Rev. Earth Planet. Sci.* **2011**, *39*, 545–576. [CrossRef]
6. Moberg, R.; Stewart, D.B.; Stackniak, D. The IEA Weyburn CO_2 Monitoring and Storage Project. In Proceedings of the 6th International Conference on Greenhouse Gas Control Technologies, Kyoto, Japan, 1–4 October 2002; pp. 219–224.
7. Baklid, A.; Korbøl, R.; Owren, G. Sleipner Vest CO_2 disposal, CO_2 injection into a shallow underground acquifer. In Proceedings of the SPE Annual Technical Conference and Exhibition, Denver, Colorado, 6–9 October 1996.
8. Paris Agreement. In Proceedings of 21st Conference of Parties (COP21) of the United Nations Framework Convention on Climate Change (UNFCCC), Le Bourget, France, 12 December 2015.

9.	CSLF. Phase III Final Report, Task force for review and identification of standards for CO_2 storage capacity estimation. In Proceedings of the Carbon Sequestration Leadership Forum (CSLF), Washington, DC, USA, 19 February 2008. Department of Energy, CSLF-T-2008-04.

10.	Saftić, B.; Kolenković Močilac, I.; Cvetković, M.; Vulin, D.; Velić, J.; Tomljenović, B. Potential for the Geological Storage of CO_2 in the Croatian Part of the Adriatic Offshore. *Minerals* **2019**, *9*, 577. [CrossRef]

11.	Koukouzas, N.; Koutsovitis, P.; Tyrologou, P.; Karkalis, C.; Arvanitis, A. Potential for Mineral Carbonation of CO_2 in Pleistocene Basaltic Rocks in Volos Region (Central Greece). *Minerals* **2019**, *9*, 627. [CrossRef]

12.	Sundal, A.; Hellevang, H. Using Reservoir Geology and Petrographic Observations to Improve CO_2 Mineralization Estimates: Examples from the Johansen Formation, North Sea, Norway. *Minerals* **2019**, *9*, 671. [CrossRef]

13.	Wasch, L.; Koenen, M. Injection of a CO_2-Reactive Solution for Wellbore Annulus Leakage Remediation. *Minerals* **2019**, *9*, 645. [CrossRef]

14.	Matter, J.; Kelemen, P. Permanent storage of carbon dioxide in geological reservoirs by mineral carbonation. *Nat. Geosci.* **2009**, *2*, 837–841. [CrossRef]

15.	Boschi, C.; Bedini, F.; Baneschi, I.; Rielli, A.; Baumgartner, L.; Perchiazzi, N.; Ulyanov, A.; Zanchetta, G.; Dini, A. Spontaneous Serpentine Carbonation Controlled by Underground Dynamic Microclimate at the Montecastelli Copper Mine, Italy. *Minerals* **2020**, *10*, 1. [CrossRef]

16.	Picazo, S.; Malvoisin, B.; Baumgartner, L.; Bouvier, A.-S. Low Temperature Serpentinite Replacement by Carbonates during Seawater Influx in the Newfoundland Margin. *Minerals* **2020**, *10*, 184. [CrossRef]

17.	Du Breuil, C.; César-Pasquier, L.; Dipple, G.; Blais, J.-F.; Iliuta, M.C.; Mercier, G. Mineralogical Transformations of Heated Serpentine and Their Impact on Dissolution during Aqueous-Phase Mineral Carbonation Reaction in Flue Gas Conditions. *Minerals* **2019**, *9*, 680. [CrossRef]

18.	Martín, D.; Flores-Alés, V.; Aparicio, P. Proposed Methodology to Evaluate CO_2 Capture Using Construction and Demolition Waste. *Minerals* **2019**, *9*, 612. [CrossRef]

19.	Kim, H.-S.; Cho, G.-C. Experimental Simulation of the Self-Trapping Mechanism for CO_2 Sequestration into Marine Sediments. *Minerals* **2019**, *9*, 579. [CrossRef]

20.	Park, J.; Yang, M.; Kim, S.; Lee, M.; Wang, S. Estimates of $scCO_2$ Storage and Sealing Capacity of the Janggi Basin in Korea Based on Laboratory Scale Experiments. *Minerals* **2019**, *9*, 515. [CrossRef]

 *minerals*

Article

Potential for the Geological Storage of CO_2 in the Croatian Part of the Adriatic Offshore

Bruno Saftić, Iva Kolenković Močilac, Marko Cvetković *, Domagoj Vulin, Josipa Velić and Bruno Tomljenović

Faculty of Mining, Geology and Petroleum Engineering, University of Zagreb, HR-10000 Zagreb, Croatia; bruno.saftic@rgn.hr (B.S.); ikolenko@rgn.hr (I.K.M.); domagoj.vulin@rgn.hr (D.V.); josipa.velic@rgn.hr (J.V.); bruno.tomljenovic@rgn.hr (B.T.)

* Correspondence: marko.cvetkovic@rgn.hr

Received: 31 July 2019; Accepted: 20 September 2019; Published: 23 September 2019

Abstract: Every country with a history of petroleum exploration has acquired geological knowledge of its sedimentary basins and might therefore make use of a newly emerging resource—as there is the potential to decarbonise energy and industry sectors by geological storage of CO_2. To reduce its greenhouse gas emissions and contribute to meeting the Paris agreement targets, Croatia should map this potential. The most prospective region is the SW corner of the Pannonian basin, but there are also offshore opportunities in the Northern and Central Adriatic. Three "geological storage plays" are suggested for detailed exploration in this province. Firstly, there are three small gas fields (Ida, Ika and Marica) with Pliocene and Pleistocene reservoirs suitable for storage and they can be considered as the first option, but only upon expected end of production. Secondly, there are Miocene sediments in the Dugi otok basin whose potential is assessed herein as a regional deep saline aquifer. The third option would be to direct future exploration to anticlines composed of carbonate rocks with primary and secondary porosity, covered with impermeable Miocene to Holocene clastic sediments. Five closed structures of this type were contoured with a large total potential, but data on their reservoir properties allow only theoretical storage capacity estimates at this stage.

Keywords: CO_2 geological storage; depleted gas fields; deep saline aquifers; Adriatic offshore; Croatia

---

1. Introduction

Regarding the distinctive characteristics of the subsurface geological setting, Croatian territory is usually subdivided into three large provinces—Pannonian basin, Dinarides and the Adriatic offshore. Only the first and the third province can offer locations with favourable conditions for the geological storage of carbon dioxide. The Dinarides can be ruled out due to several reasons. Firstly, this mountain range in Croatia is largely composed of Mesozoic carbonates that are strongly karstified to depths exceeding several kilometres. The karst hydrogeological system and its vulnerable groundwater resources effectively prevent any type of CO_2 geological storage there. The other reason is generally moderate to locally strong seismic activity [1,2], which would put both the surface installations and subsurface storage objects at risk. Thus, in prospecting for geological conditions favourable for a safe and prospective CO_2 geological storage in Croatia, one is directed both to the south-western part of the Pannonian basin and to the Adriatic offshore, the latter being far less explored but still covered by a comprehensive geological dataset, adequate for screening. This work is focused on the initial assessments of CO_2 storage potential of this extensive offshore area, based on the regional-scale knowledge of subsurface geology; i.e., the distribution and composition of lithostratigraphic units and architecture of regional-to-local structures. Why is the storage potential of the Adriatic offshore so important for Croatia? It is because almost half of the greenhouse gas (GHG) emissions from large stationary sources in the country occur along the coastline (Figure 1)—most notably in the industrial regions of Split and Rijeka, and in Istria where two large cement plants and the largest CO_2 source in Croatia, the Thermal Power Plant Plomin, are situated. Thermal Power Plant (TPP) Plomin alone is the largest single source of CO_2 in the country, exceeding 2 Mt/year according to Croatian Environmental Pollution Register [3].

Another important aspect for prospective CO_2 geological storage in the Adriatic offshore in Croatia is the decline of gas production on existing offshore gas fields in the Northern Adriatic. Consequently, these fields might be used in the future to decarbonize not only stationary CO_2 sources located along the coast, but also for inland CO_2 sources closely located or already connected by the existing pipeline network (Figure 1). Moreover, there is professional expertise of and technical potential of the otherwise declining upstream part of national petroleum industry that might be used for developing of a carbon capture and storage (CCS) system, but it will not be there for a long time. Use of this expertise for deployment of CO_2 geological storage would have unprecedented economic and environmental effects.

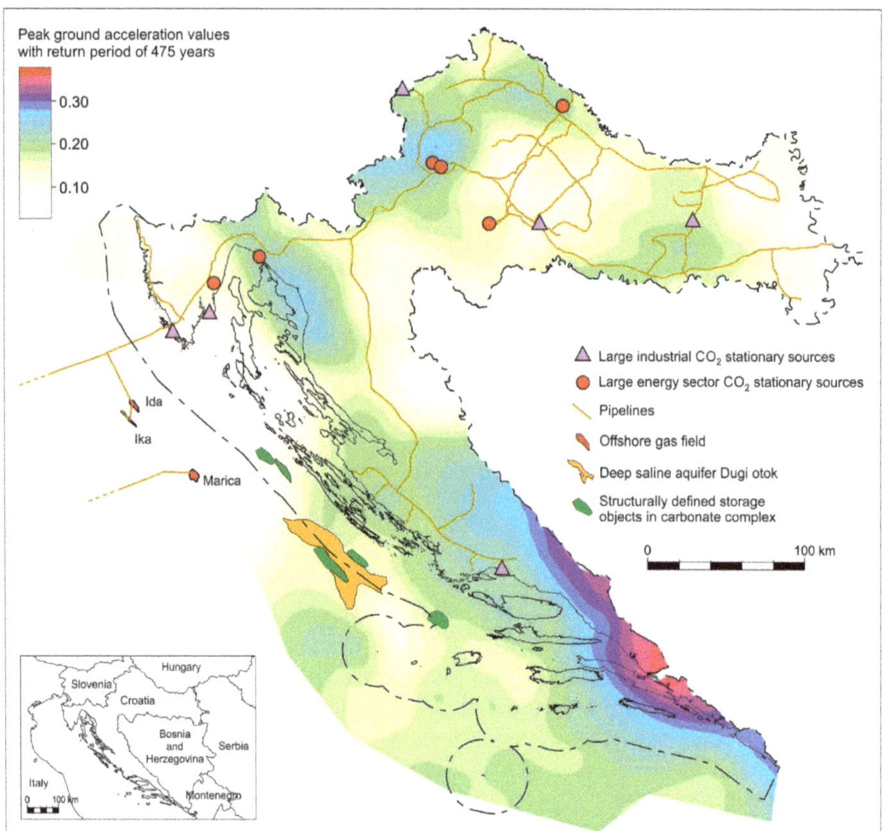

Figure 1. Location map of large stationary CO_2 sources (Croatian Environmental Pollution Register [3]), main pipeline network (after [4,5]), contours of the potential CO_2 geological storage objects in the Adriatic offshore and the peak ground acceleration values with a return period of 475 years (after [6]).

The first regional screening of CO_2 geological storage potential in Croatia was performed within the scope of the two FP6 projects—CASTOR (CO_2 from Capture to Storage) and EU GeoCapacity. This resulted in a database of the potential CO_2 storage objects, containing their geological descriptions and numerical estimates of theoretical storage capacities [7]. This database was later actualized through the FP7 project CO_2StoP [8] with the purpose of making this information uniformly structured and accessible on a European scale.

2. Geology and Petroleum Exploration of the Adriatic Offshore in Croatia

There would be no possibilities for considering the offshore CO_2 geological storage without previous HC (hydrocarbon) exploration activities that acquired data on the subsurface geological structure and lithology of rock formations in the Croatian Adriatic offshore. Interpretations evolved during five decades of intensive petroleum-geological exploration, firstly in the Northern Adriatic in the 1970s and then in other sectors southwards in 1980s. Results of initial explorations were not particularly promising [9], although some hydrocarbon shows and a few potentially economical accumulations were discovered. Major progress was made in the middle of 1990s that resulted in gas production from the Northern Adriatic offshore [10]. Several gas fields were discovered here in 1970s, first the Ivana field and later Ika and Ida fields (Figure 1) with reservoirs in Pliocene-Pleistocene clastic deposits [10–12]. Traps were formed by differential compaction, resulting in small structural closures with numerous isolated sand bodies within a progradational Plio-Pleistocene turbiditic sequence [10,13]. These thin sandy layers are characterized by intergranular porosity and markedly irregular distribution of reservoir properties [14], together with a low level of cementation. One reservoir was discovered in the underlying karstified Upper Cretaceous carbonates [10–12]. The structures are relatively shallow (from −500 to −1000 m) [10], practically meaning that only some of them might be used for CO_2 geological storage and that their storage capacities will be small. Locations of the three gas fields in the northern Adriatic offshore that were included in the EU GeoCapacity database, i.e., the Ida, Ika and Marica gas-fields, are presented in Figure 1.

The oldest rocks drilled in the Adriatic offshore are of the Permian age. According to [15], these rocks have only been drilled in two locations—one in the Italian offshore (well Amanda-1bis [16]) and one in the Croatian part (well Vlasta-1 [17]; Figures 2 and 3). Permian rocks have heterogeneous lithologic composition, comprising clastics, carbonates and evaporites [18,19]. The Lower Triassic is also characterized by mixed carbonate and clastic sediments, with both siliceous and carbonate sandstones and dolomites indicating shallow water depositional environment. Middle Triassic unit is characterized by shallow-water carbonates; however, with widespread occurrences of andesite and pyroclastics [20–23]. Evaporites can be locally found in the basal part of the Upper Triassic, more frequently in the Central and Southern Adriatic [24,25], while dolomites prevail in the Northern Adriatic area (Figure 2, with the column locations marked in Figure 4). Generally, the shallow water carbonate sedimentation in platform conditions began in the Late Triassic, on a large Southern Tethyian Megaplatform (STM) [22]. Tectonic disintegration of this megaplatform commenced by Early Jurassic rifting that resulted in formation of several smaller carbonate platforms separated by deep marine troughs and basins, giving a way to the formation of the Adriatic Basin and the Adriatic Carbonate Platform (AdCP), characterized by pelagic and platform carbonate sedimentation throughout Jurassic and Cretaceous, respectively [22]. Towards the end of Cretaceous the AdCP gradually disintegrated and emerged but carbonate sedimentation was locally restored by Paleogene transgression with the Foraminiferal limestones deposited mainly during Early to Middle Eocene when the carbonate platform sedimentation on the AdCP terminated [22]. The total thickness of the AdCP succession amounts more than 8000 m with average thickness of around 5000 m [22].

Following the lithostratigraphic subdivision generally accepted in petroleum geological exploration of the Adriatic offshore in Croatia, that is hindered by relatively scarce distribution of deep wells and seismic lines, hereafter, we will use the term "carbonate complex" for an informal lithostratigraphic unit that includes (a) Lower Jurassic (post Pliensbachian) to Middle Eocene carbonate platform succession (the AdCP succession, sensu [22]), (b) the Lower Jurassic to Middle Eocene pelagic carbonate succession of the Adriatic Basin, and (c) the underlying Upper Triassic (post Carnian) to Lower Jurassic shallow marine carbonate and clastic succession assigned by [22] to the AdCP basement or to the STM. Thus, the "carbonate complex" of the Adriatic offshore consists prevailingly of carbonate rock formations of basinal and carbonate platform origins, deposited since the Late Triassic to Middle Eocene time. In most of petroleum exploration studies (e.g., [17]) this complex is bounded on top by the Top carbonate complex horizon mapped throughout the Adriatic offshore in Croatia and shown in Figure 3.

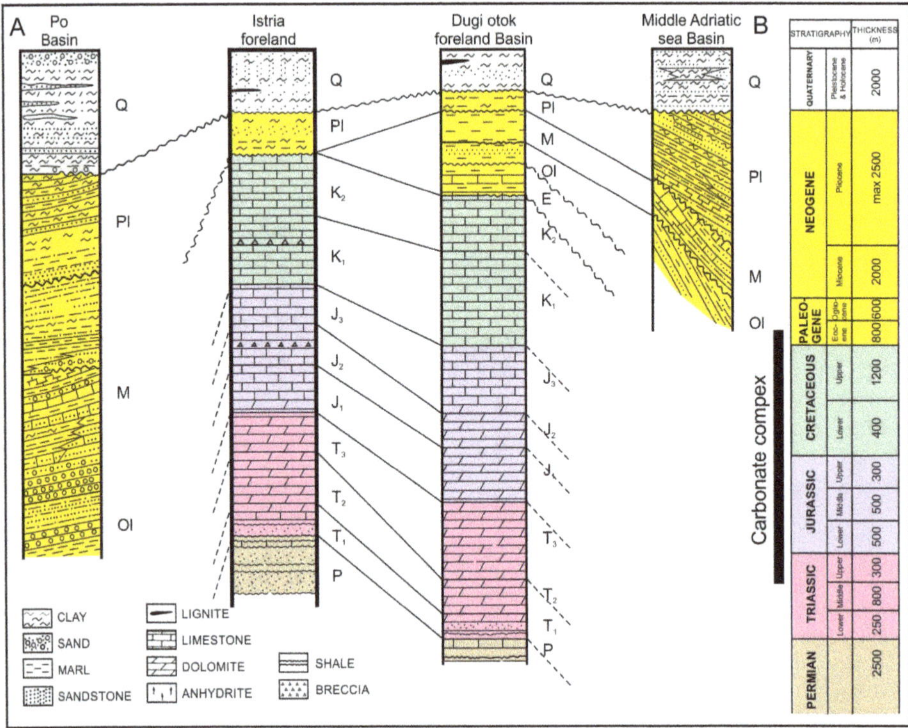

Figure 2. Schematic cross-section A–B of the Adriatic offshore (NW–SE, modified after [26–29]; locations in Figure 4).

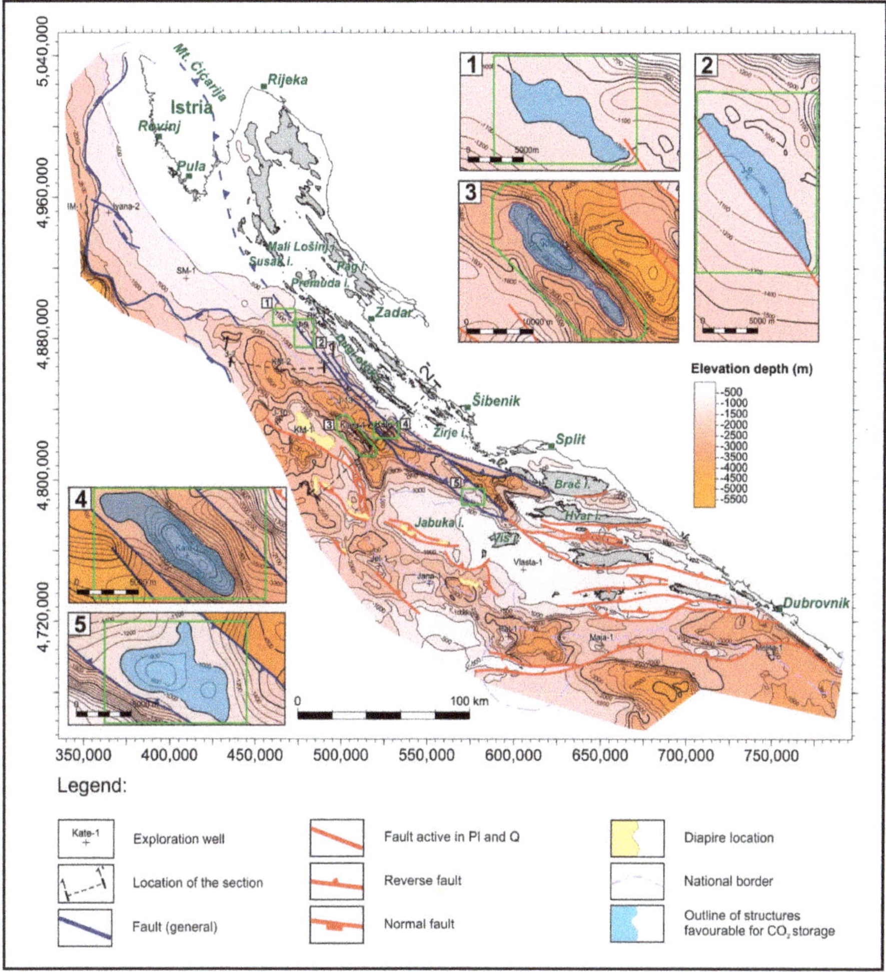

Figure 3. Structural map of the Top of carbonate complex with marked locations of potential storage objects within five structural traps. Map compiled after [25,28,30–33].

During the Middle–Late Eocene and Lower Oligocene the Adriatic offshore in Croatia was partly affected by compressional tectonics and a SW-directed propagation of thrusts that resulted in the formation of the External Dinarides fold-thrust belt, exposed along the eastern Adriatic coast and its hinterland, but also partly present in the Adriatic offshore (e.g., [30,34,35]). In the course of a SW-propagating thrust system, a large part of the AdCP succession and its basement were imbricated into a set of NW–SE striking, fault-related anticlines and synclines, that gradually led to the formation of a contemporaneous foreland basin system characterized by deposition of syntectonic flysch sediments mainly of Middle–Upper Eocene, locally of Lower Oligocene and in places, up to Lower Miocene aged sediment [22]. The continued SW-propagation of frontal thrusts locally overrode through the AdCP margin and reached up into the Adriatic basin, while more internal foreland basins gradually evolved into piggy-back basins that were filled up with a 2 km thick syntectonic clastic-carbonate succession of the Promina deposits composed of marls, calcarenites and carbonate conglomerates, at first of marine, and then of lacustrine, delta-fan and alluvial-fan origin [22,36,37]. Locally preserved Miocene deposits exposed on Pag island and in coastal hinterland of the External Dinarides, assigned to the so-called Dinaride Lake System, are exclusively of lacustrine origin. They are prevailingly composed of marls with occasional occurrences of coal seams [38,39], thus they could be considered as a post-tectonic cover in the coastal hinterland area. In contrast to these lake deposits, in contemporaneous offshore basins, Miocene deposits are represented by marine hemipelagic marls and turbidites composed of alternating marls, and calcareous and marly siltites, interbedded with sandy limestones and sandstones deposited on top of the Eocene–Oligocene marine turbidites. Based on their petrophysical characteristics, the clastic deposits of Middle Eocene to Miocene age in the Adriatic offshore are considered to have both the reservoir and sealing capabilities favourable for a regional deep aquifer formation. As a rule, the transition from Miocene to Pliocene sediments in the Adriatic offshore is marked by a regional Messinian unconformity well recognized in reflection seismic sections [17,40]. Pliocene sediments resulted from a subsequent transgression and include clays, marls and sands. In most of the offshore area there is depositional and lithologic continuity from Pliocene into Pleistocene deposits composed of sands, silts and clays with lignite interbeds, except locally where transition from Pliocene into Pleistocene is marked by transgression [41]. In the central Adriatic offshore Pleistocene and Holocene deposits can, in places, reach the thickness of 2000 m, with the total thickness of the Eocene to Holocene sequence being up to 6000 m in the deepest sub-basins. In the Northern Adriatic, the thickness of the same sequence frequently exceeds 2000 m (Figure 4).

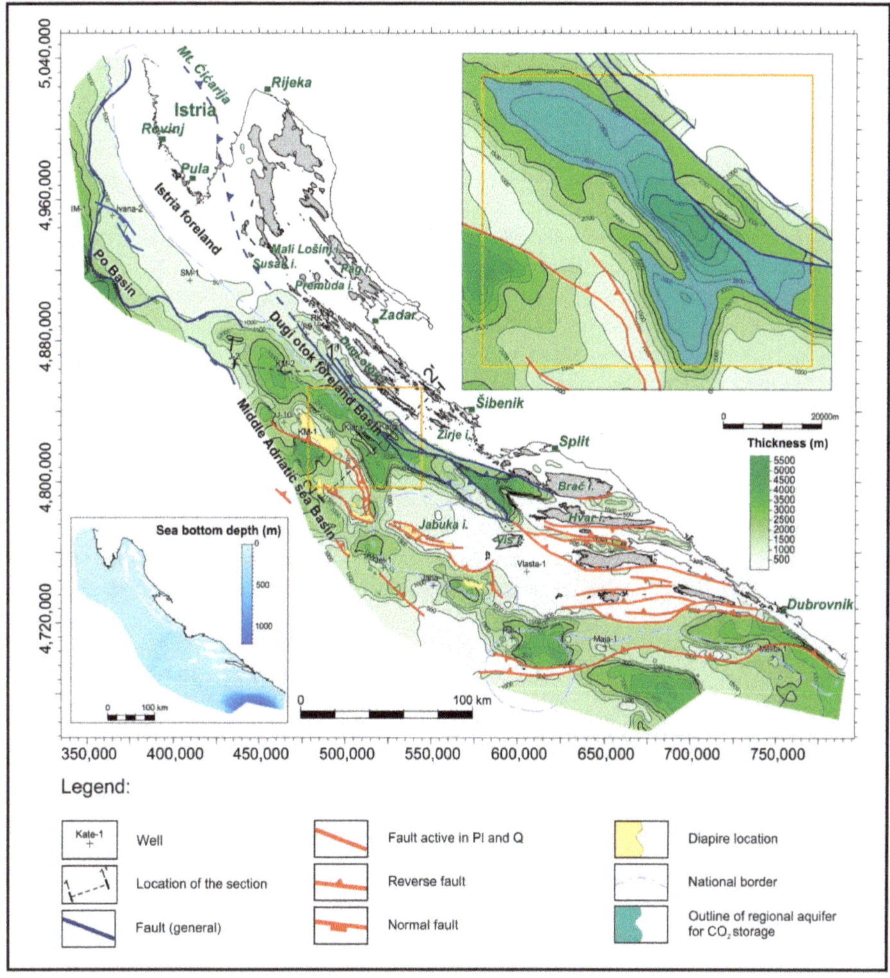

Figure 4. Thickness of an Eocene to Holocene sequence of clastic sediments with the regional, deep saline aquifer outlined as a potential CO_2 storage object. Thickness is derived based on depth of the top of the carbonate complex (Figure 3) and sea bottom depth [42].

3. First Estimates of Theoretical CO_2 Geological Storage Capacity in Gas Fields and Deep Saline Aquifers

Three different types of storage objects were found as prospective for geological CO_2 storage in the Adriatic offshore. Firstly, the Pliocene and Pleistocene sands/sandstones that have favourable petrophysical properties and are documented to be gas-tight. The second option is seen in Miocene sandstones locally present in offshore foreland basins like the Dugi otok basin, and the third is found in Upper Cretaceous limestones with primary and secondary porosity covered with impermeable Miocene or Pliocene sediments. In petroleum geological exploration terminology, these three exploration targets would be called "plays". By analogy, what is described in the following text are the three "geological CO_2 storage plays."

To evaluate the geological CO_2 storage potential of these plays, we firstly conducted regional-scale mapping of the Top carbonate complex horizon and then delineated areas that were more favourable from others. Actually, these areas are the preliminary mapped structural uplifts, which would have qualified them for the "structurally defined deep saline aquifers" if their local geological models were confirmed. These initial estimates were done based on the results of previous petroleum geological exploration, integrating them into the concept of the "theoretical storage capacity assessment," meaning that the most important properties are mapped: subsurface extension and depth range of the most important porous and permeable rock formations, thickness of their impermeable cover and zones of seismic activity that should be avoided. With regional estimates of these properties and areal extension of favourable zones, it becomes possible to make numerical estimates of storage capacity on a basin scale. This is usually called "theoretical capacity" and its only purpose is planning. It can be at first planning of land use, due to potential conflicts of interest, but the most important is planning of future targeted exploration in prospective areas. This is the way it has always been done with mineral resources, to gradually come from regional assessments to local geological models of the subsurface on locations where the exploitation (storage in this case) projects might be developed. Large capital investments in such operations dictate this procedure, which is mirror of the one used in the upstream petroleum industry, and consequently, has been proven to be the best way to substantiate the investment decisions. In that sense, a techno-economic pyramid depicting different levels of estimates of CO_2 storage capacities were developed based on the concept of energy resource pyramid introduced by McCabe [43]. Assessment of theoretical storage capacity means to make a numerical estimate of the total resource; with additional works some of it will become "effective," meaning that this is the capacity that might really be used since the uncertainties have been sufficiently reduced, and in the end the third conceived level would be the "viable" capacity that also includes economical aspects, and is by analogy equal to the "balanced reserves." In this paper the estimates of theoretical capacity are presented, based on the publicly available data. It should be noted that not all units/exploration targets are at the same level of assessment and are, therefore, described separately, in the following subchapters.

3.1. Potential Storage Objects in Depleted Gas Reservoirs

Theoretical capacity estimates were firstly performed for three gas fields in the Northern Adriatic offshore—the Ida, Ika and Marica Fields (locations in Figure 1). The capacity was calculated based on the total recoverable volume of gas under reservoir conditions, considering that CO_2 could replace the volume that was previously occupied by the gas in the reservoirs. All three assessed gas fields have multiple reservoirs of Pliocene and Pleistocene sands/sandstones, and in addition, the Ika gas field contains one reservoir in Upper Cretaceous limestones [10]. Presently, the reservoirs are not depleted. Their total potential storage capacity is estimated, and they can be converted to storage objects by making use of existing offshore installations (network of pipelines shown in Figure 1).

This theoretical storage capacity estimate has been performed based on publicly accessible data on recoverable reserves [44,45] by using the 1:1 replacement principle—the amount of CO_2 that can be stored underground into a depleted oil or gas field is equal to total oil or gas (that will be) produced:

$$m_{CO_2} = UR \times \rho_{CO_2} \times B \tag{1}$$

where B is the gas or oil formation volume factor (ratio of volume of fluid in reservoir versus volume in standard conditions); m_{CO_2} is the mass (kg) of CO_2 that can be stored; ρ_{CO_2} is the CO_2 density at reservoir conditions; and UR is the total volume of oil or gas produced; i.e., the proven ultimately recoverable recoverable oil or gas.

For calculation of CO_2 density for geosequestration, the real gas equation of state was used [46]. Formation volume factor for oil is very accurate because it was measured in laboratory. For the gas fields, assuming that the real gas volume correction, i.e., the compressibility factor Z of the gas at the surface, is 1, volume factor B_g can be expressed as:

$$B_g = 0.0034632 \times (T_r/p_r) \times Z \tag{2}$$

where T_r: reservoir Temperature (K); p_r: reservoir pressure (bar).

Table 1 shows a summary of parameters used in calculation of the CO_2 storage capacity of the three gas-fields in the Northern Adriatic. The total estimated CO_2 storage capacity for these gas-fields amounts to 32.112 Mt. Notably, most of the reservoirs of these three fields are still in production and will not be available for at least a decade.

Table 1. Summary of parameters used for CO_2 storage capacity estimation of the Northern Adriatic gas-fields.

Field Name	Stratigraphic Unit	Lithology	Depth (m)	Proven Total Recoverable Gas	B_g	CO_2 Density (t/m^3)	Estimated CO_2 Storage Capacity (Mt)
Ida	Pliocene-Pleistocene	sandstone	870	2.407	0.0051	0.3812	10.502
Ika	Pliocene-Pleistocene	sandstone/carbonates	1300	2.520	0.0045	0.5977	11.903
Marica	Pliocene-Pleistocene	sandstone	1010	1.816	0.0038	0.5396	9.707
		Total estimated CO_2 storage capacity (Mt)					32.112

3.2. Potential CO$_2$ Storage Objects in Deep Saline Aquifers

The past exploration results that were available and used in this part of the study originate from the 1971–1985 period when only few regional deep wells were drilled. These wildcat wells were not totally dry—there were gas shows in most of them and traces of heavy oil in two. With maximal depth exceeding 6000 m, this drilling campaign brought up information on the lithostratigraphy of Triassic, Jurassic, Cretaceous, Paleogene, Neogene and Quaternary sediments. Additionally, a set of correlation horizons was established then, based on the lithostratigraphy from wells and sequence boundaries observed on reflection seismics (Figure 5):

- Q—Top Pliocene;
- A—Top Miocene;
- B—Top Eocene;
- C—Top Cretaceous;
- D—Top Triassic;
- E—"Base Carbonates", Top of Permian clastic sediments.

The transversal correlation section A–B shown in Figure 2. (for location see Figure 4) is given here as an attempt to illustrate a possible reconstruction of the subsurface geology based exclusively on the vintage deep regional wells.

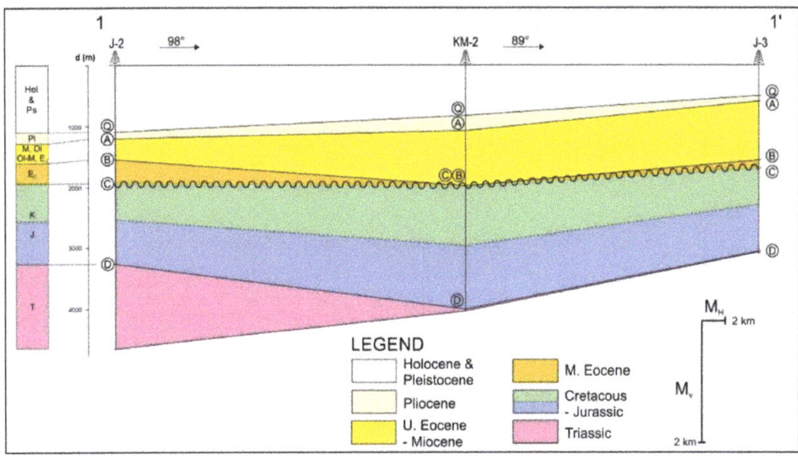

Figure 5. Schematic correlation of section 1–1′ (location in Figures 3 and 4).

The oldest E horizon delineated as the base of the carbonates was not drilled in the area of Dugi otok basin, so the thickness can only be interpreted based on the seismic data. The oldest penetrated unit is the one below the D horizon (Figure 5), composed of Triasssic dolomites and dolomitic limestones, characterized by frequent occurrence of stylolites, moldic and fenestral porosity (predominantly developed in the basal and middle part of the unit).

Jurassic and Cretaceous limestones together, build up the D–C interval. Their drilled thickness is from 900 to more than 4000 m, depending on structural position of analysed wells. The basal part of this unit is of grey to greenish dolomitic limestones with chert lenses and nodules, also with sporadic black marl intercalations. Porosity is markedly variable—from several up to even 20% based on well log interpretation [28]. The central part of this unit consists of Lower Cretaceous white limestones that have joints filled with anhydrite; limestones with stylolites; limestones with chert; and bituminous limestones and overlying dolomites. Sporadically, these sediments are characterized by increased porosities within the zones encompassing several tens of meters, interpreted to be caused by brittle tectonics. Overlying Upper Cretaceous layers are composed of dense limestones (occasionally with chert or bitumen), and bioclastic limestones (chalk), white limestones with chert and rudist limestones. Joints are not common in this unit and the existing ones are filled with organic matter—either bitumen or heavy oil, and the same goes for the stylolites. There are zones in this sub-unit where secondary porosity can be expected, but without any information about frequency and orientation of predominate joint sets. Porosities and permeabilities of Jurassic and Cretaceous limestones are strongly controlled by diagenetic and postdiagenetic processes, including dolomitization, recrystallization, dissolution, leaching, erosion and weathering. Defining the intensity and distribution of these processes in rocks would require a thorough reinterpretation of well and seismic data, which is beyond the scope of this paper.

Early to Middle Eocene limestones are the dominant lithology in the C–B interval, where basal parts of this unit also include chalk limestones and carbonate breccia. Less developed lithologies are dense laminated limestones and calcitic marls with chert nodules. Striations are observed in cores and stylolite joints as well. The youngest part of Eocene (close to B horizon) exhibits coarser carbonate bioclastics—calcarenites and calcrudites. The drilled thickness of the Eocene unit in the area is usually between 120 and 800 m, although there are wells (like Well KM-2 in Figure 5) where it is totally missing. In such cases the B horizon corresponds with the "top carbonate complex horizon".

The composite unit of the B–A interval includes clastic and carbonate sediments of Upper Eocene, Oligocene and Miocene age. Differently subsiding tectonic blocks within the Dugi otok basin are not reflected only in orientation and size of structures, but also in changes of accommodation space. Here, the Eocene-Miocene unit is found at depths between 300 and 1330 m with a variable thickness from 100 m to more than 3000 m in the central part of this structural depression. Basal part of this interval is made of Eocene flysch-like deposits—mainly marls, marly calcarenites and limestones. They are covered by Oligocene sandstones with intercalations of calcitic marl and then by the Miocene flysch-like deposits again.

The youngest unit is comprised of deposits above A and bellow Q horizon, i.e., between the Top Miocene horizon and the seabed. That unit is composed of loose, silty-sandy sediments of Pliocene, Pleistocene and Holocene age. The depth of A horizon ranges from 300 m to over 1200 m, which at the same time gives approximately the thickness of this unit. Pliocene marls, sands and clays are distributed throughout the Dugi otok basin. Quaternary sediments are transgressive in the NE part and in conformity with Pliocene in the SW region. Transgression started in early the Pleistocene age with marine sedimentation that is still ongoing. The thickness of Quaternary sediments is also variable—from 300 m to more than 1200 m.

Based on the available knowledge of subsurface geology, there is potential for geological CO_2 storage in the Adriatic off-shore in two regional units: Miocene sandstones (parts of the unit between A and B horizons, see Figure 5), and in the carbonate complex bounded on top by either C horizon or B horizon (Figure 5). The most important difference between these two units is in lithological compositions and porosity types. Miocene layers are a bit better explored and their depth range in combination with intergranular porosity appears to be more favourable, but they lack large structural closures. Underlying rocks of the carbonate complex have both the primary and secondary porosity (with locally increased permeability), and, in addition, there are numerous closed structures that are relatively easy to recognize on a structural contour map of the Top carbonate complex horizon (Figure 3). However, there is a lack of sufficient data to characterize and estimate areal distribution of reservoir properties in this unit.

The CO_2 storage potential in these two large units must, therefore, be assessed in two different ways: Miocene sandstones are studied as a regional deep saline aquifer (DSA) named "Dugi otok", because they were mapped within the Dugi otok basin, while within the carbonate complex, several structural uplifts were identified and referred to as structurally defined aquifers.

3.2.1. CO_2 Storage Potential in the Regional Deep Saline Aquifer Dugi otok (DSA Dugi otok)

Initially formed as a foreland basin ahead of SW-propagating Dinaric thrust system, sedimentation of siliciclastic deposits in the Dugi otok basin took place during the Late Eocene, Oligocene and Miocene [25,47,48]. Miocene series is made of chalk limestones with marly and sandstone interbeds. Chalk limestones and sandstones have good reservoir properties; their porosity is in 15–25% range. Based on regional seismic interpretation (seismic facies), Miocene sediments are mostly comprised of a stacked sequence of sandstone and marl layers, with some subordinate lateral lithology variations. Lower and Middle Miocene sandstone layers are the ones where CO_2 might be injected. More precisely, these are the layers and lenses of silty sandstones at depth range of 700–2100 m, regionally SW-dipping in the form of monocline unconformably covered by Pliocene marls, thus are considered as prospective for geological CO_2 storage.

In Table 2, the main characteristics of DSA Dugi otok are presented together with the parameters used to calculate its theoretical CO_2 storage capacity. An aquifer is treated as if it makes a consistent single large unit with average depth and porosity values and an estimated small proportion of pores that will eventually be filled with carbon dioxide once its plume spreads throughout the unit. This is an oversimplification of the effect of many processes that will eventually contribute to geological storage, in line with the so-called "conservative approach" taken in EU GeoCapacity CO_2 storage atlas [7]. Outline of the aquifer is shown in Figure 4. Until a detailed exploration of Miocene units in this area is made, their presence was estimated in the region where the total thickness of Eocene to Holocene sediments exceeds 3000 m. Both the storage efficiency coefficient estimate (taken as 0.02 after [49] as P50 value for clastic regional deep saline aquifers) and the way in which the unit is mapped are major contributors to the large uncertainty in the calculation of the storage capacity. This means that the number of 327.075 Mt presented in Table 2 is just a first numerical estimate of the potential and should by no means be directly compared with the numbers given in Table 1, where the potential in depleted gas fields was estimated.

Table 2. Characteristics of the regional deep saline aquifer Dugi otok.

Potential Storage Object	Average Depth (m)	Net-to-Gross	Average Porosity (%)	Pore Volume (m^3)	Initial Pore Pressure (bar)	Initial Temp. $(°C)$
	2923.5	0.2	15	16.7579×10^9	293.97	57.01
DSA Dugi otok	Pore pressure increase (%)	Pore compressibility (bar^{-1})	Pore water compressibility (bar^{-1})	CO_2 density at maximum pore pressure (kg/m^3)	Storage efficiency coefficient (-)	Total CO_2 storage capacity (Mt)
	10	3.5×10^{-5}	5.87×10^{-5}	857.75	0.02	327.075

The CO_2 storage capacity was calculated using the compressibility method (after [50]) and the volumetric method as described in [49]. The obtained capacities were than summed up, following the approach suggested by [51] that assumes that additional pore volume will be available for CO_2 storage due to compressibility of pores and initially present pore water. If pressure increase is considered, pore compressibility should be included to storage assessment:

$$c_p = 1/\delta V \times (\delta V/\delta p) \tag{3}$$

where: c_p: pore compressibility (bar^{-1}); δV: change in pore volume resulting from the change in pressure (m^3); δp: change in pressure due to injection (bar).

The maximum pore pressure was estimated to be 10% above the initial pore pressure, which is significantly less than what is estimated for structurally defined aquifers in carbonates. The reasoning behind this is that the initial pore pressure could not be expected to be intensively increased within the volume of the entire regional deep saline aquifer; the intensive pressure increase would be limited to volumes of the regional deep saline aquifer in the surroundings of the injection well, but overall average pore pressure should not increase as significantly, as it does for structurally defined deep saline aquifers; i.e., it should not be comparable to fracture pressure. The average porosity value was estimated from regional data on Miocene sandstones in the study area. Pore compressibility was estimated using the correlation of pore compressibility with the net confining pressure developed for Bandera sandstones [52], which have a similar initial porosity to the Miocene sandstones of DSA Dugi otok, amounting to 16.5%. The pore water compressibility was calculated after following equation [53]:

$$c_w = -\frac{1}{B_w}\left(\frac{\partial B_w}{\partial p}\right)_T = \frac{1}{[7.033\,p + 541.5\,C_{NaCl} - 537.0\,T + 403,300]} \tag{4}$$

where c_w is water compressibility (1/psi), B_w is water formation volume factor, p is pressure (psi), T is temperature (°F), and C_{NaCl} is salinity of pore water (g NaCl/l).

For the salinity of pore water, the value of 35 g NaCl/l, corresponding to salinity of seawater was taken. Initial pore pressure and overburden pressure were calculated using brine density of 1025 kg/m^3 and bulk wet density of overlying sediments of 2400 kg/m^3. Density of CO_2 was calculated based on the estimated values of pressure and temperature, using the equation of state after [46]. Average temperature was estimated using the geothermal gradient of 1.57 °C/100 m, which was calculated from data on temperature of the sea bottom [54] and the regional isothermal map of formation temperatures at the depth of 3000 m [31]. Pressure was calculated assuming the hydrostatic pressure gradient according to [10].

3.2.2. CO_2 Storage Potential in Anticline Structures of the Carbonate Complex

The structural map of the Top carbonate complex horizon in the Croatian part of the Adriatic off-shore (Figure 3) was constructed based on data published in one PhD thesis [28], a Master thesis [24], two graduate theses [55,56], and that in publications in scientific and professional periodicals and proceedings [29,57–59]. The regional structural model is the most important here, so characteristics of structural styles and the location of km-scale structures are described first.

In terms of the predominant structural styles and structures formed in rocks of the carbonate complex and its overlying syntectonic sediments, their orientation and distribution, the three different regions in the Adriatic offshore of Croatia can be distinguished (Figure 3):

1. Area to the WSW of the Premuda, Susak and Lošinj islands, together with the offshore west of Istrian peninsula. In this part of the Adriatic offshore, the top of the carbonate complex horizon gently dips in the WSW direction in a form of monocline (named the North-Adriatic monocline by [30]); in offshore Istria, it represents the gently WSW-dipping limb of the so called Istrian anticline [60,61] or the Istrian swell, sensu [30]. This wide and gentle anticline is bounded to the east-northeast by the frontal thrust of the External Dinarides exposed along the SW margin of the Ćićarija mountain (Figure 3), while its submerged WSW limb practically continues all the way underneath the submerged thrust front of the Northern Apennines (see in cross-section number 5 in [30]). Thus, the Istrian anticline represents a gently deformed foreland at first for the Dinarides fold-thrust belt during Middle to Late Eocene, and then for the Northern Apennines fold-thrust belt during Late Miocene to Quaternary. In the North-Adriatic monocline, at about 50 km offshore Rovinj where the core of the anticline crops out, a paleogeographic

boundary between the AdCP and the Adriatic basin is nicely preserved. According to [30,62], this boundary is interpreted as the W-dipping Early Jurassic to Paleogene normal fault that is covered by undeformed Plio-Quaternary marls and sands, that, in addition to an absence of instrumentally recorded seismicity along this boundary, suggest that it is at present, tectonically inactive. The same is true of a set of conjugate normal faults found some 20 km east of this boundary and close to the Ivana gas-field (Figure 3).

2. The area in the northern central Adriatic between the Premuda, Kornati and Žirje islands. Structurally this part of the Adriatic offshore represents presently submerged frontal part of the External Dinarides fold-thrust belt. Most, if not all, of islands in this area are fault-related anticlines formed in hangingwalls of the NE-dipping and SW verging thrust system, active during Mid-Late Eocene and Oligocene; i.e., during the main tectonic phase in the External Dinarides (see in the text above). According to interpreted reflection seismic sections available in literature (e.g., [30,62]), NE-dipping thrusts have listric geometry and sole out from two major decollement horizons: the one at approximately 5 km depth formed in Jurassic carbonates, and the other at circa 10 km depth formed in the Permo-Triassic evaporitic (salt) deposits (Figure 6). Characteristic structural styles, the morphology of reverse faults, fault-related anticlines and synclines, are nicely depicted on the transversal cross section SW of the Dugi otok island shown in Figure 6, which also shows the location of the deep well Kate-1 drilled through an anticline formed at the SW front of the Dinarides thrust system. Occasionally, NE-dipping forethrusts are associated with SW-dipping backthrusts, thus forming local pop-up structures.

3. The area in the central and southern Adriatic west of Kornati islands, and the offshore Split and Dubrovnik. Structurally, this part of the Adriatic offshore is strongly affected by salt tectonics and is comprised of numerous halokinetic structures, partly in form of salt diapirs, salt walls and salt-cored anticlines of variable size and time of formation, some of them found in cores of small and large islands like Jabuka, Brusnik, Vis, Palagruža, etc. At least in part, this area corresponds with the belt of halokinetic structures know in the Italian Adriatic offshore as the Central or Mid Adriatic Ridge (e.g., [63,64]) that we presume to extend in a SE direction all the way to the offshore Dubrovnik area and even further to offshore Montenego and Albania. A part of this belt across the Jabuka island is shown in Figure 6 and interpreted by [30] as a strike-slip corridor strongly affected by salt diapirism. In case of the Jabuka island, however, Herak et al. [65] analysed a recently recorded earthquake sequence around this island and found excellent agreement between their calculated focal mechanism and the distribution of earthquake hypocentres with the NE-dipping, Jabuka-Andrija thrust fault system. Accordingly, it is included into the list of seismogenic sources of the Adriatic offshore by Kastelic et al. [66], described there as a moderately NNE-dipping seismogenic source capable of generating earthquakes with magnitudes of 5.5. Based on this data, we have partly modified a part of the cross-section, shown in Figure 6, by proposing the NE-dipping thrust fault underneath the Jabuka island, that is supposed to splay off either from the Permo-Triassic evaporite decollement, or from an even deeper decollement within the crystalline basement, as suggested by the distribution of the Jabuka earthquake sequence. The offshore area between the Vis island and Dubrovnik shows similar structural style with a prevalence of fault-related folds associated with salt tectonics. The only difference observed there is in the prevailingly E–W strike of major faults and fault-related folds that could be controlled by variable presence of evaporites. As in case of the Jabuka island, seismic activity around major structures there is instrumentally and historically well known. Actually, in addition to the catastrophic 1667 Dubrovnik earthquake (I_0 = IX − X° EMS98; [67,68]), the most recent seismic activity here was recorded in an offshore area between the islands of Brač and Hvar (M_L = 6.1; [69]), close to the coastline in the Ston area (M_L = 6.0; [70]), and in off-shore Montenegro (M_W = 7.1; [71]). Accordingly, the ongoing tectonic activity and seismicity in this area significantly reduces its potential for CO_2 geological storage.

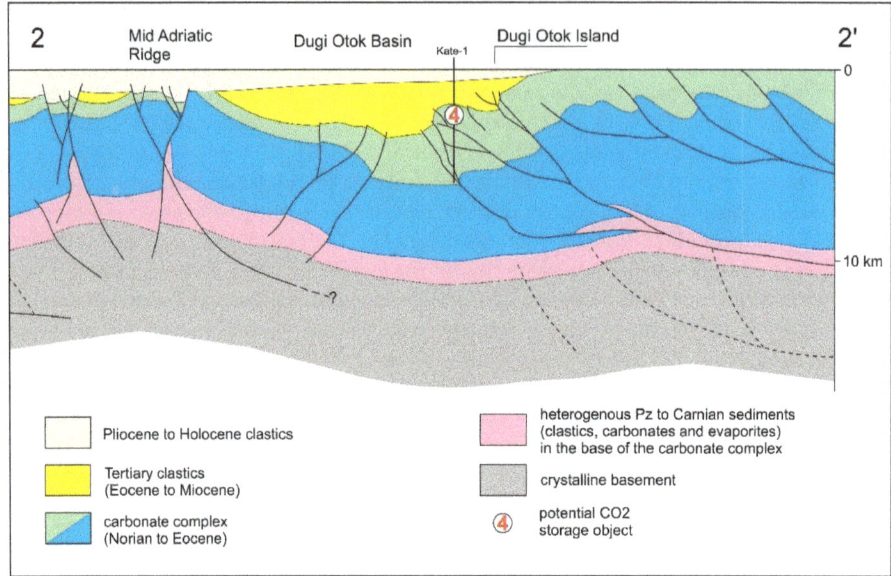

Figure 6. Transversal cross-section 2–2′ through the deep well Kate-1, south of the Dugi otok island (after [30]; stratigraphy modified after [15]). Vertical exaggeration is 2:1. Fault planes are marked with black lines. Location in Figures 3 and 4.

Altogether, five structural traps—potential underground CO_2 storage objects, are depicted, all of them identified based on the structural map of the top of the carbonate complex. Structures are shown in detail on small maps given in Figure 3. The main characteristics of the potential storage objects are given in Table 3 and main parameters used to calculate storage capacities are given in Table 4. The average effective porosity value has been extrapolated from the laboratory measurements on core samples of the Upper Cretaceous carbonates from a single well in the northern part of the Adriatic offshore [28]. To calculate pore compressibility, the correlation of pore compressibility with net confining pressure developed for carbonate rocks has been used [72]. The correlation itself was developed in Knutson and Bohor [73]. Pore water compressibility was calculated in the same manner as for DSA Dugi otok, using a correlation by Osif [53]. The maximum pore pressure is set based on criteria 90% of 0.2 bar/m fracture gradient; i.e., as 0.18 bar/m [74,75]. In that respect, maximum increase of pore pressure was averaged to 50% from initial pore pressure, which is in accordance with value of maximum pore pressure suggested to be between 1.3 and 1.8 times the initial pore pressure [76]. Storage efficiency coefficient was taken to be 0.05, which is the product of displacement efficiencies (volumetric, E_V and microscopic, E_D) and net-to-gross-ratio. For product of volumetric and microscopic displacement efficiencies, the estimated value of P_{10} in limestone formation storage objects after Goodman et al. [49] amounting to 0.1 was taken, while for the net-to-gross ratio, representing the part of the structurally defined saline aquifer having favourable petrophysical properties needed for CO_2 injection (generally corresponding to $E_{hn/hg}$ after [49], the value of 0.5 was taken. The value of net-to-gross was practically based on a rule of thumb approach, since there were not enough data to make reliable geological models of these structurally defined aquifers. The porosity data used to calculate pore volume was effective porosity, but it was extrapolated from the neighbouring well, not from the wells drilled-through the structurally defined aquifers. Also, no quantitative data of permeability were available that could be used to assess net-to-gross ratio. Temperatures were estimated using geothermal gradient that was calculated from data on temperature of the sea bottom [54] and the regional isothermal map of formation temperatures at the depth of 3000 m [31]. The calculated values were in agreement with the geothermal gradient mapped by Jelić et al. [77]. Since no data on pressure were publicly available, initial pore pressure was calculated using hydrostatic pressure gradient and this can be regarded as a reasonable assumption; i.e., no overpressure is to be expected, due to the fact that drilling operations encountered problems with total mud loss when entering the carbonate complex [19]. Densities of CO_2 were calculated based on the estimated values of pressure and temperature, using equation of state as defined in [46].

Table 3. Main characteristics of structurally defined aquifers in carbonates.

Potential Storage Object	Top Depth (m)	Average Depth (m)	Average Sea Depth (m)	Average Porosity (%)	Pore Volume (10^6 m^3)	Average Initial Pore Pressure (bar)	Average Temperature (°C)
Structure 1	891	945.5	65	18.85	209.64	95.1	26.99
Structure 2	843	921.5	65	18.85	467.14	92.7	26.61
Structure 3	1670	2085	113	18.85	4473.84	209.7	43.96
Structure 4	1772	2136	133	18.85	2066.99	214.8	44.45
Structure 5	780	890	121	18.85	778.51	89.5	23.23

Table 4. Storage capacity estimation using compressibility method for structurally defined aquifers in carbonates.

Potential Storage Object	Storage Efficiency Coefficient (-)	Pore Pressure Increase (%)	Pore Compressibility (bar^{-1})	Water Compressibility (10^{-5} bar^{-1})	CO_2 Density * (kg/m^3)	Total Storage Capacity (Mt)
Structure 1	0.05	50	10.15×10^{-5}	5.55×10^{-5}	857.68	8.99
Structure 2	0.05	50	10.15×10^{-5}	5.54×10^{-5}	856.27	20.00
Structure 3	0.05	50	6.96×10^{-5}	5.73×10^{-5}	902.58	201.90
Structure 4	0.05	50	6.96×10^{-5}	5.74×10^{-5}	904.89	93.52
Structure 5	0.05	50	11.17×10^{-5}	5.51×10^{-5}	872.18	33.95

* CO_2 density at maximum pore pressure (sum of average initial pore pressure and overpressure caused by injection).

In this way, calculated total storage capacities in five chosen structurally defined aquifers were considerably high, which makes them valid candidates for future exploration activities. Special attention should be given to the fact that in three of five potential storage objects (structures 1, 2 and 5) CO_2 is not expected to be in supercritical state, but liquid upon injection, due to low initial average temperatures that are the result of a low geothermal gradient (between 1.2 and 1.5 °C/100 m), characteristic for the studied area. This is not necessarily an issue, since according to [78], injecting CO_2 in a liquid state is energetically more efficient than in supercritical state, due to its increased density, which results in lower overpressure not only at the wellhead, but also in the reservoir, because a smaller volume of fluid is displaced.

It should be emphasized that the obtained capacities are heavily burdened by the lack of data and subsequent weaknesses of the model used for their calculation and can also be treated as theoretical values only. However, it must be noted that numbers given in Table 4 are more realistic than the estimates given for the Miocene regional aquifer (Table 2), making at least some of these objects targets for future detailed exploration.

4. Discussion

Trying to estimate the storage capacity in deep saline aquifers (DSA) always disclosed a major problem, because the available data on the subsurface geology are not detailed enough. Even in the mature petroleum provinces deep aquifers were simply not drilled through in many places and there are just a few analyses of their reservoir properties. There are frequent cases where the geometry of the reservoir rock formations can be delineated based on the regional subsurface data, but other parameters—effective thickness, porosity and temperature—need to be extrapolated from the existing hydrocarbon fields in the region, if there are any. This inevitably burdens the storage capacity estimates with a lot of uncertainties. Even more so, knowing that adequate trapping conditions in parts of these regional aquifers will only later be confirmed by targeted surveys. That is why these storage estimates are regarded as theoretical capacity only (bottom of the techno-economic resource pyramid for the capacity of CO_2 geological storage as defined in [79]).

There are the two significantly different types of formations where potential underground CO_2 storage objects might be planned and constructed in the Adriatic offshore. Firstly, there are the thick carbonate rock formations ranging in geological age from Triassic to Eocene. On the map of top of carbonate complex (Figure 3) in the more prospective zones (i.e., far from the active faults) altogether, five structures were depicted. Three (2, 3 and 4) of them were drilled by regional wells and no hydrocarbons were discovered, meaning that they can be assessed as structurally defined aquifers. Their main characteristic is the primary and secondary porosity, thus potentially high permeability, which is indicated by total mud loss during the drilling of the mentioned wells [19]. Carbonate rock formations are, in the Adriatic offshore, covered by thick succession of clastic sediments (from Eocene to Holocene age), in which most of the rocks are impermeable, most importantly the Upper Miocene and Lower Pliocene layers. The thickness of the entire clastic basin fill is given in Figure 4. Another interesting potential storage object is the deep saline aquifer—Dugi otok (DSA Dugi otok). This is a regionally defined unit of thick Miocene succession of marls and sands that filled the Dugi otok depression. Looking at the cross-section 2–2' in Figure 6, and given description of regional geology, this regional aquifer might be considered as an object worth the detailed exploration for two reasons—ample impermeable intervals (regional seals) of the Miocene and Pliocene age, and a regional dip SW by one largely undeformed structure, allowing plans to be made for the injection wells on the subsided SW part of the monocline and monitoring wells on the NE side. That is, should such general structure be confirmed by targeted exploration. The drawback for now is in the smaller proportion of permeable layers (estimated net pay of 0.2 is in the Table 2) and the same goes for the true reservoir properties, because they are also only regionally estimated. CO_2 storage capacity declared for the DSA Dugi otok is really a preliminary estimate for two reasons—its reservoir rock properties are based only on the data from three wells, and its outline follows the contour 3000 m on the map of thickness of clastic sediments (Figure 4), because that is the area where the Miocene sediments have greatest thickness, and within this area thickness of Pleistocene and Holocene sediments is the greatest, meaning that the Miocene strata are situated in depths exceeding 1000 m. It also has to be noted that the storage efficiency factor is taken to be 0.02 [49], meaning that only such a small proportion of the estimated available pore volume might retain, once being filled with carbon dioxide (at several locations that are still to be found). This storage capacity in aquifers might be prepared for use only after the deliberated exploration of these objects, not only to fully investigate their reservoir properties, but also to confirm the integrity of their cap rocks. The third option, storage objects in the three gas fields might easily be prepared for pilot injections and have upscaling potential, but this will become available only once their reservoirs are depleted and hydrocarbon exploitation licences are expired or terminated.

5. Conclusions

It is important for the Republic of Croatia to consider the possibilities of reducing their emissions by making use of CCS technology. This is the only way to achieve the Paris agreement targets in time, simply because the existing large stationary industrial sources from the energy and other industrial sectors can be cost-effectively, safely and quickly decarbonised, before the uptake of renewable technologies really starts "kicking in." Timely preparation of this will positively influence energy prices and save many jobs, not to mention that every nation should take care of its contribution to the global effort to reduce GHG emissions. Croatia is at present, still far from phasing out its fossil fuel energy sources and has a comparably high proportion of industrial emissions that will not just disappear in the near future. This can all be dealt with by evaluating the new "geological storage resource" to make use of the deep subsurface rock formations by the building of carbon capture and storage (CCS) systems.

There are favourable conditions for geological storage of CO_2 in Croatia, both in the southern part of the Pannonian basin and the Adriatic offshore [7]. The capacity declared for hydrocarbon fields is better defined than estimates for aquifer formations which still need detailed exploration in order to define the structures for storage. Regarding the Adriatic offshore, it is far less explored than Pannonian Basin, but it has a considerable dataset, enough for screening in terms of the basic characteristics of subsurface geology.

Since almost half of emissions occur in the coastal areas, the Adriatic offshore's CO_2 storage potential gains importance now that even exceeds petroleum exploration. That is simply because one can import oil or natural gas, but one cannot export CO_2. There is still the significant professional knowledge and technical potential from the otherwise declining upstream part of national petroleum industry. It simply must be put into use while the window of opportunity still exists.

The most prospective CO_2 storage objects are the small gas fields in the Northern Adriatic. Total storage potential in their reservoirs is not large (32 Mt) but it is available, and there are installations on exploitation sites which would significantly reduce investments. Another group of objects are deep saline aquifers. They offer much larger potential but with large uncertainty; that is why this is called "theoretical storage capacity". We think we have demonstrated the two most important "CO_2 storage plays" based on the regional geological data—Miocene sandstones in the Dugi otok basin (as a regional deep saline aquifer, DSA Dugi otok) and Triassic to Eocene carbonate rock formations, whose paleotopography is covered by thick impermeable layers of Miocene to Early Pliocene age, so five structurally defined aquifers were delineated, each of them representing a site where CO_2 storage capacity could be investigated on a local scale. They are all distant from sources of carbon dioxide and maybe the two of them will eventually prove to be too small to be economical, but the other three show significant potential that should not be overlooked.

Author Contributions: Conceptualization, B.S. and I.K.M.; methodology, B.S., I.K.M., M.C. and D.V.; software, I.K.M., M.C. and D.V.; validation of regional geology, J.V. and B.T.; formal analysis, I.K.M. and D.V.; investigation, B.S., I.K.M. and M.C.; resources, B.S. and M.C.; data curation, I.K.M. and M.C.; writing—original draft preparation, B.S., I.K.M. and M.C.; writing—review and editing, J.V. and B.T.; visualization, I.K.M. and M.C.; supervision, B.S.; project administration, B.S. and M.C.; funding acquisition, B.S.

Funding: The publication process was supported by the Development Fund of the Faculty of Mining, Geology and Petroleum Engineering, University of Zagreb. The authors would like to acknowledge that they have had support from European Commission (in a number of FP6, FP7 and Horizon 2020 projects), from the Croatian Environmental Protection and an Energy Efficiency Fund, from HEP (Croatian power utility who once ordered a professional study) and from the University of Zagreb with which they are affiliated.

Conflicts of Interest: The authors declare no conflict of interest.

References

1. Markušić, S.; Herak, M. Seismic zoning of Croatia. *Nat. Hazards* **1999**, *18*, 269–285. [CrossRef]
2. Herak, M.; Živčić, M.; Sović, I.; Cecić, I.; Dasović, I.; Stipčević, J.; Herak, D. Historical Seismicity of the Rijeka Region (Northwest External Dinarides, Croatia)—Part II: The Klana Earthquakes of 1870. *Seismol. Res. Lett.* **2018**, *89*, 1524–1536.
3. HAOP Croatian Environmental Pollution Register. Available online: http://roo-preglednik.azo.hr/Default.aspx (accessed on 7 June 2019).
4. Plinacro Ltd—Gas Transport System Operator. Available online: http://www.plinacro.hr/default.aspx?id=162 (accessed on 26 August 2019).
5. Hercecg, H.; Krpan, H. Optimizacija sustava dehidracije plina na platformi Ivana K. Gas dehydration optimization in the Ivana K platform. *Naft. Plin* **2019**, *39*, 84–94.
6. Herak, M. Karta Potresnih Područja. Available online: http://seizkarta.gfz.hr/karta.php (accessed on 25 August 2019).
7. Vangkilde-Pedersen, T.; Anthonsen, K.L.; Smith, N.; Kirk, K.; Neele, F.; van der Meer, B.; Le Gallo, Y.; Bossie-Codreanu, D.; Wojcicki, A.; Le Nindre, Y.-M.; et al. Assessing European capacity for geological storage of carbon dioxide–the EU GeoCapacity project. *Energy Proced.* **2009**, *1*, 2663–2670. [CrossRef]
8. Poulsen, N.; Holloway, S.; Neele, F.; Smith, N.A.; Kirk, K. Assessment of CO_2 Storage Potential in Europe. European Commission Contract No ENER/C1/154-2011-SI2.611598. CO_2StoP Final Report. 2014. Available online: http://energyx.com.au/files/56-2014%20Final%20report.pdf (accessed on 31 July 2019).

9. Cota, L.; Dalić, N.; Šikonja, Ž. INA's Experience in Hydrocarbon Exploration in Croatia. *Nafta* **2014**, *65*, 142–146.

10. Malvić, T.; Đureković, M.; Čogelja, Z.; Šikonja, Ž.; Ilijaš, T.; Kruljac, I. Exploration and production activities in northern Adriatic Sea (Croatia), successful joint venture INA (Croatia) and ENI (Italy). *Nafta* **2011**, *62*, 287–292.

11. Đureković, M.; Krpan, M.; Pontiggia, M.; Ruvo, L.; Savino, R.; Volpi, B. Geological modelling and petrophysical characterisation of turbiditic reservoirs of the Ivana gas field-R. Croatia. *Nafta* **1998**, *49*, 241–258.

12. Đureković, M.; Jovović, S.; Krpan, M.; Jelić-Balta, J. Ika gas field characterization and modeling. *Nafta* **2002**, *53*, 273–282.

13. Marić Đureković, Ž. Litofacijesne i stratigrafske značajke pleistocenskih naslaga podmorja sjevernoga Jadrana na temelju visokorazlučivih karotažnih mjerenja. In *Lithofacies and Stratigraphy of Pleistocene Deposits in North Adriatic Offshore by Using High-Resolution Well Logs*; University of Zagreb: Zagreb, Croatia, 2011.

14. Zelić, M.; Mlinarić, Ž.; Jelić-Balta, J. Croatian Northern Adriatic Ivana gas field ready for development (Reservoir characteristics and gas inflow conditions into the well). *Nafta* **1999**, *50*, 19–37.

15. Velić, J.; Malvić, T.; Cvetković, M.; Velić, I. Stratigraphy and petroleum geology of the Croatian part of the Adriatic basin. *J. Pet. Geol.* **2015**, *38*, 281–300. [CrossRef]

16. Busetti, M.; Volpi, V.; Barison, E.; Giustiniani, M.; Marchi, M.; Ramella, R.; Wardell, N.; Zanolla, C. Meso-Cenozoic seismic stratigraphy and the tectonic setting of the Gulf of Trieste (northern Adriatic). *GeoActa* **2010**, *3*, 1–14.

17. Grandić, S.; Krakatović, I.; Rusan, I. Hydrocarbon potential assesment of the slope deposits along the SW Dinarides carbonate platform edge. *Nafta* **2010**, *61*, 325–338.

18. Tišljar, J. Origin and Depositional Environments of the Evaporite and Carbonate Complex (Upper Permian) from the Central Part of the Dinarides (Southern Croatia and Western Bosnia). *Geol. Croat.* **1992**, *45*, 116–126.

19. Spaić, V. Oil and gas bearingness and structural elements of Adriatic islands and peninsulas (Outer Dinarides) with special review of anhydrite—Carbonate Mesozoic complex and diapiric belt. *Nafta* **2012**, *63*, 29–37.

20. Bahun, S. Geološki odnosi okolice Donjeg Pazariššta u Lici (trijas i tercijarne Jelar naslage). Geological relations of the surroundings of Donje Pazarište in Lika, Croatia. *Geološki Vjesn.* **1963**, *16*, 161–170.

21. Pamić, J. Triassic magmatism of the Dinarides in Yugoslavia. *Tectonophysics* **1984**, *109*, 273–307. [CrossRef]

22. Vlahović, I.; Tišljar, J.; Velić, I.; Matičec, D. Evolution of the Adriatic Carbonate Platform: Palaeogeography, main events and depositional dynamics. *Palaeogeogr. Palaeoclimatol. Palaeoecol.* **2005**, *220*, 333–360. [CrossRef]

23. Smirčić, D.; Kolar-Jurkovšek, T.; Aljinović, D.; Barudžija, U.; Jurkovšek, B.; Hrvatović, H. Stratigraphic Definition and Correlation of Middle Triassic Volcaniclastic Facies in the External Dinarides: Croatia and Bosnia and Herzegovina. *J. Earth Sci.* **2018**, *29*, 864–878.

24. Babić, K. Tektonska Kretanja i Solne Strukture u Području Vis-Biševo-Sušac. Tectonic Momvements and Salt Structures in Vis-Biševo-Sušac Area. Master's Thesis, University of Zagreb, Zagreb, Croatia, 1990.

25. Scisciani, V.; Esestime, P. The Triassic Evaporites in the Evolution of the Adriatic Basin. In *Permo-Triassic Salt Provinces of Europe, North Africa and the Atlantic Margins*; Soto, J.I., Flinch, J., Tari, G., Eds.; Elsevier: Amsterdam, The Netherlands, 2017; pp. 499–516.

26. Prelogović, E.; Kranjec, V. Geological development of the Adriatic area (Geološki razvitak područja Jadranskog mora—In Croatian). *Pomor. Zb.* **1983**, *21*, 387–405.

27. Veseli, V.; Tišljar, J.; Tadej, J.; Premec-Fuček, V. Lithofacies and Biofacies of the Cretaceous and Paleogene Carbonate Sediments in Kate-1 offshore well (Kornati Area, Croatia, Adriatic Sea). In Proceedings of the second International Symposium on the Adriatic Carbonate Platform, Zagreb, Croatia, 12–18 May 1991; p. 115.

28. Veseli, V. *Facijesi karbonatnih sedimenata mlađeg mezozoika i paleogena u pučinskim bušotinama Sjevernoga Jadrana. Late Mesosoic and Paleogene Carbonate Facies in the off-Shore Wells in the Northen Adria*; University of Zagreb: Zagreb, Croatia, 1999.

29. Grandić, S.; Veseli, V. Hydrocarbon potential of Dugi Otok basin in offshore Croatia. *Nafta* **2002**, *53*, 215–224.

30. Fantoni, R.; Franciosi, R. Mesozoic extension and Cenozoic compression in Po Plain and Adriatic foreland. *Rend. Online Soc. Geol. Ital.* **2010**, *9*, 181–196.

31. Kolbah, S.; Grandić, S. New Commercial Oil Discovery at Rovesti Structure in South Adriatic and its Importance for Croatian Part of Adriatic Basin. *Nafta* **2009**, *60*, 68–82.

32. Cazzini, F.; Zotto, O.D.; Fantoni, R.; Ghielmi, M.; Ronchi, P.; Scotti, P. Oil and gas in the adriatic foreland, Italy. *J. Pet. Geol.* **2015**, *38*, 255–279. [CrossRef]
33. Wrigley, R.; Hodgson, N.; Esestime, P. Petroleum geology and hydrocarbon potential of the adriatic basin, offshore Croatia. *J. Pet. Geol.* **2015**, *38*, 301–316. [CrossRef]
34. Grandić, S.; Boromisa-Balaš, E.; Šušterić, M. Exploration concept and characteristics of the stratigraphic and structural models of the Dinarides in Croatian offshore area PART II. Hydrocarbon Consideration. *Nafta* **1997**, *48*, 249–266.
35. Korbar, T. Orogenic evolution of the External Dinarides in the NE Adriatic region: A model constrained by tectonostratigraphy of Upper Cretaceous to Paleogene carbonates. *Earth-Sci. Rev.* **2009**, *96*, 296–312. [CrossRef]
36. Babić, L.; Zupanič, J. Laterally variable development of a basin-wide transgressive unit of the North Dalmatian Foreland Basin (Eocene, Dinarides, Croatia). *Geol. Croat.* **2012**, *65*, 1–27. [CrossRef]
37. Mrinjek, E.; Nemec, W.; Pecinger, V.; Mikša, G.; Vlahović, I.; Ćosović, V.; Velić, I.; Bergant, S.; Matičec, D. The Eocene-Oligocene Promina Beds of the Dinaric Foreland Basin in Northern Dalmatia. *J. Alp. Geol.* **2012**, *55*, 409–451.
38. Jiménez-Moreno, G.; de Leeuw, A.; Mandic, O.; Harzhauser, M.; Pavelić, D.; Krijgsman, W.; Vranjković, A. Integrated stratigraphy of the Early Miocene lacustrine deposits of Pag Island (SW Croatia): Palaeovegetation and environmental changes in the Dinaride Lake System. *Palaeogeogr. Palaeoclimatol. Palaeoecol.* **2009**, *280*, 193–206. [CrossRef]
39. De Leeuw, A.; Mandic, O.; Vranjković, A.; Pavelić, D.; Harzhauser, M.; Krijgsman, W.; Kuiper, K.F. Chronology and integrated stratigraphy of the Miocene Sinj Basin (Dinaride Lake System, Croatia). *Palaeogeogr. Palaeoclimatol. Palaeoecol.* **2010**, *292*, 155–167. [CrossRef]
40. Amadori, C.; Garcia-Castellanos, D.; Toscani, G.; Sternai, P.; Fantoni, R.; Ghielmi, M.; Di Giulio, A. Restored topography of the Po Plain-Northern Adriatic region during the Messinian base-Level drop—Implications for the physiography and compartmentalization of the palaeo-Mediterranean basin. *Basin Res.* **2018**, *30*, 1247–1263. [CrossRef]
41. Ghielmi, M.; Minervini, M.; Nini, C.; Rogledi, S.; Rossi, M. Late Miocene-Middle Pleistocene sequences in the Po Plain—Northern Adriatic Sea (Italy): The stratigraphic record of modification phases affecting a complex foreland basin. *Mar. Pet. Geol.* **2013**, *42*, 50–81. [CrossRef]
42. EMODnet Bathimetry—Understanding the Topography of the European Seas. Available online: https://portal.emodnet-bathymetry.eu/help/help.html (accessed on 2 June 2019).
43. McCabe, P.J. Energy resources—Cornucopia or empty barrel? *Am. Assoc. Pet. Geol. Bull.* **1998**, *82*, 2110–2134.
44. Velić, J. *Geologija nafte Petroleum Geology*; University of Zagreb: Zagreb, Croatia, 2007.
45. Živković, V. Proizvodne platforme eksploatacijskog polja Sjeverni Jadran. In *Production Platforms of Exploitation Field North Adriatic*; University of Zagreb: Zagreb, Croatia, 2015.
46. Span, R.; Wagner, W. A new equation of state for carbon dioxide covering the fluid region from the triple-Point temperature to 1100 K at pressures up to 800 MPa. *J. Phys. Chem. Ref. Data* **1996**, *25*, 1509–1596. [CrossRef]
47. Pavlovec, R.; Drobne, K.; Sikic, L. Upper Eocene and Oligocene in Yugoslavia. In *Developments in Palaeontology and Stratigraphy*; Pomerol, C., Premoli-Silva, I., Eds.; Elsevier: Amsterdam, The Netherlands, 1986; pp. 109–111.
48. Frixa, A.; Gorla, L.; Liverani, G.; Nini, C.; Parlov, B.; Pompadoro, G. *Eocene-Miocene Calcareous Turbiditic Play in a Dinaric Foredeep: The Dugi Otok Basin, Offshore Croatia*; AAPG: Barcelona, Spain, 2003; pp. 21–24.
49. Goodman, A.; Hakala, A.; Bromhal, G.; Deel, D.; Rodosta, T.; Frailey, S.; Small, M.; Allen, D.; Romanov, V.; Fazio, J.; et al. U.S. DOE methodology for the development of geologic storage potential for carbon dioxide at the national and regional scale. *Int. J. Greenh. Gas Control* **2011**, *5*, 952–965. [CrossRef]
50. van der Meer, L.; Egberts, P.J.P. A General Method for Subsurface CO_2 Storage Capacity Calculations. In Proceedings of the Offshore Technology Conference, Huston, TX, USA, 5–8 May 2008; pp. 889–895.
51. Vulin, D.; Kurevija, T.; Kolenkovic, I. The effect of mechanical rock properties on CO_2 storage capacity. *Energy* **2012**, *45*, 512–518. [CrossRef]
52. Zimmerman, R.W. *Chapter 8. Tubular Pores Part Two: Compressibility and Pore Structure*; Elsevier Science: Amsterdam, The Netherlands, 1991.
53. Osif, T.L. The Effects of Salt, Gas, Temperature, and Pressure on the Compressibility of Water. *SPE Reserv. Eng.* **1988**, *3*, 175–181. [CrossRef]

54. Russo, A.; Carniel, S.; Sclavo, M.; Krzelj, M. Climatology of the Northern-Central Adriatic Sea. In *Modern Climatology*; Wang, S., Gillies, R., Eds.; IntechOpen: Rijeka, Croatia, 2012; pp. 177–212.

55. Bakić, H. Strukturne značajke Jadranskog podmorja jugozapadno od Istarskog poluotoka. Structural features of Adriatic offshore Southeast of Istira Peninsula. Master's Thesis, University of Zagreb, Zagreb, Croatia, 2007.

56. Križanić, D. Strukturno-Stratigrafski odnosi i "bright-spot" anomalije u ležištima sjeverno od polja Ivana. In *Structural-Strtigraphic Relations and Bright Spor Anomalies North of IVANA Gas Field*; University of Zagreb: Zagreb, Croatia, 1999.

57. Grandić, S.; Boromisa-Balaš, E.; Šušterić, M.; Kolbah, S. Hydrocarbon possibilites in the Eastern offshore Adriatic Slope zone of Croatian area. *Nafta* **1999**, *50*, 51–73.

58. Prelogović, E.; Pribičević, B.; Ivković, Ž.; Dragičević, I.; Buljan, R.; Tomljenovic, B. Recent structural fabric of the Dinarides and tectonically active zones important for petroleum-Geological exploration. *Nafta* **2004**, *55*, 155–161.

59. Tomljenovic, B.; Herak, M.; Kralj, K.; Prelogović, E.; Bostjančić, I.; Matoš, B. Active tectonics, sismicity and seismogenic sources of the Adriatic coastal and offshore region of Croatia. In Proceedings of the Riassunti Estesi delle Comunicazioni, Trieste, Italy, 16–19 November 2009; pp. 133–136.

60. Marinčić, S.; Matičec, D. Tektonika i kinematika deformacija na primjeru Istre [Tectonics and kinematics of deformations, an Istrian Model]. *Geološki Vjesn.* **1991**, *44*, 257–268.

61. Matičec, D. Neotectonic deformations in western Istria, Croatia. *Geol. Croat.* **1994**, *47*, 199–204.

62. Grandić, S. Periplatform clastics of Croatian offshore and their petroleum geological significance. *Nafta* **2009**, *60*, 503–511.

63. Geletti, R.; Del Ben, A.; Busetti, M.; Ramella, R.; Volpi, V. Gas seeps linked to salt structures in the central adriatic sea. *Basin Res.* **2008**, *20*, 473–487. [CrossRef]

64. Casero, P.; Bigi, S. Structural setting of the Adriatic basin and the main related petroleum exploration plays. *Mar. Pet. Geol.* **2013**, *42*, 135–147. [CrossRef]

65. Herak, D.; Herak, M.; Prelogović, E.; Markušić, S.; Markulin, Ž. Jabuka island (Central Adriatic Sea) earthquakes of 2003. *Tectonophysics* **2005**, *398*, 167–180. [CrossRef]

66. Kastelic, V.; Vannoli, P.; Burrato, P.; Fracassi, U.; Tiberti, M.M.; Velensise, G. Seismogenic sources in the Adriatic Domain. *Mar. Pet. Geol.* **2013**, *42*, 191–213. [CrossRef]

67. Herak, D.; Herak, M.; Brkić, I. Great tremor, sismicity and seismic hazard of wider Dubrovnik area [Velika trešnja, seizmičnost i potresna opasnost na širem Dubrovačkom području]. *Dubrov. Čas. Književ. Znan.* **2017**, *28*, 5–18.

68. Albini, P. *The great 1667 Dalmatia Earthquake: An in-Depth Case Studdy*; Springer: New York, NY, USA, 2015.

69. Herak, M.; Orlić, M.; Kunovec-Varga, M. Did the Makarska earthquake of 1962 generate a tsunami in the central Adriatic archipelago? *J. Geodyn.* **2001**, *31*, 71–86. [CrossRef]

70. Markušić, S.; Herak, D.; Ivančić, I.; Sović, I.; Herak, M.; Prelogović, E. Seismicity of Croatia in the period 1993-1996 and the Ston-Slano earthquake of 1996. *Geofizika* **1998**, *15*, 83–102.

71. Benetatos, C.; Kiratzi, A. Finite-fault slip models for the 15 April 1979 (Mw 7.1) Montenegro earthquake and its strongest aftershock of 24 May 1979 (Mw 6.2). *Tectonophysics* **2006**, *421*, 129–143. [CrossRef]

72. Chilingarian, G.V.; Torabazdeh, J.; Robertson, J.O.; Rieke, H.H.; Mazzullo, S.J. Carbonate Reservoir Characterization: A Geologic-Engineering Analysis. In *Developments in Petroleum Science*; Chilingarian, G.V., Mazzullo, S.J., Rieke, H.H., Eds.; Elsevier: Amsterdam, The Netherlands, 1992.

73. Knutson, C.F.; Bohor, B.F. Reservoir rock behavior under moderate confining pressure. In *Rock Mechanics*; Fairhurst, C., Ed.; Pergamon: New York, NY, USA, 1963; pp. 627–658.

74. Pooladi-Darvish, M.; Moghdam, S.; Xu, D. Multiwell injectivity for storage of CO_2 in aquifers. *Energy Procedia* **2011**, *4*, 4252–4259. [CrossRef]

75. Griffith, C.A. *Physical Characteristics of Caprock Formations used for Geological Storage of CO_2 and the Impact of Uncertainty in Fracture Properties in CO_2 Transport through Fractured Caprocks*; Carnegie Mellon University: Pittsburgh, PA, USA, 2012.

76. Zhou, Q.; Birkholzer, J.T.; Tsang, C.F.; Rutqvist, J. A method for quick assessment of CO_2 storage capacity in closed and semi-Closed saline formations. *Int. J. Greenh. Gas Control* **2008**, *2*, 626–639. [CrossRef]

77. Jelić, K.; Kevrić, I.; Krasić, O. Temperatura i toplinski tok u tlu Hrvatske [Temperature and heat flow in the soil of Croatia]. In Proceedings of the First Croatian Geological Congress, Opatija, Croatia, 18–21 October 1995; pp. 245–249.
78. Vilarrasa, V.; Silva, O.; Carrera, J.; Olivella, S. Liquid CO_2 injection for geological storage in deep saline aquifers. *Int. J. Greenh. Gas Control* **2013**, *14*, 84–96. [CrossRef]
79. Bradshaw, J.; Bachu, S.; Bonijoly, D.; Burruss, R.; Holloway, S.; Christensen, N.P.; Mathiassen, O.M. CO_2 storage capacity estimation: Issues and development of standards. *Int. J. Greenh. Gas Control* **2007**, *1*, 62–68. [CrossRef]

Article

Potential for Mineral Carbonation of CO_2 in Pleistocene Basaltic Rocks in Volos Region (Central Greece)

Nikolaos Koukouzas [1,*], Petros Koutsovitis [2], Pavlos Tyrologou [1], Christos Karkalis [1,3] and Apostolos Arvanitis [4]

[1] Centre for Research and Technology, 15125 Hellas (CERTH), Greece; tyrologou@certh.gr (P.T.); karkalis@certh.gr (C.K.)
[2] Section of Earth Materials, Department of Geology, University of Patras, GR-265 00 Patras, Greece; pkoutsovitis@upatras.gr
[3] Department of Mineralogy and Petrology, Faculty of Geology and Geoenvironment, National and Kapodistrian University of Athens, Zografou, P.C. 15784 Athens, Greece
[4] Hellenic Survey of Geology and Mineral Exploration (HSGME), 13677 Attica, Greece; arvanitis@igme.gr
* Correspondence: koukouzas@certh.gr; Tel.: +30-211-106-9502

Received: 30 August 2019; Accepted: 8 October 2019; Published: 11 October 2019

Abstract: Pleistocene alkaline basaltic lavas crop out in the region of Volos at the localities of Microthives and Porphyrio. Results from detailed petrographic study show porphyritic textures with varying porosity between 15% and 23%. Data from deep and shallow water samples were analysed and belong to the Ca-Mg-Na-HCO_3-Cl and the Ca-Mg-HCO_3 hydrochemical types. Irrigation wells have provided groundwater temperatures reaching up to ~30 °C. Water samples obtained from depths ranging between 170 and 250 m. The enhanced temperature of the groundwater is provided by a recent-inactive magmatic heating source. Comparable temperatures are also recorded in adjacent regions in which basalts of similar composition and age crop out. Estimations based on our findings indicate that basaltic rocks from the region of Volos have the appropriate physicochemical properties for the implementation of a financially feasible CO_2 capture and storage scenario. Their silica-undersaturated alkaline composition, the abundance of Ca-bearing minerals, low alteration grade, and high porosity provide significant advantages for CO_2 mineral carbonation. Preliminary calculations suggest that potential pilot projects at the Microthives and Porphyrio basaltic formations can store 64,800 and 21,600 tons of CO_2, respectively.

Keywords: basalts; carbonation; CO_2 storage; hydrochemistry; regional heat flow

---

1. Introduction

The use of fossil fuels (coal and oil) in the industries has increased the CO_2 emissions in the atmosphere. Anthropogenic CO_2 is a major greenhouse gas that contributes to the change of climate [1,2]. To mitigate the problem of global warming, several technologies have been developed. CO_2 Capture and Storage (CCS) is one of the most advanced technologies that mediates the increase of the CO_2 contents in the atmosphere [3]. Basaltic rocks exhibit appropriate physicochemical properties for the implementation of carbonate mineral precipitation, through interaction of the Ca-Mg-Fe rich minerals with carbonic acid, derived from the dissolution of the injected CO_2 in water [4]. The newly formed minerals mostly consist of calcite, magnesite and siderite [5,6], which provide the potential for long-term and safe storage. Selection of the appropriate type of basalt and region for implementing CO_2 storage techniques via mineral carbonation requires detailed mineralogical and petrophysical (porosity, permeability) studies. The nature of the injected CO_2 affects the integrity and trapping potential of the rock material [7]. CO_2 is present in the supercritical form (sCO_2) at pressure and temperature conditions that correspond to depths greater than 1 km. In such environments, sCO_2 can give rise to various geochemical reactions, causing the dissolution/precipitation of primary and secondary minerals, as well as changes in porosity and permeability properties [8,9]. Successful CCS pilot injection projects have been implemented, including the sites Ferrybridge (UK), Aberthaw Pilot Plant (UK), Puertollano (Spain), Ketzin (Germany), and Hvergerdi (Iceland; CarbFix project) [10].

Alkali basaltic rocks with the potential of CO_2 storage are relatively restricted but widespread throughout mainland Greece [10]. Main areas of basalt appearance are located in the regions of Pindos (NW Greece; [11]), Central and Southern Aegean islands [12–14], Koziakas [15], Othris [16], Evia island [17], and Argolis [18]. The present study focuses on studying the Porphyrio and Microthives alkali basaltic outcrops for their mineralogical potential of CO_2 sequestration. The study areas are located 8 and 12 km south-southwest of the industrial city of Volos, respectively (Figure 1). These volcanic rocks formed along with other adjacent scattered volcanic centers that were active during the Late Pleistocene–Quaternary period, including the islands of Lichades, Achilleio, and Agios Ioannis between the gulfs of Pagasitikos and North Evoikos. Their formation is attributed to back-arc extensional volcanism and affected by the activity of the Northern Anatolian fault [19–21]. They comprise massive lavas and pyroclastic rocks that include basaltic rock fragments and pumice. These volcanic rock formations are located in the Pelagonian Zone and the Eohellenic tectonic nappe [22,23]. The Pelagonian Zone is part of the Internal Hellinides, and it can be distinguished into two metamorphic and non-metamorphic groups, respectively [24,25]. It was over-thrusted by the Eohellenic nappe during the Late Jurassic to Early Cretaceous period [24,26]. In the studied regions, the Pelagonian Zone mostly consists of clastic sedimentary rocks, limestones, and ophiolitic occurrences [22,23]. The Eohellenic tectonic nappe consists mainly of metamorphosed sedimentary rocks, serpentinites, and ophicalcites [22,23], as well as gneissic formations of the Volos Massif [21], composed of gneiss, muscovite, and mica-chlorite schists.

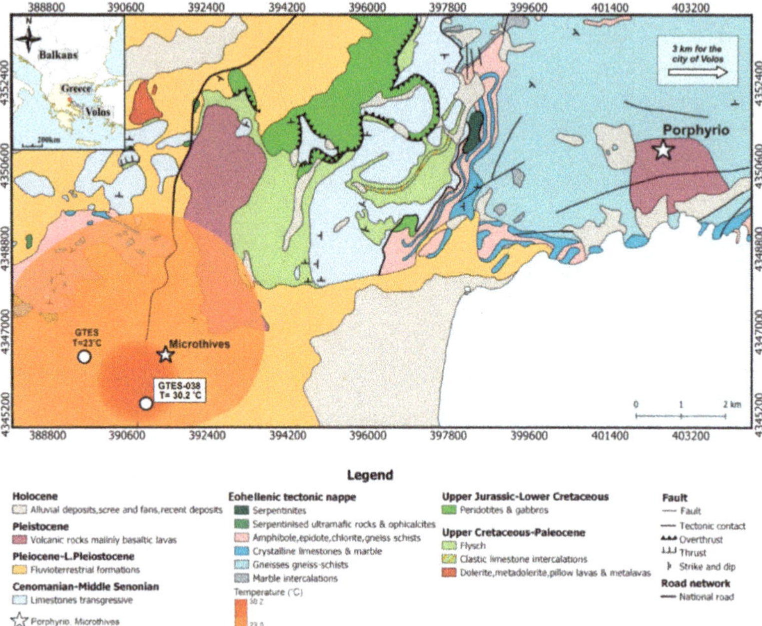

Figure 1. Geological map of Microthives locality and calculated water temperatures, EGSA87.

This study presents new mineralogical, mineral chemistry, and petrographic data of the volcanic rocks from the localities of Microthives and Porphyrio, coupled with hydrochemical data of groundwater samples from irrigation wells, to estimate their potential for the development of geological carbon capture and storage (CCS) [27]. The present study focuses on examining the physicochemical features necessary for applying CCS technologies focusing on: (i) degree of alteration, (ii) nature and geochemistry of the basalts, (iii) presence of Ca-bearing minerals, (iv) porosity (v), indications of enhanced heat calculated in groundwater samples from irrigation wells, and (vi) locality advantages.

2. Materials and Methods

This study includes the investigation of rocks that have been collected from the region of Volos (Central Greece; SE Thessaly), focusing on the volcanic occurrences of the Porphyrio and Microthives localities. Sampling was carried out to select the most appropriate rocks regarding their porosity and mineralogical assemblage. Modal composition of pores was calculated by applying ~500 counts on each thin section. Calculations were cross-correlated with image analysis techniques performed on the same thin sections. More than 50 rock samples were examined through detailed petrographic observations upon polished thin sections with the use of a Zeiss Axioskop-40 (Zeiss, Oberkochen, Germany), equipped with a Jenoptik ProgRes CF Scan microscope camera at the Laboratories of the Center for Research and Technology, Hellas (CERTH). Mineral chemistry analyses were conducted at CERTH using a SEM–EDS JEOL JSM-5600 scanning electron microscope (Jeol, Tokyo, Japan), equipped with an automated energy dispersive analysis system ISIS 300 OXFORD (Oxford Instruments, Abington, UK), with the following operating conditions: 20 kV accelerating voltage, 0.5 nA beam current, 20 s time of measurement, and 5 μm beam diameter. SEM-EDS facility was calibrated to obtain accurate quantitative results using standard reference materials. In order to perform standardised quantitative analyses, thin sections were flat, polished, and carbon coated. XRD analyses were conducted at CERTH using a Philips X'Pert Panalytical X-ray diffractometer (Malvern Panalytical, Malvern, UK), operating with Cu radiation at 40 kV, 30 mA, 0.020 step size, and 1.0 sec step time. For the evaluation of the XRD patterns, DIFFRAC plus EVA software v.11 was deployed (Bruker, MA, USA) based on the ICDD Powder Diffraction File (2006). Physicochemical data (from the Hellenic Survey of Geology and Mineral Exploration (HSGME)) [28], including temperature and pH values, are also provided for two groundwater samples from local irrigation wells (sample GTES-038; 250 m depth, sample GTES-040; 180 m depth).

3. Results

3.1. Petrography and Mineral Chemistry

Basalts from Microthives and Porphyrio localities exhibit porphyritic, vesicular textures. The groundmass is fine-grained holocrystalline, being either trachytic or aphanitic (Figure 2a–f) and often enriched in oxide minerals (ilmenite and magnetite). The porosity, after the examination of an extended number of thin sections of basaltic rock samples ($n > 50$), varies highly from 5 to 40% in the more massive and porous samples, respectively. The vast majority, though, were determined to have porosity that ranges from 15% to 23% (Avg. 18%). Vesicles are in cases filled with secondary calcite.

Their mineralogical assemblage is predominantly composed by prismatic subhedral and rarely euhedral clinopyroxene (15–30%) and olivine (10–20%) phenocrysts (450–700 μm diameter), enclosed within a clinopyroxene and plagioclase-rich groundmass (50–60%). Plagioclase is mostly restricted in the groundmass, appearing in the form of needle- to lath-shaped crystals that compositionally are either bytownite and labradorite ($An_{68.9–71.6}$). Accessory minerals (<5%) include alkali-feldspar, quartz, calcite, amphibole, orthopyroxene, apatite, opaque minerals (ilmenite, titanomagnetite, and magnetite), and pyrite.

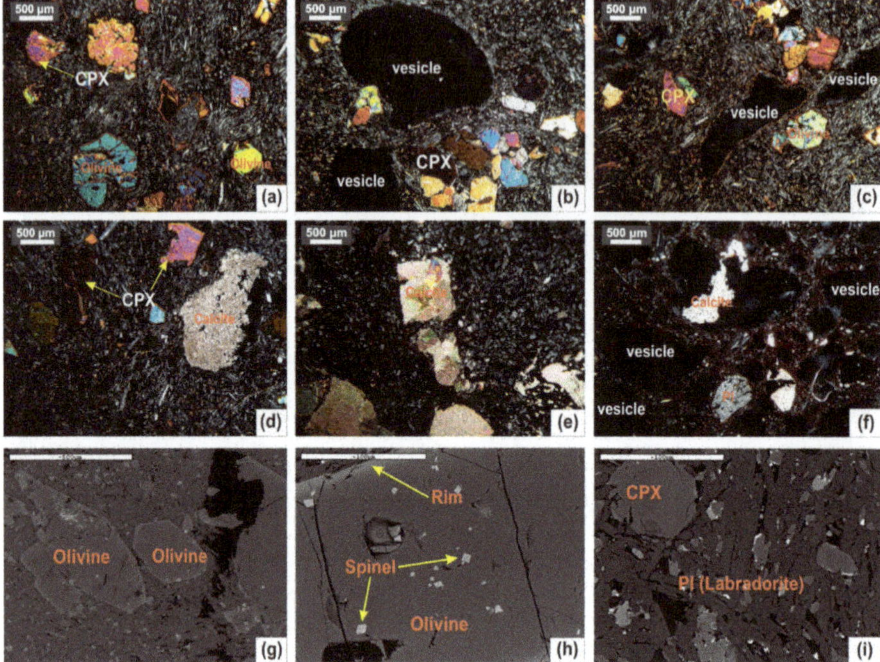

Figure 2. (**a**) Olivine and clinopyroxene phenocrystals in a hypo-crystalline trachytic groundmass mostly consisting of lath-shaped plagioclase but also K-felspar (Sample M3). (**b,c**) Vesicular basaltic lava samples M1 and M8, mainly consisting of clinopyroxene and also olivine phenocrystals, exhibiting glomeroporphyritic textures. It includes a hypo-crystalline trachytic groundmass, as well as vesicular textures. (**d**) Vesicular basaltic lava sample M2, with clinopyroxene and olivine phenocrystals in a trachytic groundmass. Vesicles are occasionally filled with secondary calcite-forming amygdaloidal textures. (**e**) Vesicular basaltic lava sample M5, within a microcrystalline vesicular groundmass, filled with secondary calcite. (**f**) Pyroclastic tuff with a high percentage of vesicles. Groundmass locally aphanitic with rare feldspar phenocrysts. (**g–i**) BSE (Back Scattered Electron) images with olivine and clinopyroxene phenocrysts.

Clinopyroxene is mainly classified as augite and less often as diopside, displaying highly variable TiO_2 and Al_2O_3 contents (0.55–2.94 wt.% and 2.22–7.69 wt.%, respectively). SiO_2 contents range between 44.52 and 51.34 wt.%. Representative compositions of olivine are presented in Table 1. They contain 38.10–40.55 wt.% SiO_2 and variable FeO and MgO contents (10.30–24.90 wt.% and 36.58–48.50 wt.%, respectively). Mg# ranges between 72.78 and 89.36 wt.%.

The mineralogical composition of the studied basaltic rocks was also investigated by powder X-ray diffraction (XRD). In accordance with petrographic observations and mineral chemistry results, the main mineral phases were confirmed with XRD patters, based upon the DIFFRACplus EVA software (version11, Bruker, MA, USA) recommendations. In particular, the peaks at ~51.0° 2θ correspond to the olivine porphyroblasts, whereas clinopyroxene corresponds to peaks at 29.80–30.80° 2θ. The presence of magnetite in small amounts is characterised by small peaks at ~30° 2θ, ~52° 2θ and 62.20–62.80° 2θ. The plagioclase crystals of the basaltic groundmass were recognised by the peaks at ~28.0° 2θ, ~22.0° 2θ, and ~24.30° 2θ.

Table 1. Representative mineral chemistry analyses. (Abbreviations: Ol: olivine, Cpx: clinopyroxene, Plg: plagioclase, K-fs: K-feldspar, Amph: amphibole, Opx: orthopyroxene, Spl: spinel, n: number of analysis, $Mg\# = 100 \times$ molar $MgO/(MgO + FeO_t)$, $Cr\# = 100 \times$ molar $Cr_2O_3/(Cr_2O_3 + Al_2O_3)$.)

Min.	Ol	Ol	Ol	Ol	Cpx	Cpx	Cpx	Cpx	Cpx	Plg	Plg
Sample	M3	M3	M3	M7	M3	M3	M7	M7	M3	M3	M7
n:	7	1	5	2	3	4	1	1	1	2	4
SiO_2	40.55	39.04	38.1	39.61	50.23	51.34	49.96	49.06	44.52	50.34	49.15
TiO_2	0.03	0.04	0.14	0.14	1.27	0.55	1.17	1.58	2.94	0.12	0.14
Al_2O_3	0.02	0.04	0.03	0.06	4.16	2.22	5.22	4.46	7.69	31.6	32.57
FeO	10.3	17.62	24.39	14.14	5.64	5.79	5.37	6.05	8.17	0.46	0.57
MnO	0.16	0.13	0.2	0.13	21.66	21.77	21.58	23.16	23.57	13.84	14.34
MgO	48.5	42.33	36.58	45.48	15.9	17.07	15.32	14.4	11.88	-	-
CaO	0.15	0.38	0.45	0.13	21.66	21.77	21.58	23.16	23.57	13.84	14.34
Na_2O	-	-	-	-	0.5	0.62	0.68	0.66	0.74	3.3	2.92
K_2O	-	-	-	-	0.09	0.03	0.07	0.06	0.74	3.3	2.92
Cr_2O_3	0.03	0.04	0.05	0.04	0.22	0.41	0.38	0.49	0.02	-	-
NiO	0.29	0.13	0.12	0.24	0.03	0.07	0.02	0.05	0.05	-	-
Total	100.02	99.75	100.07	99.94	99.93	100.03	99.9	100.08	99.81	99.88	100.02
Mg#	89.36	81.07	72.78	85.15							

Min.	K-fs	K-fs	Amph	Amph	Opx	Glass
Sample	M3	M5	M3	M3	M3	M7
n:	4	1	1	1	5	2
SiO_2	65.38	64.5	51.59	53.78	53.02	57.09
TiO_2	0.07	0.06	0.47	0.3	0.7	1.65
Al_2O_3	18.75	18.7	23.35	27.68	0.78	17.53
FeO	0.23	0.54	2.34	0.57	17.85	4.27
MnO	-	-	0.04	0.11	0.85	0
MgO	-	-	4.32	0	23.95	2.25
CaO	0.27	0.15	11.66	9.81	2.46	6.84
Na_2O	4.03	3.57	2.5	3.76	0	3.02
K_2O	11.02	12.02	0.49	1.47	0.1	6.91
Cr_2O_3	-	-	0.31	0.02	0.15	0.36
NiO	-	-	0.26	0.07	0.13	0.08
Total	99.75	99.54	97.33	97.55	99.99	100

Min.	Spl	Spl	Spl
Sample	M3	M3	M3
n:	3	1	2
Cr_2O_3	35.56	23.72	30.36
Al_2O_3	31.27	43.51	22.35
TiO_2	0.93	0.5	4.21
FeO	16.82	14.14	33.27
MgO	14.56	16.72	8.91
MnO	0.05	0.09	0.23
NiO	0.19	0.18	0.13
SiO_2	0.43	0.7	0.39
Total	99.81	99.56	99.82
Mg#	62.08	67.56	37.6
Cr#	43.26	26.77	47.67

3.2. Rock Classification and Geochemistry

Volcanic rocks from the region of Volos correspond to small scattered outcrops with an age range from 0.5 to 3.4 Ma [21,29]. Their formation was attributed to Pleistocene back-arc extension in the Aegean Sea [19,30,31]. Based upon the total alkali–silica (TAS) diagram (Figure 3a), the extensional-related volcanic rocks from the Volos plot formed within the basaltic trachyandesite and trachyandesite fields. Pleistocene basalts from the adjacent regions of Kamena Vourla, Lichades islands, Psathoura, and Achilleio also plot in the same compositional fields (Figure 3a). Chondrite-normalised REE patterns of the Volos basaltic rocks (Figure 3b) are highly enriched in LREE ($La/Yb_{CN} = 0.34$–0.44) and also present notable negative Eu anomalies ($Eu_{CN}/Eu^* = 0.73$–0.80), with the later implying plagioclase fractionation. These are additionally characterised from trace element ratios that account for a clear OIB (Ocean Island Basalt)-signature: $Zr/Nb = 4.66$–19.82, $La/Nb = 0.75$–10.38, and $Ba/Th = 39.4$–100.95 [32]. Basalts from the aforementioned adjacent regions exhibit lower LREE enrichments (Figure 3b), indicating higher degrees of partial melting and/or differentiation processes.

The Pleistocene extensional-related basaltic rocks from Volos and the adjacent regions differ from other recent age (Pliocene–present) volcanic rocks in Greece. The latter are subduction-related volcanics from the South Aegean (Methana [26,33], Nisyros [34], and Santorini [2,35,36]), associated with the subduction of the African plate beneath Eurasia [37–39]. These compositionally correspond to subalkaline basalts and andesites (Figure 3a), which possess significantly lower LREE and also higher HREE contents (Figure 3b). From this comparison, it is evident that the basaltic rocks from Volos are among the very few alkaline basaltic rocks of recent age that are compositionally suitable for considering mineral carbonation of CO_2.

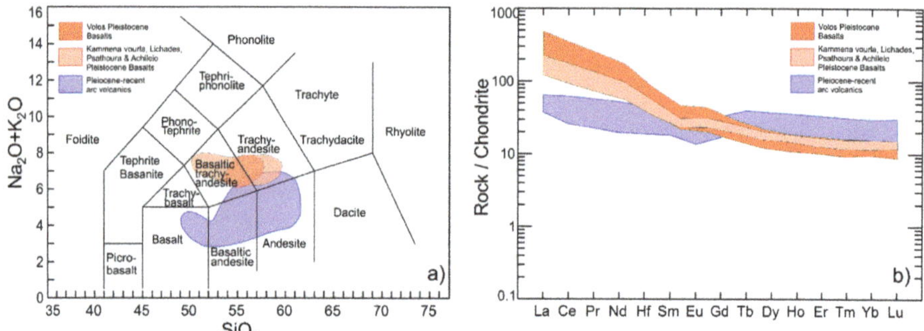

Figure 3. (**a**) Total alkali–silica (TAS), $Na_2O + K_2O$ vs. SiO_2 [40], and (**b**) chondrite-normalised REE patterns [41,42] of volcanic rocks from Volos, Kamena Vourla, Psathoura, Achilleio, Lichades [21], Methana [26,33], Nisyros island [34], and Santorini island [2,35,36].

3.3. Water Chemistry and Temperature Data

Geothermal data from the Almyros–Microthives basin [28] indicate that the north part of the basin is characterised by Pleistocene volcanic activity. Deep groundwater (sample GTES-038; >250 m depth) exhibits a temperature of 30.2 °C and a pH of 7.30, whereas shallow groundwater (sample GTES-040; probably 170–180 m depth) presents a temperature of 23.0 °C and a pH of 7.40. The elevated water temperatures appear in the vicinity of the basaltic dominated areas. Based on the Castany classification [43], the analysed groundwaters belong to hypothermal, neutral-to-alkaline types. Their total dissolved solids (TDS) content is 660 mg/L (Table 2). TDS calculation was based on the cations and anions sum, including HCO_3^- ($0.49 \times (HCO_3^-)$) and B. The total hardness values are 309 and 363 mg/L for the deep and shallow groundwaters, respectively. Non-carbonated hardness values are 22 and 0 mg/L for the deep and shallow groundwater samples, respectively.

Table 2. Hydrochemical analyses of groundwater samples from Microthives locality [28]. T (°C); conductivity (µS/cm); concentrations (mg/L); total dissolved solids (TDS) (mg/L).

Sample	T	pH	Cond.	TDS	Ca	Mg	Na	K	CO_3	HCO_3	Cl	SO_4	NO_3	SiO_2
GTES-038	30.2	7.70	989.0	660	55.30	41.70	80.50	2.25	0.00	287.0	165	24.60	0.00	34.90
GTES-040	23.0	7.60	693.0	460	38.70	65.10	17.60	1.30	0.00	437.0	19.50	9.80	3.72	56.0

From the Hem [44] and Sawyer and McCarty [45] classifications, the analysed groundwater samples are classified as very hard. Deep and shallow groundwater samples belong to the Ca-Mg-Na-HCO_3-Cl and the Mg-HCO_3 hydrochemical types, respectively (Figure 4).

Hydrogeochemical comparisons between the two water samples from Microthives and Aegean seawater [47], are discussed below (see Discussion paragraph).

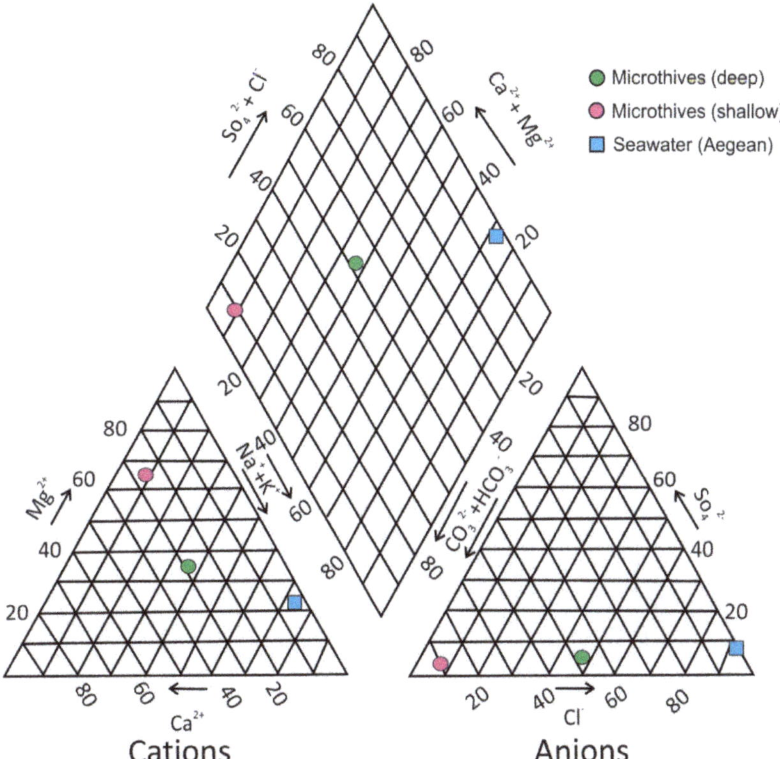

Figure 4. Piper diagram [46] for the water samples of Microthives and Aegean regions.

4. Discussion

4.1. Mineral Reactions

Despite the high availability of basalts on the Earth's surface [48–50], only few basaltic types have the appropriate petrophysical and chemical properties [3,6,50–52] to serve as host rocks suitable for CO_2 mineral carbonation. The basaltic rocks from the localities of Porphyrio and Microthives possess proper mineralogical, chemical, and textural features to apply CO_2 sequestration techniques. These features include the high abundance of Ca-bearing minerals, as well as their silica-undersaturated alkaline composition and high porosity. Mineral chemistry reactions that result from this interaction can be modeled based on data provided from this study.

The physicochemical properties of water strongly affect the formation of carbonate minerals during the interaction of basalts with the CO_2 injected fluids. Carbonation with the presence of water can lead to higher amounts of sequestered CO_2 compared to the dry carbonation processes [53,54]. The dissolution of CO_2 in water further affects the liquid reactivity, due to the high amounts of the released H^+ [3,4,6,50]. The concentration of Mg in water can affect the crystal growth of calcite, whereas, at high temperatures, Mg can precipitate in the form of solid mineral phases [3]. In addition, the water saturation reflected from the water/rock ratio (W/R) determines the dissolution of basaltic rocks, which is higher in CO_2 saturated solutions compared to the undersaturated ones (W/R: 10/1 and 2/1 respectively; atmosphere [3]).

The underground water analysed from the region of Microthives is classified as neutral to alkaline (pH = 7.30). Dissolution of CO_2 in water produces carbonic acid. The gradual mixing of the alkaline groundwater with the acidic injection fluids starts up with the entrance of the injected fluid into the storage formation and ends up with the entrance of the fluid in the monitoring wells [55]. After the mixing process, the formation fluids become more acidic, presenting lower pH values [55]. This acidic pH is characterised by a high concentration of dissolved inorganic carbon (DIC), making the water reactive with the basaltic rocks, due to the high H^+ contents [50].

Addition of CO_2 is expected to lower the pH of water due to the release of H^+ ions, according to the following chemical reactions [50]:

$$CO_2 + H_2O = H_2CO_3 \tag{1}$$

$$H_2CO_3 = HCO_3^- + H^+ \tag{2}$$

Basaltic rocks are rich in Ca, Mg, and Fe, providing the potential for CO_2 mineralisation in the form of carbonate minerals. The released H^+ ions (chemical reaction-2) increase the reactivity of water, resulting in dissolution of the primary basalt minerals and the precipitation of Ca^{2+}, Mg^{2+}, and Fe^{2+} in the form of carbonate minerals [4,50], according to the following chemical reaction:

$$(Ca,Mg,Fe)^{2+} + H_2CO_3 \rightarrow (Ca,Mg,Fe)CO_3 + 2H^+ \tag{3}$$

Carbonation of olivine is described by mineral reaction-4. The high MgO contents (MgO: 36.58–48.50%) of the studied olivine phenocrystals will produce high amounts of magnesite. This reaction is developed with slow rates in the natural systems, suggesting that the carbonation of olivine must be enhanced by a large-scale storage method for CO_2 mineralisation [56,57]. The formation of hydromagnesite is favoured at low temperatures and can be described by reaction-5 [58]. At low temperatures ($T < 60\ °C$), indirect precipitation of magnesite can occur via hydromagnesite dehydration [58]. This process is described through the two-way reaction-6 [59].

$$Mg_2SiO_{4(s)} + 2CO_2 \rightarrow 2MgCO_{3(s)} + SiO_{2(s)} \tag{4}$$

$$5Mg_2SiO_{4(s)} + 8CO_{2\ (gas)} + 2H_2O_{(liq.)} \rightarrow 2\ (4MgCO_3 \cdot Mg(OH)_2 \cdot 4H_2O)_{(s)} + 5H_4SiO_{4(aq)} \tag{5}$$

$$4Mg(CO_3) \cdot Mg(OH)_2 \cdot 4H_2O \leftrightarrow 4MgCO_3 + Mg(OH)_2 + 4H_2O \tag{6}$$

The studied olivine crystals of Microthives and Porphyrio basalts are mostly composed by forsterite. In that case, the olivine carbonation can be further described by the following mineral reaction:

$$Mg_2SiO_{4(s)} + 4H^+_{(aq)} \rightarrow 2Mg^{2+} + SiO_{2(s)} + 2H_2O \tag{7}$$

Dissolution of clinopyroxene is developed according to the following mineral reaction:

$$MgCaSi_2O_6 + 4H^+ \rightarrow Mg^{2+} + Ca^{2+} + 2H_2O + 2SiO_{2(aq)} \tag{8}$$

The release of Ca^{2+} cations is described by the dissolution of anorthite rich plagioclase according to the chemical reaction-9:

$$CaAl_2Si_2O_8 + 8H^+ \rightarrow Ca^{2+} + 2Al^{3+} + 4H_2O + 2SiO_{2(aq)} \tag{9}$$

Orthopyroxene appears in the form of accessory enstatite crystals. Dissolution of enstatite is described by mineral reaction-10 [60]:

$$MgSiO_3 + 2H^+ \rightarrow Mg^{2+} + SiO_2 + H_2O \tag{10}$$

Precipitation of calcite (reaction-11 [50]) during hydrothermal alteration of basaltic rocks is strongly associated with temperature and depth. The Ca^{2+} required for calcite precipitation is mostly derived from the primary calc–silicate minerals and the glass matrix of the basaltic protolith. These minerals mostly include clinopyroxene (CaO: 21.58–23.57%), plagioclase (CaO: 13.84–14.34%), and amphiboles (CaO: 9.81–11.66%).

$$Ca^{2+} + CO_2 + H_2O \rightarrow CaCO_3 + 2H^+ \tag{11}$$

Calcite formation is not favoured at temperatures higher than 290 °C [61] and depths between 200 and 400 m [62]. The time required for carbonate minerals precipitation strongly depends on the abundance of divalent cations, the fluid *P–T*, the liquid chemistry, the CO_2 saturation, and the pore surface area [4]. Diagrams of basalt dissolution rates vs. pH (Figure 5a,b) were designed using data from the literature [50,63–65]. The aforementioned diagrams indicate that during the mixing of the background water with the CO_2 injected fluids, the pH decrease enhances the dissolution rate of forsterite (Mg-olivine) and augite (clinopyroxene). The crystalline basalts in Microthives and Porphyrio localities are mostly composed by clinopyroxene and olivine phenocrystals within a glass-rich matrix. Clinopyroxene is mostly classified as augite, whereas olivine is characterised by relatively high MgO contents (Table 1). The glass-rich basalts are characterised by relatively constant dissolution rates for pH values between 4 and 7.3, whereas their dissolution rates increase with further pH decrease. For pH values lower than 4, the dissolution rate will be rapidly increased and become similar with that of forsterite. This indicates that during the initial stages of the CO_2 injection, more glass-rich basalts will be dissolved with lower rates compared to the crystalline ones. During the interaction of Microthives and Porphyrio basalts with CO_2 injected fluids, clinopyroxene-olivine porphyroblasts [3,6,50,66] and the anorthite-glass rich matrix will be dissolved with similar rates against their pH [3,6,50,67]. The aforementioned results indicate that clinopyroxene and olivine porphyroblasts will be the first mineral phases to be dissolved during the CO_2 injection.

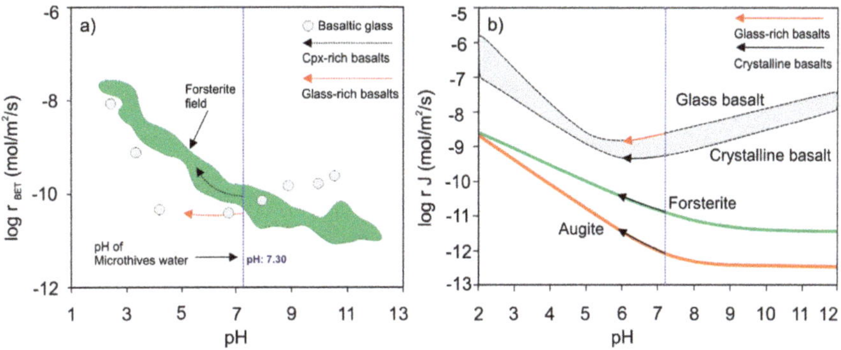

Figure 5. (a) Modified diagram of the dissolution rate of the forsterite (*T*: 25 °C; [64]) and basaltic glass (*T*: 30 °C) [65] vs. pH. The dissolution rate is normalised to the BET surface area of the dissolving mineral and glass grains. (b) Modified diagram [68] of the dissolution fluxes (mol m^{-2} s^{-1}) at *T*: 25 °C of crystalline and glassy basalts. Forsterite and augite dissolution rates taken from [63].

Based on the experimental results from Gislason et al. [50] (Figure 5a), the dissolution rate of olivine increases from 10^{-10} to $10^{-8.5}$ (mol/m^2/s) for pH values ranging from 7.3 (Microthives water pH) to 1.5. These results are in agreement with the experiments of Palandri and Kharaka [63] that indicate a comparable increase of forsterite dissolution rate from $10^{-10.5}$ to $10^{-8.5}$ (mol/m^2/s) for pH values ranging from 7.3 to 2. Dissolution of augite vs. pH follows similar trends, ranging from 10^{-12} to $10^{-8.5}$ (mol/m^2/s) for the same pH range with augite (Figure 5b). Experimental results suggest that dissolution rate of diopside will be three orders of magnitude slower compared to other silicate minerals, such as olivine at 25 °C [69,70]. Data provided by Palandri and Kharaka [63] point to an increase of the plagioclase dissolution rate from $10^{-11.5}$ to 10^{-10} (mol/m^2/s) for pH values ranging from 7 to 2.

Dissolution rate of CO_2 in water strongly depends on the water temperature, the partial pressure of CO_2, and the salinity of the medium [50]. Carbonation rate of secondary minerals is strongly associated with the acidic or alkaline nature of the water. Experimental results at a temperature of 25 °C under acidic and neutral conditions show that the carbonation rate of calcite, magnesite, and siderite ranges are $10^{-0.3}$–$10^{-5.81}$ mol/(m^2/s) [63], $10^{-6.38}$–$10^{-9.34}$ mol/(m^2/s) [63], and $10^{-3.74}$–$10^{-8.90}$ mol/(m^2/s) [71], respectively. This further suggests that precipitation of carbonate minerals is mostly favoured during the late stages of the CO_2 injection, characterised by lower pH values compared to the formation groundwater (pH: 7.3 for Microthives groundwater). Availability of divalent cations is the main limiting step during CO_2 mineralisation in basalts [5]. Basalts of 8% average MgO correspond to 0.087 CO_2 g/g basalt converted to magnesite [3]. Abundance of Mg-olivine in the studied basalts from the regions of Microthives and Porphyrio support their high potential for magnesite precipitation. The relatively low alteration grade of the studied basalts provides additional advantages regarding their potential for mineral carbonation, due to their higher carbonation grades compared to the altered ones [72].

4.2. Groundwater Chemistry from Irrigation Wells

Chemical comparison between the groundwater samples from the studied localities indicate that shallow groundwater is more depleted in Cl$^-$ and Na$^+$ compared to the deep one. Cl$^-$ is a relatively mobile element that does not incorporate into secondary minerals after being released from the dissolution of the basaltic protolith [51]. The different Cl$^-$ contents between the analysed borehole groundwater samples are attributed to their distance from seawater [73], origin, and circulation. In particular, the sampling site of the deep groundwater is located closer to the Aegean Sea compared to the shallow one. The aforementioned difference is attributed to the mixture between the deep groundwater from Microthives and seawater and confirmed by the ionic ratios (Table 3), coupled with the Langelier–Ludwig [74] and Piper plots [46] (Figure 6a,b). The shallow groundwater presents similar Na–Cl contents compared to those of the groundwater from Iceland (Figure 6a). This suggests that both water samples were not affected by mixture processes with seawater.

Table 3. Ionic ratios of water samples from the region of Microthives (mg/L) [28].

Ionic Ratio	Mg/Ca	Na/K	Na/Cl	SO$_4^{2-}$/Cl$^-$	HCO^{3-}/Cl$^-$	Cl$^-$/F	Cl$^-$/Br	Cl/Li	Na/Li
Deep sample	0.75	35.78	0.49	0.15	1.74	369	77.0	6600	3220
Shallow sample	1.68	13.54	0.90	0.50	22.41	57.0	57.0	2167	62,583

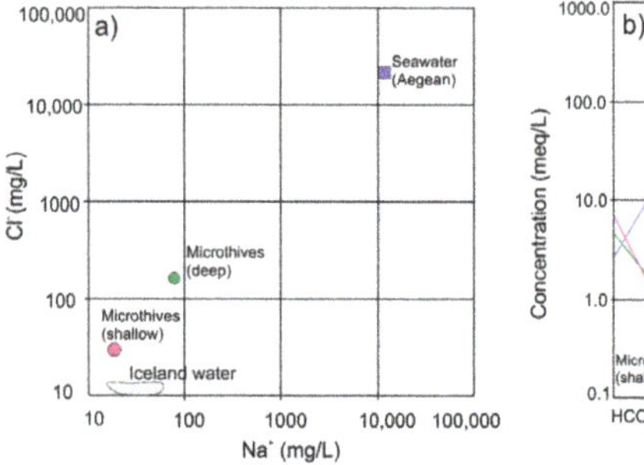

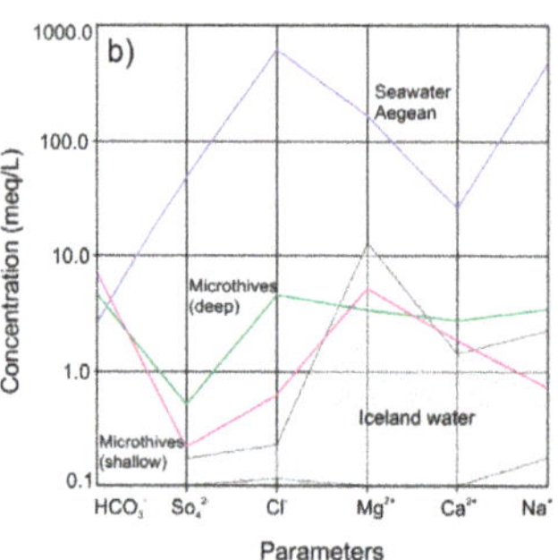

Figure 6. (a) Ludwig–Langelier [74] diagram for the water sample of Microthives and the seawater sample from the Aegean Sea. (b) Schoeller diagram [75] for the Microthives water sample and the Aegean seawater [28,51].

The groundwater composition is also affected by the water–rock interaction during the circulation of rainwater through basalts [51].

4.3. Indications of Enhanced Heat

Volcanic rocks in the Porphyrio and Microthives localities were developed in an extensional back-arc geotectonic setting affected by the activity of the Northern Anatolian Fault [19–21]. This back-arc extension was evolved with respect to the active volcanic arc of the South Aegean [21] and gave rise to the generation of Late Pleistocene basalts. The age of the magmatism is very crucial to the determination of the heat source [76]. In particular, the active magmatism is indicative of elevated heat sources, compared to the inactive or extinct magmatism that are associated with heat remnants and/or additional radioactive-heat [76,77].

The back-arc extension developed in Porphyrio and Microthives localities indicates a recent-inactive enhanced heat, characterised by the development of relatively shallow and young magma chambers [76]. These systems are mainly developed in divergent plate margins [76], usually including two distinct zones of different T and pH conditions [76,78–80]. In particular, the outflow zone has a lower T and neutral-to-alkaline pH groundwaters [80] compared to the upflow, which is more acidic [81]. In the cases of inactive magmatic sources, the produced heat is strongly associated with crystallised, but still-cooling, magmatic bodies [76]. According to this model, the main heat source is provided by the Pleistocene magmatic melts, whereas the presence of faults further enhances the recharge of meteoric waters [76]. A similar heating source was developed in Hungary as a result of a Miocene extension that caused a high thermal attenuation of the lithosphere [82,83]. In the current study, the elevated water temperatures were mainly observed close to the basaltic rock occurrences (T = 30.2 and 23.0 °C for GTES-038 and GTES-040 groundwater irrigations wells, respectively). Enhanced water temperatures are also recorded in the adjacent regions of Kamena Vourla (Central Greece; East Thessaly) and Lichades islands (Central Greece; North Evoikos Gulf), corresponding to 25–41.3 °C and 41 °C, respectively [84]. These regions are related to scattered volcanic centers, which were active during the Late Pleistocene–Quaternary period, similarly to those of the Porphyrio and Microthives localities. The above data suggest that this activity is associated with the extensional back-arc tectonic setting. Based upon the geological mapping of the Porphyrio and Microthives localities, coupled with the elevated temperatures of the groundwater samples (irrigation wells GTES-038 and GTES-040), the elevated temperatures in the studied region are strongly associated with the basalt occurrence underneath the Neogene alluvial sediments. The water pH in the Microthives locality (pH: 7.20–7.30; [28]) and the adjacent regions of Kamena Vourla and Aidipsos (pH: 6.28 and 6.80 respectively; [85]) indicate that these waters are derived from the outflow zone, which is characterised by a neutral-to-alkaline pH [81].

4.4. A Case Scenario for Mineral Carbonation in the Micothives Basalts

The Microthives and Porphyrio basaltic occurrences are potential sites for CO_2 storage [86]. The research area is located 10 km away from the industrial zone of Volos, a significant source of CO_2 emissions. The case study scenario presented in this study is based on the results of the CarbFix project [50,55]. Carbon storage through injection of water dissolved CO_2, is a potential applicable CCS scenario for the volcanic rocks of Microthives and Porphyrio localities.

The CarbFix method does not require the presence of a cap rock, since the dissolved CO_2 is not buoyant [55]. The process of CO_2 dissolution during the injection into basaltic rocks [55] of the Microthives and Porphyrio localities, can be enhanced due to the higher porosity that these rocks present (average porosity: 18%). There is a strong association between the porosity and permeability of the basaltic rocks and their alteration grade [55]. Thus, the younger and less-altered basalts are more appropriate for CO_2 storage compared to the older types. Basaltic rocks of the current study belong to the relatively young extensional Pleistocene volcanic activity, and, hence, they were not affected by a high alteration grade. The pH value in the groundwater from the Microthives locality is

7.3, which is similar to that of the target zone prior to the injection of CO_2 in the CarbFix project [10,50]. After the initial pH decrease, due to the mixing of the groundwater fluids with the hydrous injected CO_2, the reaction paths of basaltic glass at 25 °C [51] indicate that pH becomes more alkaline due to the P_{CO_2} decrease during the water–rock interaction.

Regarding the diffusivity of water-dissolved CO_2 in basalts, we provide preliminary calculations with Equation (12) [87]:

$$D = D_0 \cdot \varphi^{m} \tag{12}$$

where D is diffusion coefficient; D_0 is diffusion of the water dissolved CO_2, ($1.92 \cdot 10^{-5}$ cm^2/s [88]; φ is porosity of basalt (0.18–0.23 for our studied basalts); and m is Archie's coefficient, (m: 2.3 [89]). By applying the aforementioned equation, it is estimated the diffusion coefficient ranges from $38 \times \cdot 10^{-8}$ cm^2/s to 65×10^{-8} cm^2/s, respectively.

One of the major parameters in the CarbFix project is the substantial quantities of water for the dissolution of CO_2 during injection [50]. Basaltic outcrops of Microthives–Porphyrio localities are in proximity with the Aegean Sea, giving the potential for high storage capacities, due to the unlimited seawater supply [6,50,90,91].

We provide preliminary calculations that estimate the CO_2 that could be stored in the frames of pilot projects for the two basalt locations of Microthives and Porphyrio. For this purpose, we apply the function below:

$$\text{Storage Capacity} = \sum(V \times \varphi \times \rho \times \varepsilon) \tag{13}$$

where V is the volume of the basaltic outcrop; φ is the average porosity = 18%; ρ is the specific gravity of the sCO_2; and ε is the sCO_2 storage ratio.

The Microthives basaltic outcrop has a surface of ~8 km^2; therefore, the potential pilot project can be realised at an estimated volume of 300 m (length) × 200 m (width) × 300 m (depth) = 18×10^6 m^3. Taking into consideration the average porosity of basalts from our studied site (18%), the specific gravity of the scCO$_2$ (400 kg/m^3; at 10 MPa and 50 °C [92,93]), and the scCO$_2$ storage ratio of basalts (5% [94]), the Microthives basaltic outcrop could store an amount of 64,800 tons of CO_2. The Porphyrio basaltic formation is smaller, and, therefore, by assuming an estimated volume of 200 m (length) × 100 m (width) × 300 m (depth) = 6×10^6 m^3, it could store a calculated amount of 21,600 tons of CO_2. The maximum capability of CO_2 storage, considering the highest porosity of the studied suite (23%), corresponds to 82,800 tons and 27,600 tons for the Microthives and Porphyrio basalts, respectively. The size of these outcrops could serve for storage of much larger amounts of CO_2 after deployment of pilot tests.

The charged water can significantly increase the energy consumed for the CO_2 injection. From the CarbFix experience, it is evident that the cost of storage and transport corresponds to $17/ton of dissolved CO_2 injected [50,95], which doubles the cost compared to the classic CO_2 injection in sedimentary basins [50,96]. This cost is balanced by the lower monitoring after the injection period, due to the non-buoyant nature of the mineralised CO_2 [50]. The development of a cost-effective scenario is further enhanced by the relatively short distance of the basaltic dominated areas (~10 km) from the industrial area of Volos, reducing the cost of transport.

5. Conclusions

Pleistocene volcanic rocks are present in the region of Volos (Central Greece) and in the specific localities of Microthives and Porphyrio. They are classified as basaltic and trachyandesitic lavas and were formed due to back-arc extension of the Aegean Sea. Their geochemical affinities suggest that these are alkaline basalts of OIB affinity. Results from detailed petrographic examination show that their porosity ranges between 5% and 40% with vesicles, which, in a few rock samples, partly host calcite. The vast majority of the studied samples exhibit porosity that ranges between 15% and 23%.

A recent-inactive magmatic heating source present in the Microthives basaltic vicinity, affected the groundwater temperature regime. Enhanced groundwater temperatures are also recorded in adjacent regions with basalts of similar composition and age, suggesting that this activity is associated with the extensional back-arc tectonic setting. Deep and shallow groundwater samples are classified as Ca-Mg-Na-HCO_3-Cl and the Mg-HCO_3 hydrochemical types respectively. Measured groundwater temperatures from irrigation wells, at depths between 170 and 250 m, reach up to ~30 °C.

Basalts from the region of Volos have the necessary appropriate physicochemical features to be considered as potential sites for implementing carbon capture and storage (CCS) technologies due to (i) low alteration grade, (ii) silica-undersaturated alkaline composition, (iii) presence of Ca-bearing minerals, (iv) high porosity, and (v) indications of enhanced heat. The proximity of the basaltic rocks to the sea gives the opportunity for exploitation of the unlimited water sources during the CO_2 injection. Furthermore, these outcrops are in close distance to the industrial area of Volos, providing the potential for the development of a financially feasible scenario. Preliminary calculations suggest that potential pilot projects at the Microthives and Porphyrio basaltic formations can store 82,800 and 27,600 tons of maximum CO_2, respectively, although their size could serve for storage of much larger amounts of CO_2 after deployment of pilot tests. Further and detailed petrological, petrophysical, geochemical, hydrochemical, geothermal, and financial research studies are needed prior to deployment of pilot tests in the region of Volos.

Author Contributions: All authors actively participated in a balanced manner at all stages of the research presented in this paper. This involved participation of all authors in sample collection in the field, performing laboratory work and manuscript writing.

Funding: This research received no funding.

Acknowledgments: We would like to express our sincerest thanks to the Reviewers and the Editor for their constructive comments and useful suggestions that have substantially helped to improve this paper.

Conflicts of Interest: The authors declare no conflict of interest.

References

1. IPCC. Intergovernmental Panel. In *Climate Change 2013 the Physical Science Basis*; Cambridge University Press: New York, NY, USA, 2013.
2. Davis, W.J. The Relationship between Atmospheric Carbon Dioxide Concentration and Global Temperature for the Last 425 Million Years. *Climate* **2017**, *5*, 76–110.
3. Rosenbauer, R.J.; Thomas, B.; Bischoff, J.L.; Palandri, J. Carbon sequestration via reaction with basaltic rocks: Geochemical modeling and experimental results. *Geochim. Cosmochim. Acta* **2012**, *89*, 116–133. [CrossRef]
4. Adam, L.; Otheim, T.; van Wijk, K.; Batzle, M.; McLing, T.L.; Podgorney, R.K. CO_2 Sequestration in Basalt: Carbonate Mineralization and Fluid Substitution. In *SEG Technical Program Expanded Abstracts 2011*; Society of Exploration Geophysicists: Tulsa, OK, USA, 2011; pp. 2108–2113.
5. Oelkers, E.H.; Gislason, S.R.; Matter, J. Mineral carbonation of CO_2. *Elements* **2008**, *4*, 333–337. [CrossRef]
6. Gislason, S.R.; Wolff-Boenisch, D.; Stefansson, A.; Oelkers, E.H.; Gunnlaugsson, E.; Sigurdardottir, H.; Sigfusson, B.; Broecker, W.S.; Matter, J.M.; Stute, M. Mineral sequestration of carbon dioxide in basalt: A pre-injection overview of the CarbFix project. *Int. J. Greenh. Gas Control* **2010**, *4*, 537–545. [CrossRef]
7. Koukouzas, N.; Kypritidou, Z.; Purser, G.; Rochelle, C.A.; Vasilatos, C.; Tsoukalas, N. Assessment of the impact of CO_2 storage in sandstone formations by experimental studies and geochemical modeling: The case of the Mesohellenic Trough, NW Greece. *Int. J. Greenh. Gas Control* **2018**, *71*, 116–132. [CrossRef]
8. Black, J.R.; Carroll, S.A.; Haese, R.R. Rates of mineral dissolution under CO_2 storage conditions. *Chem. Geol.* **2015**, *399*, 134–144.
9. Gaus, I. Role and impact of CO_2—Rock interactions during CO_2 storage in sedimentary rocks. *Int. J. Greenh. Gas Control* **2010**, *4*, 73–89. [CrossRef]
10. Kelektsoglou, K. Carbon capture and storage: A review of mineral storage of CO_2 in Greece. *Sustainability* **2018**, *10*, 4400. [CrossRef]
11. Saccani, E.; Photiades, A. Mid-ocean ridge and supra–subduction affinities in the Pindos ophiolites (Greece): Implications for magma genesis in a forearc setting. *Lithos* **2004**, *73*, 229–253. [CrossRef]

12. Stouraiti, C.; Pantziris, I.; Vasilatos, C.; Kanellopoulos, C.; Mitropoulos, P.; Pomonis, P.; Moritz, R.; Chiaradia, M. Ophiolitic remnants from the upper and intermediate structural unit of the Attic-Cycladic Crystalline Belt (Aegean, Greece): Fingerprinting geochemical affinities of magmatic precursors. *Geosciences* **2017**, *7*, 14. [CrossRef]

13. Mortazavi, M.; Sparks, R. Origin of rhyolite and rhyodacite lavas and associated mafic inclusions of Cape Akrotiri, Santorini: The role of wet basalt in generating calcalkaline silicic magmas. *Contrib. Miner. Pet.* **2004**, *146*, 397–413. [CrossRef]

14. Bachmann, O.; Deering, C.D.; Ruprecht, J.S.; Huber, C.; Skopelitis, A.; Schnyder, C. Evolution of silicic magmas in the Kos-Nisyros volcanic center, Greece: A petrological cycle associated with caldera collapse. *Contrib. Miner. Pet.* **2012**, *163*, 151–166. [CrossRef]

15. Pomonis, P.; Tsikouras, V.; Hatzipanagiotou, K. Geological evolution of the Koziakas ophiolitic complex (W. Thessaly, Greece). *Ofioliti* **2005**, *30*, 77–86.

16. Koutsovitis, P. Gabbroic rocks in ophiolitic occurrences from East Othris, Greece: Petrogenetic processes and geotectonic environment implications. *Miner. Pet.* **2012**, *104*, 249–265. [CrossRef]

17. Pe-Piper, G.; Panagos, A.G. Geochemical characteristics of the Triassic volcanic rocks of Evia: Petrogenetic and tectonic implications. *Ofioliti* **1989**, *14*, 33–50.

18. Saccani, E.; Beccaluva, L.; Photiades, A.; Zeda, O. Petrogenesis and tectono-magmatic significance of basalts and mantle peridotites from the Albanian—Greek ophiolites and sub-ophiolitic mélanges. New constraints for the Triassic—Jurassic evolution of the Neo-Tethys in the Dinaride sector. *Lithos* **2011**, *124*, 227–242. [CrossRef]

19. Fytikas, M.; Innocenti, F.; Manetti, P.; Mazzuoli, R.; Peccerillo, A.; Villari, L. Tertiary to Quaternary evolution of volcanism in the Aegean region. In *The Geological Evolution of the Eastern Mediterranean*; Dixon, J.E., Robertson, A.H.F., Eds.; Geological Society Special Publications: London, UK, 1984; Volume 17, pp. 687–699.

20. Pe–Piper, G.; Piper, D.J. *Neogene Backarc Volcanism of the Aegean: New Insights into the Relationship between Magmatism and Tectonics*; Special Papers; Geological Society of America: Boulder, CO, USA, 2007; p. 17.

21. Innocenti, F.; Agostini, S.; Doglioni, C.; Manetti, P.; Tonarini, S. Geodynamic evolution of the Aegean: Constraints from the Plio-Pleistocene volcanism of the Volos–Evia area. *J. Geol. Soc. Lond.* **2010**, *167*, 475–489. [CrossRef]

22. Katsikatsos, G.; Mylonakis, J.; Vidakis, M.; Hecht, J.; Papadheas, G.; Dimou, E.; Papazeti, E.; Skourtsi-Koroneou, V.; Hadjicostanti–Tsalachouri, I.; Karamicahlou–Kavali, A.; et al. *Geological Map of Greece, Volos Sheet*; Institute of Geology and Mineral Exploration of Greece: Athens, Greece, 1978.

23. Katsikatsos, G.; Mylonakis, J.; Triantaphyllis, E.; Papadheas, G.; Psonis, C.; Staila-Monopolis, S.; Skourtsi–Coroneou, V.; Hadjicostani-Tsalachouri, I.; Georgiou-Nikolaidi, A.; Benaki-Dragoumanou, E.; et al. *Geological Map of Greece, Velestino Sheet*; Institute of Geology and Mineral Exploration of Greece: Athens, Greece, 1978.

24. Katsikatsos, G.H. *Geology of Greece*; University Publications: Patra, Greece, 1992; p. 451.

25. Papanikolaou, D. Timing of tectonic emplacement of the ophiolites and terrane paleogeography in the Hellenides. *Lithos* **2009**, *108*, 262–280. [CrossRef]

26. Pe–Piper, G.; Piper, D.J.W. *The Igneous Rocks of Greece: The Anatomy of an Orogen*; Gebrüder Borntraegen: Berlin/Stuttgart, Germany, 2002.

27. Rigopoulos, I.; Vasiliades, M.A.; Ioannou, I.; Efstathiou, A.M.; Godelitsas, A.; Kyratsi, T. Enhancing the rate of ex situ mineral carbonation in dunites. *Adv. Powder Technol.* **2016**, *27*, 360–371. [CrossRef]

28. Vakalopoulos, P.; Efthimiopoulos, T.; Arvanitis, A.; Xenakis, M.; Vougioukalakis, G.; Galamakis, D.; Gkagka, M.; Lachana, G.; Kanellopoulos, C.; Fragkogiannis, G.; et al. Geothermal exploration in Eastern Thessaly. I.G.M.E., NSRF (National Strategic Reference Framework) 2007–2013/Operational Programme "Competitiveness and Entrepreneurship"/Project "Geothermal energy exploration in selected areas, in order to reduce energy dependency and environmental impact. Assessment of hot groundwater and geothermal resources (GEOTHEN)". Athens, Greece, 2016; p. 238. Available online: http://igme.gr/index.php/proionta-ypiresies/23-erga/erga-espa-2007-2013 (accessed on 15 May 2016).

29. Pe–Piper, G.; Piper, D.J.W. Plio–Pleistocene ages of high-potassium volcanism in the northwestern part of the Hellenic arc. *Tschermaks Mineral. Petrogr. Mitt.* **1979**, *26*, 163–165. [CrossRef]

30. Ninkovich, D.; Hays, J.D. Mediterranean island arcs and origin of high potash volcanoes. *Earth Planet. Sci. Lett.* **1972**, *16*, 331–345. [CrossRef]

31. Pe, G.; Panagos, A. Comparative geochemistry of the Northern Eubeoecos lavas. *Bull. Geol. Soc. Greece* **1976**, *9*, 95–130, (In Greek with English Abstract).

32. Weaver, B.L. The origin of ocean island basalt end-member compositions: Trace element and isotopic constraints. *Earth Planet. Sci. Lett.* **1991**, *104*, 381–397. [CrossRef]

33. Pe, G.G. Petrology and geochemistry of volcanic rocks of Aegina, Greece. *Bull. Volcanol.* **1973**, *37*, 491–514. [CrossRef]

34. Wyers, G.P.; Barton, M.D. Geochemistry of a transitional Ne-trachybasalt-Q-trachyte lava series from Patmos (Dodecanesos), Greece: Further evidence for fractionation, mixing and assimilation. *Contrib. Miner. Pet.* **1987**, *97*, 279–291. [CrossRef]

35. Huijsmans, J.P.P.; Barton, M.D.; Salters, V.J.M. Geochemistry and evolution of the calc-alkaline volcanic complex of Santorini, Aegean Sea. *J. Volcanol. Geotherm. Res.* **1988**, *34*, 283–306. [CrossRef]

36. Druitt, T.H.; Mellors, R.A.; Pyle, D.M.; Sparks, R.S.J. Explosive volcanism on Santorini, Greece. *Geol. Mag.* **1989**, *126*, 95–126. [CrossRef]

37. Nicholls, I.A. Petrology of Santorini volcanic rocks. *J. Pet.* **1971**, *12*, 67–119. [CrossRef]

38. Nicholls, I.A. Santorini volcano, Greece. Tectonic and petrochemical relationships with volcanics of the Aegean region. *Tectonophysics* **1971**, *11*, 377–385. [CrossRef]

39. Pe, G.G.; Piper, D.J.W. Vulcanism at subduction zones; The Aegean area. *Bull. Geol. Soc. Greece* **1972**, *9*, 113–144.

40. Le Maitre, R.W. *Igneous Rocks—A Classification and Glossary of Terms*; Cambridge University Press: Cambridge, UK, 2002.

41. Mcdonough, W.F.; Sun, S.S. The composition of the Earth. *Chem. Geol.* **1995**, *120*, 223–253. [CrossRef]

42. Sun, S.S.; McDonough, W.F. Chemical and isotopic systematics of oceanic basalts; implications for mantle composition and processes. In *Magmatism in the Ocean Basins*; Saunders, A.D., Norry, M.J., Eds.; The Geological Society by Blackwell Scientific Publications: London, UK, 1989; Volume 42, pp. 313–345.

43. Castany, G. *Traité Pratique des Eaux Souterraines*; Dunod: Paris, France, 1963.

44. Hem, J.D. *Study and Interpretation of the Chemical Characteristics of Natural Water*, 2nd ed.; (Water Supply Paper); U.S. Government Printing Office: Washington, WA, USA, 1970.

45. Sawyer, C.N.; McCarty, P.L. *Chemistry and Sanitary Engineers*; McGrawHall: New York, NY, USA, 1966.

46. Piper, A.M. A graphic procedure in the geochemical interpretation of water analyses. *Trans. Am. Geophys. Union.* **1944**, *25*, 914–928. [CrossRef]

47. Xenakis, M.; Kavouridis, T.; Vrellis, G.; Vakalopoulos, P. *Exploration and Identification of Geothermal Fields on Chios Island*; Institute of Geology and Mineral Exploration of Greece (I.G.M.E.), Department of Geothermal Energy and Thermometallic Waters: Athens, Greece, 2007.

48. Dessert, C.; Dupré, B.; Gaillardet, J.; François, L.M.; Allegre, C.J. Basalt weathering laws and the impact of basalt weathering on the global carbon cycle. *Chem. Geol.* **2003**, *202*, 257–273. [CrossRef]

49. McGrail, B.P.; Schaef, H.T.; Ho, A.M.; Chien, Y.J.; Dooley, J.J.; Davidson, C.L. Potential for carbon dioxide sequestration in flood basalts. *J.Geophys. Res. Solid Earth* **2006**, *111*. [CrossRef]

50. Gislason, S.R.; Broecker, W.S.; Gunnlaugsson, E.; Snæbjörnsdóttir, S.; Mesfin, K.G.; Alfredsson, H.A.; Aradottir, E.S.; Sigfusson, B.; Gunnarsson, I.; Stute, M.; et al. Rapid solubility and mineral storage of CO_2 in basalt. *Energy Procedia* **2014**, *63*, 4561–4574. [CrossRef]

51. Alfredsson, H.A.; Oelkers, E.H.; Hardarsson, B.S.; Franzson, H.; Gunnlaugsson, E.; Gislason, S.R. The geology and water chemistry of the Hellisheidi, SW-Iceland carbon storage site. *Int. J. Greenh. Gas Control* **2013**, *12*, 399–418. [CrossRef]

52. Schaef, H.T.; McGrail, B.P.; Owen, A.T. Carbonate mineralization of volcanic province basalts. *Int. J. Greenh. Gas Control* **2010**, *4*, 249–261. [CrossRef]

53. Rigopoulos, I.; Petallidou, K.C.; Vasiliades, M.A.; Delimitis, A.; Ioannou, I.; Efstathiou, A.M.; Kyratsi, T. Carbon dioxide storage in olivine basalts: Effect of ball milling process. *Powder Technol.* **2015**, *273*, 220–229. [CrossRef]

54. Gerdemann, S.K.; O'Connor, W.K.; Dahlin, D.C.; Panner, L.R.; Rush, H. Ex situ aqueous mineral carbonation. *Environ. Sci. Technol.* **2007**, *41*, 2587–2593. [CrossRef]

55. Snæbjörnsdóttir, S.Ó.; Wiese, F.; Fridriksson, T.; Ármansson, H.; Einarsson, G.M.; Gislason, S.R. CO_2 storage potential of basaltic rocks in Iceland and the oceanic ridges. *Energy Procedia* **2014**, *63*, 4585–4600.

56. Teir, S.; Kuusik, R.; Fogelholm, C.-J.; Zevenhoven, R. Production of magnesium carbonates from serpentinite for long–term storage of CO_2. *Int. J. Miner. Process.* **2007**, *85*, 1–15. [CrossRef]

57. Koukouzas, N.; Ziogou, F.; Gemeni, V. Preliminary assessment of CO_2 geological storage opportunities in Greece. *Int. J. Greenh. Gas Control* **2009**, *3*, 502–513. [CrossRef]

58. Hähnchen, M.; Prigiobbe, V.; Storti, G.; Seward, T.M.; Mazzotti, M. Dissolution kinetics of fosteritic olivine at 90–150 degrees C including effects of the presence of CO_2. *Geochim. Cosmochim. Acta* **2006**, *70*, 4403–4416. [CrossRef]

59. Zhang, P.C.; Anderson, H.L.; Kelly, J.W.; Krumhansl, J.L.; Papenguth, H.W. *Kinetics and Mechanisms of Formation of Magnesite from Hydromagnesite in Brine United States*; Department of Energy, US Department of Energy, Sandia National Labs.: Albuquerque, NM, USA; Livermore, CA, USA, 2000; pp. 1–26.

60. Clark, D.E.; Gunnarsson, I.; Aradóttir, E.S.; Arnarson, M.Þ.; Þorgeirsson, Þ.A.; Sigurðardóttir, S.S.; Sigfússon, B.; Snæbjörnsdóttir, S.Ó.; Oelkers, E.H.; Gíslason, S.R. The chemistry and potential reactivity of the CO_2-H_2S charged injected waters at the basaltic CarbFix2 site, Iceland. *Energy Procedia* **2018**, *146*, 121–128. [CrossRef]

61. Franzson, H. Reservoir Geology of the Nesjavellir High-Temperature Field in SW-Iceland. In Proceedings of the 19th Annual PNOC—EDC Geothermal Conference, Manila, Philippines, 5–6 March 1998; pp. 13–20.

62. Tómasson, J.; Kristmannsdóttir, H. High temperature alteration minerals and thermal brines, Reykjanes, Iceland. *Contrib. Miner. Pet.* **1972**, *36*, 132–134. [CrossRef]

63. Palandri, J.L.; Kharaka, Y.K. *A Compilation of Rate Parameters of Water-Mineral Interaction Kinetics for Application to Geochemical Modeling*; U.S. Geological Survey Open File Report (of 2004-1068); National Energy Technology Laboratory—United States Department of Energy: Menlo Park, CA, USA, 2004; p. 71.

64. Pokrovsky, O.S.; Schott, J. Forsterite surface composition in aqueous solutions: A combined potentiometric, electrokinetic, and spectroscopic approach. *Geochim. Cosmochim. Acta* **2000**, *64*, 3299–3312. [CrossRef]

65. Oelkers, E.; Gislason, S. The mechanism, rates and consequences of basaltic glass dissolution: I. An experimental study of the dissolution rates of basaltic glass as a function of aqueous Al, Si and oxalic acid concentration at 25 °C and pH = 3 and 11. *Geochim. Cosmochim. Acta* **2001**, *65*, 3671–3681. [CrossRef]

66. Knauss, K.; Nguyen, S.; Weed, H. Diopside dissolution kinetics as a function of pH, CO_2, temperature, and time. *Geochim. Cosmochim. Acta* **1993**, *57*, 285–294. [CrossRef]

67. Gudbrandsson, S.; Wolff-Boenisch, D.; Gislason, S.; Oelkers, E. Experimental determination of plagioclase dissolution rates as a function of its composition and pH at 22 °C. *Geochim. Cosmochim. Acta* **2014**, *139*, 154–172. [CrossRef]

68. Pollyea, R.M.; Rimstidt, J.D. Rate equations for modeling carbon dioxide sequestration in basalt. *Appl. Geochem.* **2017**, *81*, 53–62. [CrossRef]

69. De Paolo, D.J.; Cole, D.R.; Navrotsky, A.; Bourg, I.C. Geochemistry of Geologic CO_2 Sequestration. *Mineral. Soc. Am.* **2013**, *77*, 536.

70. Kaszuba, J.; Yardley, B.; Andreani, M. Experimental perspectives of mineral dissolution and precipitation due to carbon dioxide-water-rock interactions. *Rev. Mineral. Geochem.* **2013**, *77*, 153–188. [CrossRef]

71. Golubev, S.V.; Bénézeth, P.; Schott, J.; Dandurand, J.L.; Castillo, A. Siderite dissolution kinetics in acidic aqueous solutions from 25 to 100 °C and 0 to 50 atm p CO_2. *Chem. Geol.* **2009**, *265*, 13–19. [CrossRef]

72. Liu, D.; Agarwal, R.; Li, Y.; Yang, S. Reactive transport modeling of mineral carbonation in unaltered and altered basalts during CO_2 sequestration. *Int. J. Greenh. Gas Control* **2019**, *85*, 109–120. [CrossRef]

73. Sigurdsson, F.; Einarsson, K. Groundwater resources of Iceland—Availability and demand. *Jökull* **1988**, *38*, 35–53.

74. Langelier, W.F.; Ludwig, H.F. Graphical Methods for Indicating the Mineral Character of Natural Waters. *J. Am. Water Works Assn.* **1942**, *34*, 335–352. [CrossRef]

75. Schoeller, H. Geochimie deseaux souterraines. *Rev. Inst. Franc. Pètrole. Paris* **1955**, *10*, 181–213.

76. World Bank. *Best Practices Guide for Geothermal Exploration*; Bochum University of Applied Sciences: Bochum, Germany, 2014; p. 196.

77. McCoy-West, A.; Milicich, S.; Robinson, T.; Bignall, G.; Harvey, C.C. Geothermal resources in the Pacific Islands: The potential of power generation to benefit indigenous communities. In Proceedings of the 36th Workshop on Geothermal Reservoir Engineering, Stanford University, Stanford, CA, USA, 31 January–2 February 2011.

78. Williams, C.F.; Reed, M.J.; Anderson, A.F. Updating the classifi cation of geothermal resources. In Proceedings of the hirty-Sixth Workshop on Geothermal Reservoir Engineering Stanford University, Stanford, CA, USA, 31 January–2 February 2011; p. 7.

79. Giggenbach, W.F. Magma degassing and mineral deposition in hydrothermal systems along convergent plate boundaries. *Econ. Geol. Soc. Explor. Geol. Bull.* **1992**, *97*, 1927–1944.

80. Hochstein, M.P. Assessment and modelling of geothermal reservoirs (small utilization schemes). *Geothermics* **1988**, *17*, 35. [CrossRef]

81. Moeck, I.S. Catalog of geothermal play types based on geologic controls. *Renew. Sustain. Energ. Rev.* **2014**, *37*, 867–882. [CrossRef]

82. Horváth, F. Towards a mechanical model for the formation of the Pannonian basin. *Tectonophysics* **1993**, *226*, 333–357. [CrossRef]

83. Békési, E.; Lenkey, L.; Limberger, J.; Porkoláb, K.; Balázs, A.; Bonté, D.; Vrijlandt, M.; Horváth, F.; Cloetingh, S.; van Wees, J.-D. Subsurface temperature model of the Hungarian part of the Pannonian Basin. *Glob. Planet. Chang.* **2018**, *171*, 48–64. [CrossRef]

84. Andritsos, N.; Arvanitis, A.; Papachristou, M.; Fytikas, M.; Dalambakis, P. Geothermal Activities in Greece During 2005–2009. In Proceedings of the World Geothermal Congress 2010, Bali, Indonesia, 25–29 April 2010.

85. Baba, A.; Bundschuh, J.; Chandrasekharam, D. *Geothermal Systems and Energy Resources: Turkey and Greece*, 1st ed.; CRC Press/Balkema: Boca Raton, FL, USA, 2014; p. 291.

86. Koutsovitis, P.; Koukouzas, N.; Magganas, A. Carbon Storage Potential in Pleistocene Volcanic Rocks of the Magnesia Area (Central Greece). In Proceedings of the 19th EGU General Assembly EGU2017, Vienna, Austria, 23–28 April 2017; p. 3942.

87. Navarre–Sitchler, A.; Steefel, C.I.; Yang, L.; Tomutsa, L.; Brantley, S.L. Evolution of porosity and diffusivity associated with chemical weathering of a basalt clast. *J. Geophys. Res. Earth Surf.* **2009**, *114*. [CrossRef]

88. Cussler, E.L. *Diffusion: Mass Transfer in Fluid Systems*; Cambridge University Press: New York, NY, USA, 1997; p. 580.

89. Peng, S.; Hu, Q.; Hamamoto, S. Diffusivity of rocks: Gas diffusion measurements and correlation to porosity and pore size distribution. *Water Resour. Res.* **2012**, *48*. [CrossRef]

90. Wolff–Boenisch, D.; Wenau, S.; Gislason, S.R.; Oelkers, E.H. Dissolution of basalts and peridotite in seawater, in the presence of ligands, and CO_2: Implications for mineral sequestration of carbon dioxide. *Geochim. Cosmochim. Acta* **2011**, *75*, 5510–5525. [CrossRef]

91. Goldberg, D.S.; Takahashi, T.; Slagle, A.L. Carbon dioxide sequestration in deep-sea basalt. *Proc. Natl. Acad. Sci. USA* **2008**, *105*, 9920–9925. [CrossRef] [PubMed]

92. Span, R.; Wagner, W. A new equation of state for carbon dioxide covering the fluid region from the triple-point temperature to 1100 K at pressures up to 800 MPa. *J. Phys. Chem. Ref. Data* **1996**, *25*, 1509–1596. [CrossRef]

93. Spycher, N.; Pruess, K. CO_2-H_2O mixtures in the geological sequestration of CO_2. II. Partitioning in chloride brines at 12–100 °C and up to 600 bar. *Geochim. Cosmochim. Acta* **2005**, *69*, 3309–3320. [CrossRef]

94. Gislason, S.; Oelkers, E. Carbon Storage in Basalt. *Science* **2014**, *344*, 373–374. [CrossRef]

95. Ragnheidardottir, E.; Sigurdardottir, H.; Kristjansdottir, H.; Harvey, W. Opportunities and challenges for CarbFix: An evaluation of capacities and costs for the pilot scale mineralization sequestration project at Hellisheidi, Iceland and beyond. *Int. J. Greenh. Gas Control* **2011**, *5*, 1065–1072. [CrossRef]

96. Global CCS Institute. *Economic Assessment of Carbon Capture and Storage Technologies*; Global CCS Institute: Canberra, Australia, 2011.

Article

Using Reservoir Geology and Petrographic Observations to Improve CO$_2$ Mineralization Estimates: Examples from the Johansen Formation, North Sea, Norway

Anja Sundal * and Helge Hellevang

Department of Geosciences, University of Oslo, 0371 Oslo, Norway; helge.hellevang@geo.uio.no
* Correspondence: anja.sundal@geo.uio.no; Tel.: +47-22-85-66-52

Received: 9 September 2019; Accepted: 24 October 2019; Published: 31 October 2019

Abstract: Reservoir characterization specific to CO$_2$ storage is challenging due to the dynamic interplay of physical and chemical trapping mechanisms. The mineralization potential for CO$_2$ in a given siliciclastic sandstone aquifer is controlled by the mineralogy, the total reactive surface areas, and the prevailing reservoir conditions. Grain size, morphologies and mineral assemblages vary according to sedimentary facies and diagenetic imprint. The proposed workflow highlights how the input values for reactive mineral surface areas used in geochemical modelling may be parameterized as part of geological reservoir characterization. The key issue is to separate minerals both with respect to phase chemistry and morphology (i.e., grain size, shape, and occurrence), and focus on main reactants for sensitivity studies and total storage potentials. The Johansen Formation is the main reservoir unit in the new full-value chain CO$_2$ capture and storage (CCS) prospect in Norway, which was licenced for the storage of CO$_2$ as of 2019. The simulations show how reaction potentials vary in different sedimentary facies and for different mineral occurrences. Mineralization potentials are higher in fine-grained facies, where plagioclase and chlorite are the main cation donors for carbonatization. Reactivity decreases with higher relative fractions of ooidal clay and lithic fragments.

Keywords: CCS; CO$_2$ storage; mineralization; carbonatization; mineral trapping; mineral sequestration; Johansen Formation; North Sea; sedimentary facies

---

1. Introduction

Saline aquifers hold the largest potential for geological CO$_2$ storage considering total volume, economic and environmental factors [1]. CO$_2$ storage is considered one important measure for the imminent reduction of greenhouse gas emissions and climate change mitigation [2]. Most suitable reservoir candidates, pilot projects, and commercial operations utilize siliciclastic deeply buried sandstones [3,4]. In evaluating the suitability of saline aquifers for CO$_2$ storage, geological characterization is of crucial importance in estimating the reservoir property distribution and reactivity under prevailing reservoir conditions. Sedimentary facies and burial diagenesis control the petrophysical properties and mineralogical composition of the reservoir host rock, and to some extent the chemistry of pore water. These factors must be specified when evaluating the relative effect of various trapping mechanisms for CO$_2$ (i.e., structural, residual, solubility, ionic, and mineral trapping [5]). The physical and chemical immobilization of CO$_2$ are important controls in risk assessments.

Predictions of the CO$_2$ trapping potential of a storage reservoir over hundreds to thousands of years requires a sound understanding of the geochemical reactions that will come about when CO$_2$ is injected and the thermodynamic system is perturbed [6]. Such predictions ideally require detailed knowledge about the mineralogy, formation water chemistry, mineral surface reactivities, and reaction rates. These data are then used as input in the geochemical batch or reactive transport numerical simulations (e.g., [7–12]). This is, however, not trivial for several reasons. First, there is no simple way to accurately estimate reactive surface areas of the various reactive mineral phases without careful sediment analyses and theoretical models to relate reactive and total surface areas [13–16]. This may lead to corresponding orders-of-magnitude uncertainties in the rates of CO_2 mineral trapping [17]. Second, the most commonly used rate models, i.e., based on transition state theory (TST), have been suggested to largely overestimate the growth rates of secondary carbonates at low temperatures and in the shorter time scales (<100–1000 years) [10,17,18]. Third, data on the mineralogy may in many cases be available only as crude XRD data, without details on the individual mineral morphologies, grain size, sediment maturity, etc.

The relative importance of the various trapping mechanisms for injected CO_2 in aquifers has been discussed ever since Gunther and co-workers published their geochemical simulations on solubility, ionic, and mineral trapping in the early nineties [19,20]. This relates especially to how fast these reactions are, and if they will impose porosity/permeability changes. This has implications if true complex multiphase reactive flow simulations are needed, or if flow and reactions can be partly separated. Most commonly, reservoir simulations of CO_2 storage only include the dissolved CO_2 in contact with separate phase CO_2, and disregard the mineral-formation water reactions due to slow reaction rates. Furthermore, the heterogeneity of reservoirs with respect to mineralogy and grain size has seldom been taken into account (e.g., [7,16,21–23]). However, some mineral phases and occurrences do seem to react and contribute to mineralization in shorter time scales (100's of years) (e.g., [10,24]) and are thus valid for consideration in sensitivity studies of storage reservoir performance.

We show how to more accurately estimate input parameter values for reactive mineral surface areas, as used in the geochemical modelling of long-term mineralization potential for CO_2. Reservoir models can be improved by upscaling from pore- and grain-scale to sedimentary facies distributions with the associated reactive mineral characteristics [25]. A general workflow is outlined, which can be applied to improve facies and mineral specific estimates of reactive surface areas and mineralization potential for CO_2 in sandstone aquifers.

Case Study: The Johansen Formation, North Sea (NORWAY)

Simulation examples with input from the Johansen Formation are provided. The Johansen Formation is part of the Northern Lights full scale storage prospect offshore Norway (Figure 1), which is highly relevant at this time due to imminent drilling and plans for CO_2 injection [26]. The first formal license for injecting and storing CO_2 as part of full-value chain carbon capture and storage (CCS) was approved by Norwegian authorities as of 2019 (Exploitation Licence EL001, by the Norwegian Petroleum Directorate) [26].

The Johansen Formation (Dunlin Group) is a sandstone of early Jurassic age [27]. This prospective reservoir is located offshore of the city of Bergen on the western Norwegian coast (Figure 1). It displays thicknesses in the order of 100–180 m and is located at burial depths of 2–3 km. The saline aquifer is in parts underlying the operating Troll Gas Field in the North, and as a premise for storing CO_2, there is to be no risk of interference with on-going production [4]. Thus, the potential injection area considered in evaluations of storage potential for CO_2 is located approximately 20 km south of Troll, at top formation depths in the order of 3 km. The Cook Formation is likely to be in contact with the Johansen Formation and provide as a secondary reservoir unit. The main sealing unit is the Drake Formation mudstone [4,28,29].

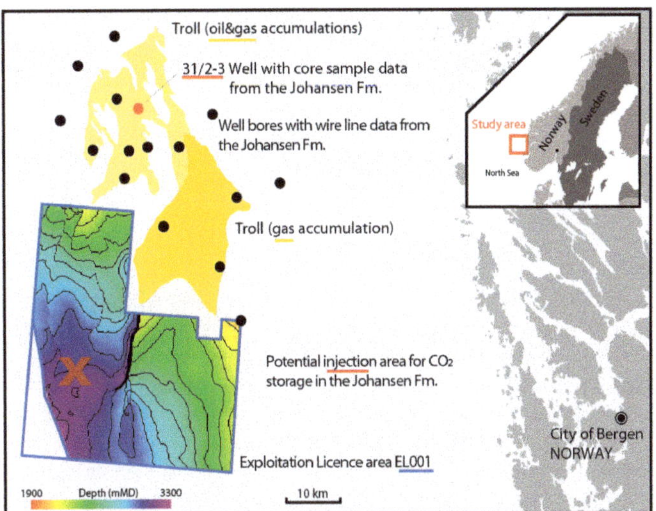

Figure 1. The Johansen Formation is a prospective CO_2 storage reservoir offshore of Norway, located at burial depths of ca. 2100–3200 m. The operating hydrocarbon field "Troll" (yellow) is located north of the licensed CO_2 injection area "EL001". The cored well 31/2-3 is marked in red, and additional wells with wire line data from the Johansen Formation are marked as black dots. There are no well data available from EL001 as of yet, while an appraisal well is planned [26]. Source data are available at factmaps.npd.no, with suggested injection area and depth maps as shown in [29].

The Johansen Formation is interpreted as a progradational to retrogradational sequence of shallow marine sandy deposits sourced from the east [28,29]. The depositional environment in the licenced injection area (Figure 1) has been interpreted to comprise lower to upper shoreface deposits based on seismic data and extrapolation (across distance and depth) of well data from the Troll area [28]. However, an accurate facies description of sandstone in the injection area is not feasible until an appraisal and/or injection well is drilled and sample material becomes available. The shallow marine facies and mineral assemblages of the Johansen Formation appear analogous to several other CO_2 reservoir candidates on the Norwegian Shelf, e.g., the Sognefjord, Fensfjord, Krossfjord, Cook, and Gassum formations [29–35].

2. Estimating CO_2 Mineralization Potential (I): Model Parameterization

In CO_2 storage, the chemical characteristics of the sediment are of particular importance with respect to estimating mineralization potential. Thorough, descriptive petrographic studies using optical- and scanning electron-microscopy (SEM) methods in addition to quantitative bulk mineralogy analysis such as X-ray diffraction data (XRD) are necessary to characterize the reservoir rock with respect to reactivity.

2.1. Qualitative and Quantitative Reservoir Mineralogy

It is useful to define reactivity and make separate geochemical categories within the sedimentological framework. Changes in grain size and mineralogy (phase and occurrence) are particularly important. Bed stacks, or para-sequences (sensu Van Wagoner et al. [36,37]), may serve as a scale of reservoir subdivision, depicting depositional trends; e.g., upwards coarsening or fining trends in grain sizes, indicating changes in depositional regime with time. In the case of the Johansen Formation, reservoir grade sandstones recognised in wells are subdivided in lower shoreface (very fine-grained) and upper shoreface (medium-grained) deposits, interbedded with mudstones and/or carbonate cemented layers [28,29] (Figure 2).

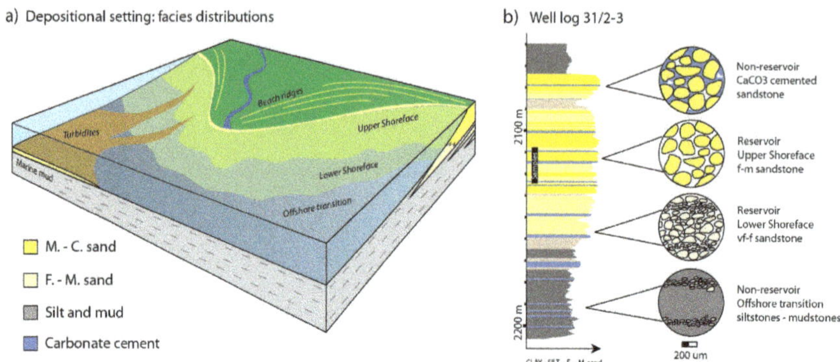

Figure 2. (**a**) Simplified sedimentary facies distributions for the Johansen Formation. Mineralization potential for the reservoir intervals are given for upper and lower shoreface sandstones; (**b**) an interpreted lithological log (vertical section through the sandy Johansen Formation with the muddy over- and underlying Amundsen Formation) from well 31/2-3: based on wire line log data (available at npd.factpages.no), cuttings, and a short cored section (2116–2134 m) from which rock samples were collected. The succession consists of prograding and aggrading parasequences of upper and lower shoreface deposits, with mudstones representing flooding events. Carbonate cemented sandstone layers may form within or on the top of beds due to dissolution and re-precipitation of calcareous material (e.g., shells). These layers are close to impermeable, and provide barriers to fluid flow in otherwise permeable reservoir sandstone. Generally, micro-scale observations from the different facies settings show that total grain surface area in contact with pore water (white void) increases with decreasing grain size.

Though desirable, geological cores through the entire reservoir zone from saline aquifers are rarely available. Usually data from shorter core sections or sidewall cores must be interpolated with respect to vertical and lateral facies changes. In the case of the Johansen Formation, available core data are collected from a well tens of kilometres away from potential injection areas, and at shallower burial depths (Figure 1). Thin sections provide means for 2D porosity estimations, grain size, and mineral content (vol. %). Porosity and permeability plug test data are available from side wall cores [28].

One of the most common means for mineralogical quantification is X-ray diffraction (XRD) (e.g., the Rietveld method), as it is inexpensive, fast, and requires little sample material. The method may be crude or specific with respect to mineral phases, depending on the effort and knowledge put into interpretation of the results and treatment of sample material [38]. Analyses of grain size specific fractions are more suited for reactivity estimates—e.g., clay separation in fluid suspension.

Identification of main cation donors for mineralization in a given reservoir can be performed using the bulk mineralogy. Chlorite is a major constituent in the clay fraction of the Johansen Formation [29], and geochemical studies find chlorite to be a significant cation donor (i.e., Fe2+ and Mg2+ supply through rapid dissolution) in CO_2 carbonatization [9,10,18,39]. Feldspars also provide a significant reactant, as plagioclase (albite and oligoclase) dissolve within relatively short time-scales (100 s of years), contributing Na^+ and Ca^{2+} to solution [10,18,19]. This study will focus on the characterization of chlorites and feldspars, while the same kind of analyses should be undertaken in case of other or more reactive constituents (e.g., mafic minerals).

2.1.1. Characterization of Chlorites

Chlorite is a phyllosilicate mineral, with Fe-rich chamosite $(Fe,Mg)_5Al(Si_3Al)O_{10}(OH)_8$ and Mg-rich clinochlore $(Mg,Fe)_5Al(Si_3Al)_{10}(OH)_8$ as common varieties. Detrital chlorites derived from mafic volcanic or metamorphic terrains are commonly Mg-rich clinochlores, whereas chlorites sourced from peralkaline granites tend to generate Fe-rich chamosites [40]. Diagenetic chlorite is a significant constituent in many siliciclastic reservoirs in the Norwegian North Sea [33]. Autigenic chlorite may form by recrystallization of precursor clay minerals during early burial; e.g., from smectite (<70 °C) [41,42] or from berthierine (90 °C) [33,34,43]. Chlorite may also form as an alteration product from degradation of mafic minerals (e.g., biotite, pyroxene, amphibole). In addition to provenance and detrital mineralogy, depositional environment also exerts a control on chlorite occurrence. As summarized in a literature review by Maast [44] (and references therein), Fe-rich chlorite coating is associated with sediments deposited in marine environments, near river mouths and under tropical conditions, whereas Mg-rich chlorite coating is commonly found in continental sediments, deposited under arid- to semi-arid conditions. Chemical speciation with respect to Fe/Mg ratios is important in the selection of suitable kinetic parameters for geochemical simulations, as chamosite and clinochlore display different reaction potentials. Chamosites with Fe/(Fe+Mg) values between 0.57 and 0.91 are the most common in studied North Sea reservoirs [45]. Most geochemical studies implement kinetic data from [46], which only provide kinetic constants for Mg-rich clinochlore. It seems that also recent investigations into chlorite kinetics focus mainly on clinochlore (e.g., [47–49]). The effect of varying chemical composition of chlorite on dissolution rates is uncertain, as no thorough studies have been performed in this realm. It has been claimed to have little effect [50]. However, some experimental studies indicate significantly higher rate constants for Fe-chlorite [33]. However, the precipitation rates for siderite ($FeCO_3$), magnesite ($MgCO_3$), and Fe-Mg-Ca solid solutions are not the same, which provides another argument for differentiation. It is likely that more kinetic data will become available and include more detailed solid solution speciation in the future. As part of the geological characterization, XRD-spectra may be modelled for estimation of element ratios, as shown for a typical Fe-rich chlorite (chamosite) from the Johansen Formation (Figure 3a).

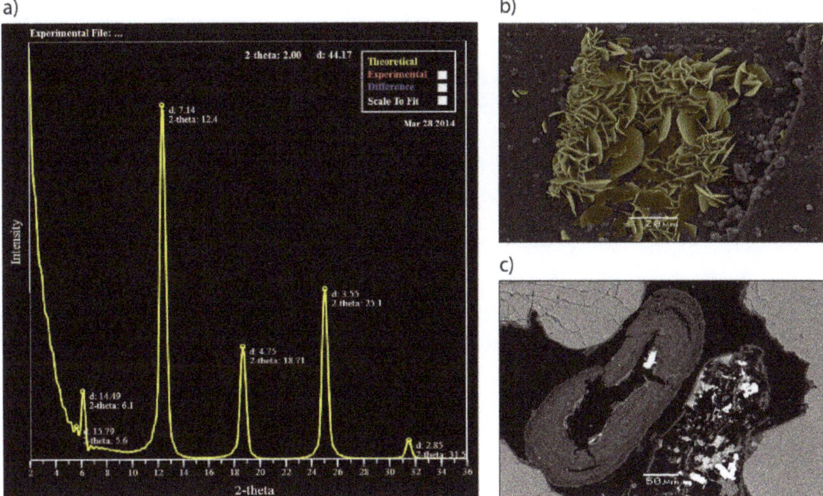

Figure 3. Chlorite solid solutions and occurrence: (**a**) Modelled X-ray diffraction pattern by use of Newmod II, showing a fit with typical chlorites found in a potential CO_2 reservoir in the North Sea; the Johansen Formation. The best fit was found for a Fe2.34 chamosite, with a Fe/(Fe+Mg) ratio of 0.93; (**b**) Scanning electron microscope image of grain coating chlorite from laboratory experiments of daphnite growth (yellow colour applied for reference). Individual crystals are half disk-shaped, growing perpendicular to the host grain surface, ranging in size from 2–15 μm; (**c**) Scanning electron microscope image of ooidal chlorite. Crystal growth occurs in dense, concentric layers around a nucleus grain, which has been dissolved in this case. Grain coating chlorite covers the surface of framework quartz grains. Ghost rims of chlorite coats remain where the framework grain has been dissolved.

In a disaggregated sample, e.g., separated in clay (<2 µm), fine (2–250 µm), and medium (>250 µm) grain size classes, chlorite may be present in all fractions as quantified by XRD. The clay fraction would comprise pore-filling chlorites from diagenetic degradation of detrital, percolated clay and/or diagenetic chlorite from disassembled pseudomorphs of altered grains (e.g., degraded biotite) or mud-clasts. The sand fraction classes could comprise grain coating chlorite, from precursor clay coats. These appear as platy clay-fraction crystals growing tangential or perpendicular on the host grain surface (Figure 3b), and may be more or less resistant to mechanical sample treatment. Another common chlorite occurrence, ooidal, may also be included in the sand fraction. Ooids are spherical grains, with concentric layers of a coating mineral (e.g., clays, carbonates, phosphates) adsorbing on and accumulating around a nucleus-grain. Ooidal chlorite (Figure 3c) forms by recrystallization of precursor clay. These grains have a dense structure with low permeability. Other examples are chloritic pellets and dense diagenetically altered pseudomorphs. Thus, reaction potentials calculated as surface area per wt% mineral from XRD analysis, assuming a uniform clay fraction, would be overestimated in ooid-rich sediments. Additionally, pore-filling clays may not be accessible for intruding reactive fluid [22,23,51], which may cause overestimation of the clay fraction reactivity.

Petrographic studies (e.g., modal mineralogy or point counting) of thin sections in optical microscopes provide a volumetric estimate (vol %) of the mineral assemblage and porosity, which in combination with a description of grain shapes, sizes, micro-porosity from SEM, pore connectivity, and extent of coating translates directly to 2D specific surface areas. As with sieving before XRD analysis, point-counting methods may be used to separate mineral occurrences in grain size classes, in combination with descriptions of grain shape. In addition to chlorite, several reactants may appear in different grain size classes, representative of different reaction potentials.

2.1.2. Characterization of Feldspars

It is relevant to quantify the relative contributions and occurrences of feldspars (i.e., microcline/orthoclase/sanidine, albite, anorthite, and their solid solutions), as kinetics and dissolution potentials in the presence of CO_2 vary significantly [46]. Anorthite is rarely preserved in clastic rocks, as it is chemically unstable and weathers easily [52]. Generally the feldspar assembly varies according to provenance and hinterland geology (i.e., felsic or mafic, igneous or methamorpic rock), and relative feldspar/quartz contents are higher in finer grained facies.

In the Johansen Formation (and aforementioned siliciclastic reservoirs of the North Sea) K-feldspar and Na-plagioclase are abundant (Figure 4). Plagioclase occurs as monocrystalline, diagenetically etched grains, partly dissolved and/or severely altered to sericite. The plagioclase fraction is less than the original detrital composition, but the overall reactive surface area is probably higher than the direct relation to average grain size, due to the diagenetic, secondary porosity within individual grains. The chemical composition in single grains is closer to the albite endmember, with Na >> Ca (determined with electron microprobe). Albitization of K-feldspar grains is common in siliciclastic reservoirs at temperatures >65 °C [53], which would add to the more reactive fraction of feldspars compared to assemblages at shallower depths. If the reservoir conditions in the injection area differ from the sample site it is necessary to extrapolate such diagenetic alterations, or perform sensitivity studies. In the available data set from the Johansen Formation microcline is the most abundant phase, and occurs as monocrystalline grains, some with authigenic overgrowths. K-feldspar is less corroded than plagioclase, and the reactivity is thus likely more directly proportional to average grain size. In perthitic grains, one constituent may be more corroded than the other (Figure 4b), increasing the proportion of reactive surface areas. The feldspar component in lithic fragments (e.g., gneiss and granite) is less corroded and exposes smaller mineral surface areas relative to the absolute volume fraction.

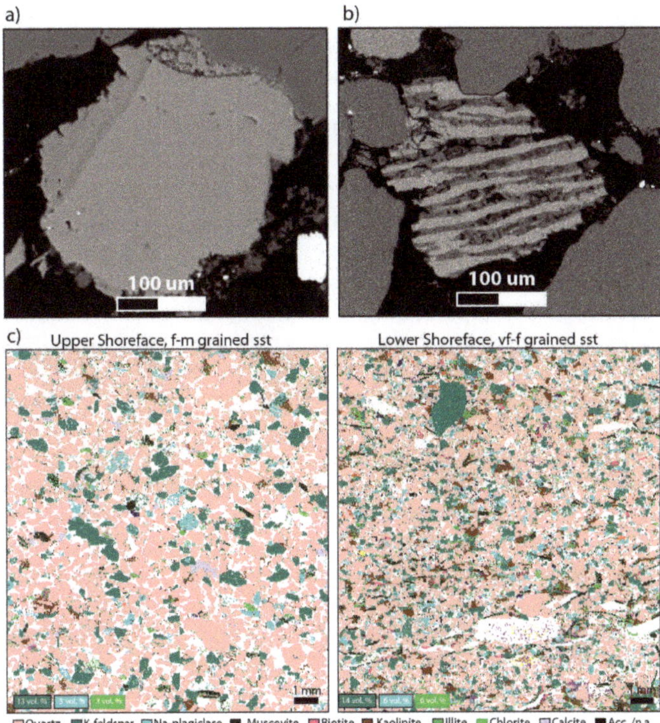

Figure 4. Feldspar occurrences and elemental mapping: (**a**) Scanning electron micrograph of K-feldspar grain with spiky, euhedral overgrowths. Note the compositional change across the outline of the original, detrital grain. The autigenic component is pure microcline, compared to the detrital K-feldspar grain with more heavy elements; (**b**) feldspar perthite grain (scanning electron micrograph). The albite component (dark colour) is partly dissolved, while microcline (light colour) is preserved. This grain is likely to display large mineral surface area compared with average grain sizes; (**c**) mineral maps from Scanning Electron Microscopy (QemScan analysis, Equinor—by C. Kruber), showing the relative volume fractions of the main mineral constituents. Corresponding wt% from XRD (Rietveld) are: 10 wt% K-felspar, 6 wt% Albite, and 2 wt% chlorite in this f-m grained sandstone (2125.4 m), and 12 wt% K-felspar, 5 wt% Albite and 9 wt% chlorite in this very fine grained sandstone (2129.3 m).

2.2. Reaction Potential

The mineralization potential is given by the amount of available cations per given rock volume, but the term reaction potential is more useful for summarizing the geochemical processes. In the case of the Johansen Formation, plagioclase (5–8 wt% in samples [29]) is the most reactive phase in fine- and medium-grained sand fractions, while K-feldspar (9–12 wt% in samples [29]) is more abundant. Fe-rich chlorite (1–9 wt% in samples [29]) is the most reactive clay phase. There is generally more clay in the finer grained, lower shoreface facies (e.g., Figure 4).

The potential for CO_2 to be mineralized, i.e., trapped in solid state, depends firstly on the amount of CO_2 added to the system and less on the solubility in formation water, considering salinity, pressure, temperature, and thermodynamic constraints. CO_2 is transported through the aqueous phase during mineralization [18]. The solution composition applied in simulations (Table 1) was selected based on analogous reservoirs in the North Sea [54]. There is currently no data available on detailed water composition from the Johansen Formation.

Table 1. Aqueous solution input for kinetic simulation.

T	pH	Na	K	Ca	Mg	Fe	Al	Cl	Alk	Si	O	O_2
°C		*ppm*	*ppm*	*ppm*	*ppm*	*ppm*	*ppm*	*ppm*	*meq*	*ppm*	*ppm*	*Log P*
96	5.9	7544	113	890	53	0	1×10^{-8}	13,187	293	1×10^{-4}	1×10^{-3}	−50

Aqueous trapping capacity; CO_2 + HCO_3^- + CO_3^{2-}, is relatively small and in the order of a few percent [55]. Further dissolution of residual CO_2 adds to the dissolution potential [56]. Carbonate precipitation reactions consume bicarbonate and cations from solution, lowering the pH as H^+ is produced. Dissolution of silicate minerals consumes protons, and release cations, HCO_3^-, aqueous silica and/or secondary clay minerals to solution (e.g., [19]). Carbonate stability is enhanced by increased pH, and the cation supply drives further carbonate precipitation. Dissolution and precipitation are interconnected through these feedback mechanisms [56], and the rate of either will be controlled by the slowest reaction [18]. Carbonate precipitation, most often considered a more rapid reaction compared to silicate dissolution (e.g., [54]), may in some settings provide the rate limiting reaction, such as for low temperature settings [18].

As a first approximation of reactivity and identification of primary reactants, initial geochemical batch simulations including the full mineral assemblage are adequate. For example PHREEQC, TOUGHREACT, and other numerical tools may be applied for batch geochemical modelling in combination with thermodynamic databases such as llnl.dat, phreeqc.dat, or equivalents, including kinetic expressions with nucleation growth rate equations (e.g., [10]).

Based on previous geochemical studies of siliciclastic reservoirs from the North Sea and elsewhere, it may be concluded that a few percent of scattered carbonate equilibrates instantly, that quartz is close to chemically inert, and that reactive accessory minerals present in small amounts (<<1 wt%) are insignificant on reservoir scale. One sedimentary facies may be represented by several samples, which in turn should be averaged with respect to grain size distributions, porosity and mineral content. Subsequently, cases for geochemical simulations, may be defined. If the sediment sorting is poor, it may be relevant to divide the sand fraction in two or more classes. Each mineral is assigned a representative wt% within each class according to petrographic studies of occurrence.

2.3. Reactive Surface Areas

Estimation of reactive surface area (m^2/liter pore water) must relate weight or volume percent of mineral to grain size, shape, porosity, and mineral density. Aged, coated, diagenetically altered and/or weathered grain surfaces are expected to display lower reactivity compared to crushed sample material commonly used in laboratory studies of kinetics.

The mineral content given as wt% from XRD must be translated to the specific surface area by relating mineral density and grain shape (e.g., spheres or circular disks) in geometric formulas. Spherical grains are an appropriate assumption if grain sizes are adjusted according to appearance, e.g., 0.1 μm diameter for the clay fraction if assuming spherical grains, rather than 2 μm diameter if measuring more realistic clay appearances such as flakes (Figure 3b). Porosity is a characteristic of the sedimentary facies and diagenetic imprint, which must be estimated for the associated sample and/or interpolated to the study area. The geometric surface area may be described as:

$$\bar{S}_i = \frac{3x_i \rho_{solid}}{\rho_i}\left[\frac{1}{\varphi} - 1\right]\sum_j \frac{x_j}{r_j} \tag{1}$$

where S_i is the average specific surface area of mineral i for the appropriate facies (m^2/L pore water). x is the mass fraction of mineral i, and ρ_{solid} (g/L) and ρ_i (g/m^3) are the average density of the total solid and density of mineral i respectively, φ_i is porosity, r is the mean radius of grains belonging to the discrete size group j, and x is the fraction of grains belonging to the same discrete grain size group. As a next step in detailed studies of separate mineral phases, the reactive surface area may be further adjusted according to petrographic observations. The true reactive surface area S_t, differs from S_i, as only some parts of the surface is reacting at any given time (e.g., [10,17]). Grain roughness may increase S_t by up to several orders of magnitude compared to S_i, while grain coats and "aged" surfaces have the opposite effect. Diagenetic processes may provide inaccessible (−) or accessible (+) micro-porosity within grains or mud aggregates. Appropriate fractions may be estimated qualitatively and/or quantitatively by elemental analysis and microscopy.

For example, the reactive surface area of plagioclase in lithic fragments may be assigned a lower reactive surface area, $S_r < S_i$, compared to plagioclase as monocrystalline grains, where $S_r = S_i$. In the case of etched plagioclase grains with additional internal porosity, $S_r > S_i$. Using spheres as proxy for geometric grain shapes is sufficient in most cases, as long as the true morphology is considered. Needle-like crystals (e.g., illite) may for example be represented as a series of small spheres, and must be accounted for by reducing grain size. Clay minerals with flaky occurrence (e.g., chlorite in Figure 3b) are most reactive at the edges, and thus, $S_r << S_i$, as shown in Figure 5a. Assuming spherical grain shapes is therefore not necessarily a drastic simplification. For shales, where the connected porosity is low, reactive surface areas have to be estimated from a geometric model of the pore space rather than the solid phase [17].

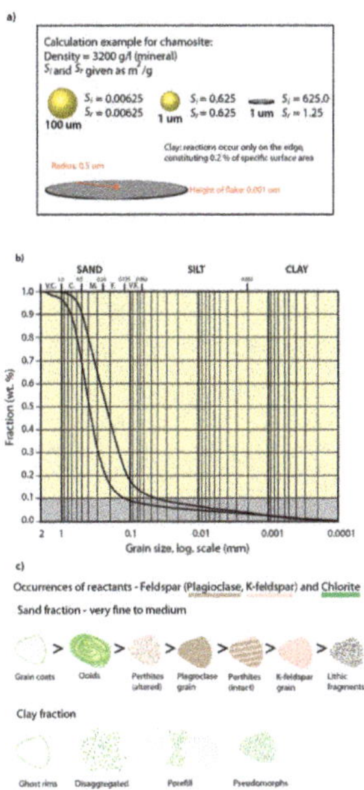

Figure 5. Parameterization of reactive surface areas: (**a**) Examples of grain geometries in relation to specific and reactive surface areas. For platy clay minerals $S_r << S_i$, as reactions only occur along the edges; (**b**) Grain size distribution curves for typical lower (finer) and upper (coarser) shoreface facies. The small amounts of clay in reservoir sandstone and small amounts of available sample material makes separate clay analysis for detailed speciation difficult. Note that if disaggregating samples to analyse the clay fraction separately in XRD, a large part of the reactive phase may sort as sand (e.g., chamosite grain coats and ooids). Bulk XRD for the same samples show 6–9 wt% chlorite; (c) Reactivity quantification may be performed on bulk XRD data if typical mineral occurrences are described (vol %), and reactive surface area assigned. Example from the Johansen Formation: plagioclase, K-feldspar, chamosite, sorted from right to left according to reactivity in sand and clay fractions, respectively. Occurrence affecting reactive surface area is related both to sedimentary facies and diagenetic alterations (e.g., altered perthites).

3D imagery of crystal habits in SEM is useful in quantification of reactive surfaces (Figure 3b). By use of 2D element-mapping of thin sections, mineral distributions may be efficiently estimated. These methods are of particular importance when variations in solid-solution chemistries relate to different reaction potentials; e.g., for feldspars, chlorites, smectites, carbonates, sometimes by orders of magnitude (e.g., [46]). Micro-porosity and fluid access is challenging to quantify, and total porosity is likely to be underestimated by microscopy methods. The total surface within connected pores may be measured in 3D by use of micro-tomography or Hg intrusion. In combination, these methods may be applied to estimate reactive surface areas, as described in [23]. However, available sample material or budget is often limiting 3D characterization.

3. Estimating CO_2 Mineralization Potential (II): Example Modelling

Batch reaction models were performed with the geochemical software PHREEQC v3 using the built-in phreeqc.dat thermodynamic database, allowing robust estimates of the CO_2 fugacity coefficient using the Peng–Robinson EOS model [57]. CO_2 solubilities are however slightly high because a Poynting (pressure) correction for the gas solubilities is not included in PHREEQC. For the Johansen Formation (e.g., 96 °C, 300 bar, 0.5M NaCl, from [28]) the true solubility is 1.29 mol/Kgw (using the SAFT v1 method as described in [56]), whereas the PHREEQC solubility is 1.52 mol/Kgw. Nevertheless, the high solubility does not alter the prediction of carbonatization potential as simulations were run at a constant CO_2 pressure and pH rapidly approaches a value close to 5 (4.9–5.4) at calcite saturation and over a large range of CO_2 pressures (100–300 bar) [10,17].

To model the kinetics of mineral reactions the rate equations presented in [10] were used, with a transition state theory (TST) based rate law for dissolution and a nucleation-growth equation for growing mineral phases. The exception was growth of dawsonite, which had to be estimated using a local-equilibrium assumption (forming at equilibrium) because of convergence problems when it was included in the kinetic assemblage. This has been demonstrated to cause an overestimate of the amount of dawsonite that forms at short time-scales, whereas the models are less sensitive at the longer time-scales [17]. Kinetic data (rate constants for dissolution, nucleation and growth, and apparent activation energies), were taken from [17] (Table 2).

Table 2. Kinetic data.

Mineral	log k+ $^{T = 25\ pH = 0}$	E_{a+} [1]	N [2]	log k- [3]	E_{a-}
Albite	−10.2	65.0	0.46	log k+ −2	E_{a+}
Oligoclase	−9.7	65.0	0.46	log k+ −2	E_{a+}
K-feldspar	−10.1	52.0	0.5	log k+ −2	E_{a+}
Chamosite 7A	−9.8	22.0	0.53	log k+ −2	E_{a+}
Quartz/Chalcedony	EQUIL				E_{a+}
Calcite	EQUIL				E_{a+}
Kaolinite	EQUIL				E_{a+}
Dawsonite[4]	−4.5	63.8	0.98	log k+ −2	E_{a+}
Siderite	−7.5	48.0	0.94	log k+ −2	E_{a+}
Ankerite [5]	−3.2	56.7	0.5	log k+ −2	E_{a+}

(1) Apparent activation energy (kJ/mol) for dissolution, listed in [46]; (2) Reaction order with respect to protons [46]; (3) Growth rate constants at 25 °C (mol/m^2 s). pH dependencies are unknown and neglected for growth; (4) Rate coefficient at 25 °C and pH = 0 (mol/m^2 s), apparent activation energy and reaction order with respect to protons from [58]; (5) Lacks data and set to the same as dolomite [46].

The true reactive surface area (S_t) differs from the geometric values (S_i) because of grain shape, surface roughness, and because only parts of the surface are taking part in the reaction at a given time. Roughness may increase the surface area by 1–2 orders of magnitude, whereas the fraction of the total surface that is reactive may be 1% or less. These two effects therefore partly cancel each other, but the extent of this is difficult to assess and depends on several factors. Generally, aged sediments may have orders of magnitude lower reactive surface area than activated crushed materials. The sensitivity of mineral carbonatization on the reactive surface area has earlier been demonstrated in [10,17], and we will here simply use the geometric model (Equation (2)) and focus the sensitivity study on sedimentological features (e.g., chlorite and feldspar morphologies and mean sediment grain sizes).

$$\overline{S}_i = \frac{3x_i\rho_{solid}}{\overline{r}\rho_i}\left[\frac{1}{\varphi} - 1\right]$$

(2)

For the case studies, we divided simulations into very fine sand (r = 0.05 mm) and medium sand (r = 0.2 mm), representing typical lower and upper shoreface facies of the Johansen Formation (Figure 2) [28,29]. In these simulations quartz (nucleation surface), feldspar, Fe-chlorite ooids, and rock fragments were considered to be of the same size, whereas clay particles (kaolinite, chlorite, smectite) were considered to have a mean radius of 1 μm. Sensitivity of feldspar occurrences were simulated for 4.8 wt% perthitic K-feldspar, 1 wt% lithic K-feldspar, 2 wt% perthitic Na-feldspar, and 3 wt% plagioclase. Chamosite input was 4 wt% porefill and 1 wt% ooid. The porosity was set at 25% and reservoir conditions (300 bar, 96 °C) were not varied between scenarios. Model sensitivity studies for temperature and nucleation growth are on-going.

3.1. Carbonatization of Chlorite

Two of the chlorite occurrences observed in the Johansen Formation, i.e., in ooids and as pore filling and grain coating cements, display very different reactivity. Because large parts of the chlorite in ooids are inside the grain and prevented from being in contact with the reactive solutions, ooid-chlorite is assumed to have about two orders of magnitude lower specific reactive surface area than the pore-filling and grain-coating chlorites. The abundance of ooids varies in the cored interval of well 31/2-3 in the Johansen Formation, but is generally not dominant relative to more accessible pore-filling, pseudomorph alterations, and grain coats. We therefore varied the fraction of ooid-chlorite from 0 to 20 vol %. Chlorite morphology is also expected to show significant lateral and stratigraphic variations. Chloritic ooids are recognised also in other potential storage formations such as the overlying Cook Formation [34] and in the Gassum Formation [35].

The simulations show that the amount of chlorite dissolved over short to medium time spans (<1000 years) very much depends on the amount that is high-reactive, i.e., the pore-filling and grain-coating chlorite (Figures 6 and 7). The time it takes to completely dissolve the ooidal chlorite is approximately 10,000 years, also in the 20% ooid-chlorite case, and it is therefore no difference in the dissolved amount at this time scale (Figure 7a). Figure 7b shows pH changes and the amount of secondary carbonates (siderite, ankerite, and dawsonite) that form in the 20% chlorite-ooid case over 100 years. Siderite is the only Fe-carbonate to form, and the amount is proportional to the amount of chlorite that dissolved (1.8 moles of siderite formed for each mole of chlorite dissolved). The short delay of four years before onset of growth (Figure 7b) was due to the nucleation induction time. As kaolinite was defined to be at equilibrium with the formation water and dawsonite formed according to the local-equilibrium assumption, dawsonite formed immediately from the dissolved CO_2 and the formation water Na^+ and Al^{3+}, but the growth rapidly stopped as no further Na^+ was supplied (Figure 7b).

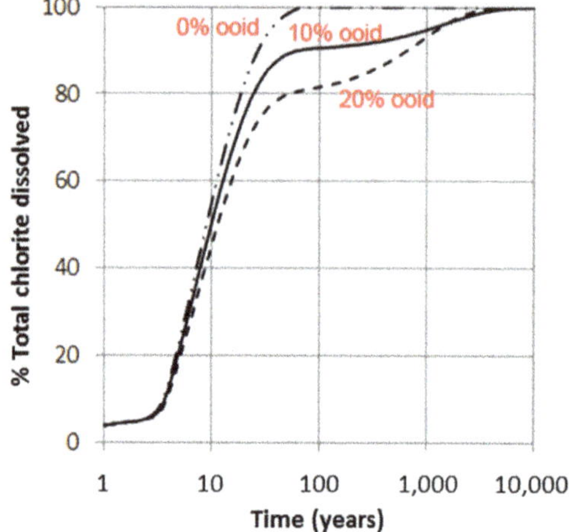

Figure 6. Simulated dissolution of chlorite (chamosite) (e.g., Johansen Formation, 96 °C, 300 bar CO_2); where chlorite was separated into two parts: (1) highly reactive pore filling/grain coating chlorite with large reactive surface area (1 μm grains); and (2) low-reactive ooids (125 μm aggregates where reactions are only assumed on the aggregate surface). Up to 20% chlorite in ooids was simulated.

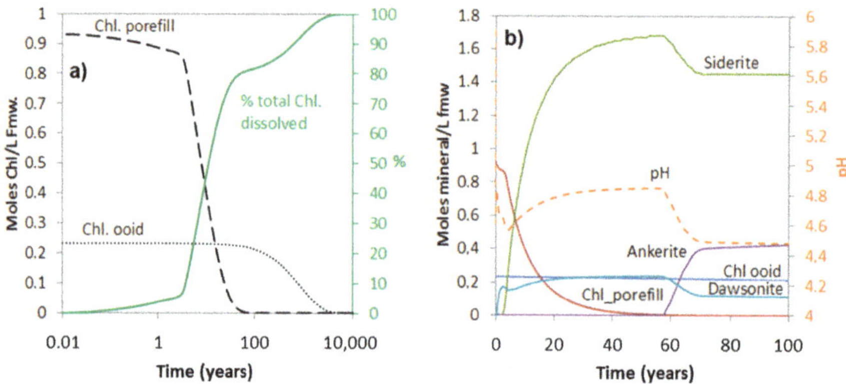

Figure 7. Simulated chlorite dissolution (**a**), and corresponding secondary carbonate formation and pH (dotted curve) evolution (**b**), for 100 years of CO_2-chlorite interactions with initial materials consisting of 20% chlorite as ooids, and the remaining fraction as high-reactive pore-filling or grain-coating materials.

3.2. Carbonatization of Feldspars

The feldspars in the Johansen Formation have been divided into five different types based on their chemistry and morphology. The lithic feldspars (plagioclase and K-feldspar) were assumed to be in a mineral mixture inside spherical fragments, and have reactive surface areas corresponding to the mineral fraction in the fragment. K-feldspar was also found along albite in perthitic fragments, and some plagioclase occurred as larger, preserved detrital grains. In these simulations we assumed that all grains (minerals and lithic) where in the same size, and we simulated two different settings: very fine grained sand (0.1 mm) corresponding to lower shoreface deposits, and medium grained sand (0.4 mm) being representative for upper shoreface deposits [29]. The difference in grain size leads to a four times larger reactive surface area for the very fine sandstone, and correspondingly faster dissolution of the feldspar grains and faster formation of the secondary dawsonite (Figure 8). In the very fine sand, detrital plagioclase dissolved completely within 50 years, leading to a corresponding dawsonite growth. In the medium-grained sandstone the same reaction takes four times longer, indicating that the dissolution occurs at far-from-equilibrium under-saturation and that the rate therefore is proportional to the reactive surface area. The difference in carbonatization potential is mainly seen on the short term, with the very fine sandstone having a potential of about 50% more CO_2 bound and immobilized in dawsonite after 100 years (Figure 8). The difference is smaller in the long term (1000–10,000 years) as all Na-feldspars are eventually replaced by dawsonite. Including minor calcite contents (e.g., 1 wt%) causes some minor recrystallization of calcite to ankerite, but the overall storage potential does not change much. The K-feldspars were not completely dissolved after 10,000 years, but K-feldspars have earlier not been regarded as a source for dawsonite ([59] and references therein). Some recent data may, however, indicate that K-dawsonite may also form, but this is still highly uncertain [60]. The feldspars in intact lithic fragments are much less reactive compared to individual plagioclase grains, and it takes thousands of years to dissolve the lithic plagioclase even in the very fine sandstone.

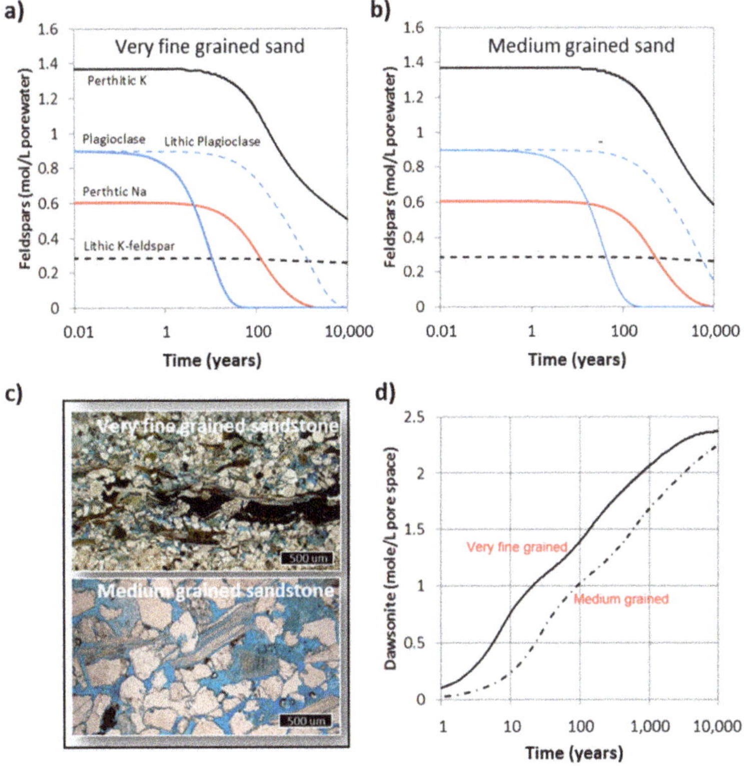

Figure 8. Simulated feldspar dissolution of equal wt% feldspars for (**a**) very fine grained sand, and (**b**) medium-grained sand. Feldspars were divided into perthitic K- and Na-feldspars, K-feldspar and plagioclase in lithic fragments, and detrital plagioclase grains; (**c**) micrographs from corresponding facies in the Johansen Formation: very fine grained and medium grained sandstones. Pore space is filled with blue epoxy; (**d**) amount of dawsonite that forms from feldspar dissolution for very fine grained and medium grained sand lithologies. The two cases correspond to (**a**) and (**b**) (this figure).

If available mineral data are XRD analyses only and no petrographic information is available, we cannot distinguish the lithic plagioclase component (as long as the chemical compositions of the two are similar). The rates of plagioclase dissolution and dawsonite formation may in such cases be highly uncertain. Although we know quite well the local composition (well 31/2-3) of the Johansen Formation rocks, there may be spatial variations and we illustrated this by simulating three cases with different fractions of plagioclase bound in lithic fragments (Figure 9). If all plagioclase is monocrystalline and detrital (0% in lithic fragments), nearly 70% of the plagioclase has dissolved after only 100 years and some significant amounts of dawsonite forms (Figure 9b). On the other hand, if most (90%) of plagioclase is within lithic fragments, less than 20% has dissolved after 100 years and much less dawsonite forms. The difference is, however, smaller at longer time scales and quite small when approaching 10,000 years (Figure 9). We have so far simulated chlorites and feldspars and their carbonatization potentials separately. Because these mineral groups share elements such as Al and Si, dissolution of one may affect the other. We therefore simulated the combined chlorite-feldspar assemblage and compared the carbonatization potential with the individual mineral-group simulations (Figure 10).

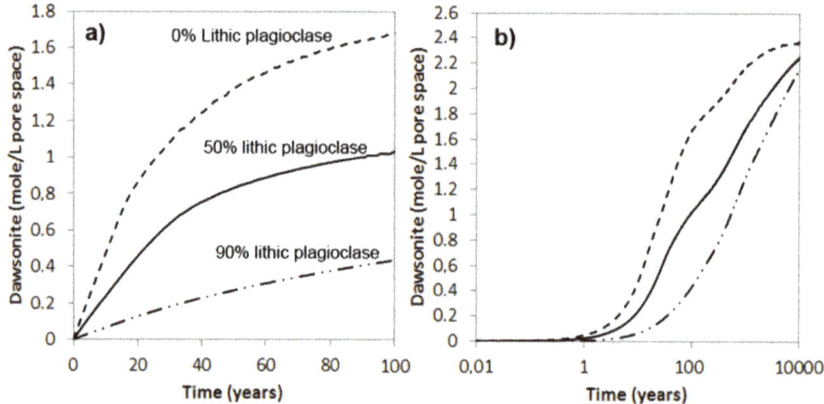

Figure 9. The amount of dawsonite formed with varying amount of plagioclases lithic rock fragments (low-reactive) or as separate detrital crystals (high-reactive) over (**a**) 100 years, and (**b**) 10,000 years.

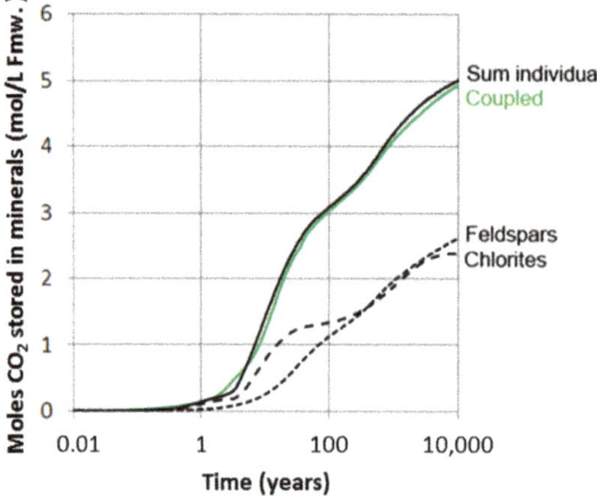

Figure 10. Amount of CO_2 stored in secondary carbonates (mol/L fmw) in simulations only taking into account either feldspars or chlorites, and compared to the results in coupled simulations taking into account both feldspars and chlorites. No/little difference suggests that pH is the same in both cases. The motivation to run separate simulations is to save time in more complex large-scale reactive transport simulations.

4. Discussion

The reactivity depends on the mineral assemblage, grain size and morphology, which all vary according to sedimentary facies within one sandstone reservoir unit, as well as on the diagenetic imprint and in-situ reservoir conditions. Characterizing and describing a potential storage candidate with respect to the spatial distribution of sedimentary facies is an efficient means for separating zones with different mineralization potentials for CO_2 in siliciclastic reservoirs.

4.1. Workflow

The workflow for parameterizing input for geochemical models described above is summarized in Figure 11. The main challenge with respect to geological characterization of saline aquifers is that hard data are often scarce. Thus, interpolation across large distances and/or burial depths from areas where data are available, to a less mapped, potential injection site are carried out. Absolute reactivity may often not be estimated, but taking into account facies changes (e.g., more clay and feldspar in finer grained facies) and diagenetic imprint (such as albitization with increasing temperature), some scenarios may be defined for initial geochemical bulk modelling including all phases and identification of main reactants. For the Johansen Formation the main cation donors were identified in bulk simulations as albite and Fe-chlorite [61]. Subsequently, estimating the facies-specific reaction potential through geochemical simulations takes a detailed description of the reactants into account. Solid-solutions must be specified and parameterized with suitable kinetic parameters [10,46] and assigned to one or more grain size/shape class occurrences (Figure 5).

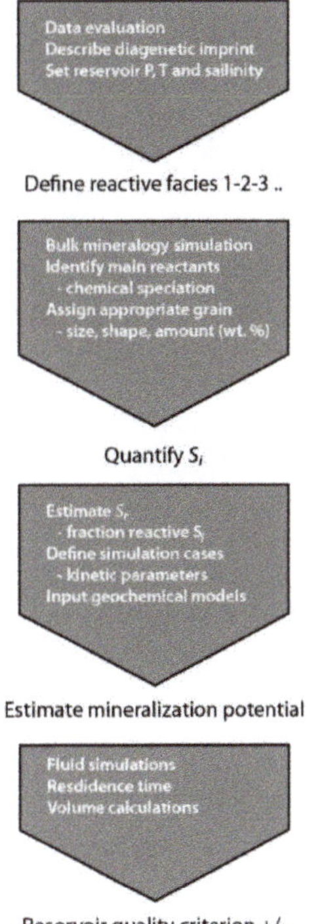

Figure 11. Workflow for estimating mineralization potentials specific to sedimentary facies and mineral occurrence in reservoir characterization.

The total input reactive surface area (S_t) may be defined through petrographic studies and quantification of mineral assemblage as described here and/or according to other methodologies (e.g., [22,23,62]). Depending on the magnitude and kinetics of pH-changes during simulations (i.e., not too large fluctuations), simulations for each reactant (dissolving phase) may be run separately.

4.2. Input Data

XRD quantification procedures provide cheap and frequently available data. Direct use of bulk rock XRD data (sometimes lacking specification of sample type and treatment) as input for geochemical modelling of mineral trapping, with uniform grain sizes and associated reactive surface areas, is common procedure. Evaluating the effect of phase occurrence and grain shape/size may change the time scale and magnitude of mineralization potential significantly, as shown here. Without complementary geological knowledge about the depositional environment and burial history of siliciclastic rocks, reservoir scale estimations of reaction potentials may be grossly wrong. Appreciating the large uncertainty due to natural heterogeneity and lack of data, a range of geochemical models can be constructed to illustrate effects of alternative interpretations and data interpolation. Effects of system variability may also be explored by means of stochastic analysis: for example by Monte Carlo sampling from experimental surface area measurements [16].

In the Johansen Formation the content of reactive minerals varies in well 31/2-3, and until the assembly is confirmed in the actual injection area, facies related scenarios may be defined for sensitivity studies of trapping potential.

4.3. Upscaling

Having defined reactive surface areas and mineral assemblages specific to each facies represented in the reservoir, bulk reservoir reaction potentials (dissolution + mineralization) may be estimated. By fluid flow simulations of CO_2 injection and migration (e.g., Eclipse, TOUGH, and similar numerical tools), total dissolved volumes of CO_2 within specific layers or in each defined facies setting may be quantified. The current facies model for the Johansen Formation (Figure 12) is based on acoustic impedance data, and must be verified and calibrated with well data before reactivity distributions can be predicted. Scenario based modelling of fluid distributions and dissolution potentials, however, indicate that the dissolution potential (e.g., 29% of 160 Mt injected CO_2 after 1000 y) could be less than the mineralization potential (Figure 10) on the same time scales (100's of years), which would cause a catalytic effect of mineralization on further dissolution of the residual phase, where supply of CO_2 would be the rate limiting factor. This is highly dependent also on reservoir pressure and temperature conditions, which still are to be measured in the proposed injection area.

Decoupling of models is an advantage in that fluid simulations may include realistic topography, fine grid sizes, proper equations of state (EoS) [63] and relative permeability curves (e.g., [51]) assigned to facies [28], all within the limits of computing capacity (CPU). Considering the residence time of CO_2 at the dispersive plume front, as well as the dissolution of residually trapped CO_2 left behind a migrating plume, the reaction potential in the injection area and along the predicted migration path may be evaluated. The formation water salinity has negligible effect on the mineralization potential, as the moles of cations in the solid phases are several orders of magnitude larger than in the water at any time, and the aqueous phase can be regarded as merely a transporter of CO_2 during the mineralization [18]. Significant mineral precipitation may retard plume migration, increase the dissolution potential, immobilize CO_2 and is considered the safest trapping mechanism [2,55,56]. The reaction potential is higher in fine grained facies (because of larger reactive surface areas and higher relative fractions of cation donor minerals), which in combination with higher fractions of residual CO_2 would support more efficient immobilization. With a porosity of 25%, the estimated volumetric long term trapping potential would be 55 kg CO_2 per m^3 reservoir.

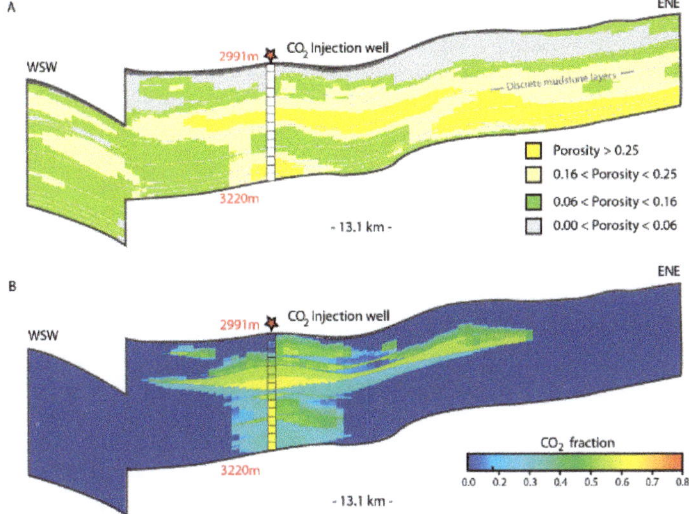

Figure 12. Fluid distribution relative to sedimentary facies with different mineralization potentials: (**A**) Interpreted facies distribution related to porosity class and interpreted from seismic attribute analysis (acoustic impedance). Upper shoreface (high porosity and permeability) in dark yellow, lower shoreface (intermediate porosity and permeability) in light yellow. Non-reservoir siltstones and mudstones in green and grey. Model described in [28]; (**B**) fluid distribution at 100 years, after 50 years of injection 3 Mt/y CO_2 through a well perforated in the lower half of the reservoir. In this case 16% of CO_2 was dissolved, 38% residually trapped [28].

4.4. Hydrogeochemical Trapping

Mineral and ionic trapping reactions for CO_2 are often disregarded in reservoir characterization and modelling schemes, due to slow kinetics. Geochemical studies indicate, however, that the dissolution of clay and silicate minerals and the subsequent precipitation of carbonates may be significant also on time scales less than 100 years (e.g., [10,24]), and approach immobilization potentials in the same size order as dissolution in pore waters [7]. Thus, it may be argued that mineralization potential and the associated increase in dissolution potential should be taken into account in general reservoir characterization schemes.

Immobilization of CO_2 enhances storage security [2]. Between two otherwise suitable reservoir candidates (e.g., high injectivity, safe cap-rock, appropriate temperature and pressure conditions), the safer option would be the more reactive reservoir setting (e.g., mineralogy, salinity, pH, temperature) providing permanent storage of CO_2 through dissolution-, ionic- and mineral- trapping, disregarding near-well pore-clogging by rapid salt precipitation in this context. Geological heterogeneity may cause plume spreading, increase dissolution and immobilization potential [28,64].

De-coupling of models for transport and reactions, as proposed here, introduce challenges with regards to timing and linked processes such as aqueous speciation of CO_2 and pH-changes during silicate dissolution and carbonate precipitation [56]. Furthermore, the dynamics of porosity changes due to mineral precipitation may not be incorporated [65]. However, in coupled geochemical and transport models, reservoir geometries and geological heterogeneities are not accounted for, e.g., [7], which impose the most important control factors with respect to fluid distributions (Figure 12). Fluid flow models may highlight preferential flow paths within distinct facies, bypass zones, and plume separations due to layered heterogeneities. The reactive surface area is expected to be highest in rocks of sedimentary facies with smaller average grain size and higher clay content, as well as increasing with associated lower porosities and smaller pore throats. This relation is valid only down to effective porosities <6% and associated permeabilities <100 mD [66], at which point it is no longer realistic that all pores are swept with CO_2. Absolute estimations of mineralization potential by decoupled methods are not possible due to constrains of present day CPU capacity, but reservoir scale relative evaluations may be made by bulk volume calculations of residual and dissolved CO_2 present in a given sedimentological facies setting at different time steps during fluid migration (Figure 12). The implementation of geological models in sensitivity studies may provide insight towards long-term reservoir behaviour.

The result shows that there is no significant difference between batch simulations using the complete reactive mineral assemblage, and the results of running separate simulations for the feldspars and chlorite and summing up the carbonatization potential (Figure 10). This indicates that pH of the simulations are also very similar, as pH strongly affects reaction rates. Such simplifications may be beneficial for running reactive-transport simulations of larger and more complex CO_2 storage systems as the addition of kinetic reactions to flow simulations adds a substantial CPU load. All simulations were run at constant CO_2 pressure, implying that the batch system is in communication with a CO_2 source with sufficient CO_2 to feed the necessary five moles required for a complete carbonatization of the feldspar-chlorite assemblage. The amount of CO_2 required for a complete carbonatization at time scales when CO_2 is still dominantly mobile (<100–1000 years) is about 2.5 to 3.8 moles/L Fmw (Figure 10). With a solubility of 1.28 moles/Kg Fmw in the Johansen Formation (SAFT v1 calculations: [59]), we need about 1.2 to 2.5 additional moles of CO_2 per liter formation water for a complete carbonatization. This can be fed from CO_2 trapped residually. The minimum volume of residual CO_2 and percent residual required per liter pore water was estimated using a CO_2 density of 680.13 kg/m^3 [63]. The estimated amounts of residually trapped CO_2 for 1.2 and 2.5 moles are then 7.2% and 13.9% respectively.

4.5. Kinetic Rate Uncertainties for Chlorite

Kinetic data used for most CO_2 storage simulations are taken from Palandri and Kharaka [46]. There has, however, been generated more recent data and some experiments have also been performed at conditions more relevant to CO_2 storage (i.e., relevant CO_2 pressures). Because the Palandri and Kharaka [46] review has incorporated data for all feldspars of interest to CO_2 storage in sedimentary basins (anorthite, Na-rich plagioclases, albite, K-feldspar), and there are no more recent studies that change the rate constants or temperature dependencies, we have here focused on the variation in data for chlorites, and the few data of Fe-rich chlorites and the total lack of data for the Fe-endmember chlorite. Only rate data from experiments at acidic conditions will be compared here as they are most relevant for CO_2 storage. A summary of the compiled chlorite data is given in Table 3.

Table 3. Kinetic data for chlorites.

Study	CO_2 [(1)]	pH	T	log k [(2)]	E_a	N [(3)]	log k	log k
		range	*°C*		*(kJ/mol)*		*at 37 °C, pH 5*	*at 75 °C, pH 5*
Clinochlore-14A	0	Acidic	–	−11.1	88.0	0.5	−13.0	−10.6
Fe-rich (Mg/Fe = 1.4)	0	Acidic	25–95	−9.8	94.3	0.49	−11.6	−9.1
Clinochlore-14A	0.1–0.5 M	3.0–5.7	100–275	−9.9	25.1	0.49	−12.2	−11.5
Clinochlore-14A	120–200 M	3.5–5.4	50–120	−12.0	16.0	0.076	−12.3	−11.9

(1) Experimental CO_2 molar concentration (denoted with an 'M' after the value) or CO_2 pressure. (2) Rate constant k (mol/m^{2}s) at pH = 0 and 25 °C, assuming a rate equation of the form R = kSaH$^+$n(1 − Ω) (see Palandri and Kharaka, 2004 [46]). (3) Rate order with respect to the proton activity.

Average values from Palandri and Kharaka [46] have been estimated from the published data prior to 2004. This compilation suggests a dissolution rate constant of the Mg-endmember chlorite (clinochlore-14A) of 7.76×10^{-12} mol/m^2 s at pH = 0 and 25 °C (all rate constants will from here be discussed at this reference point), and with an apparent activation energy of 88.0 kJ/mol. Lowson et al. [50,67] examined the dissolution rates of an Fe-rich chlorite (molar Mg/Fe = 1.4) and found a rate constant of 1.62×10^{-10} mol/m^{2}s, significantly larger than the average value listed by Palandri and Kharaka [46] for the Mg-endmember, but with a similar and even larger apparent activation energy (94.3 kJ/mol). Smith et al. [48] examined the clinochlore-14A end-member and found a rate constant comparable with [50] (1.23×10^{-10} mol/m^2·s), but with a much smaller apparent activation energy (25.1 kJ/mol). Finally, Black and Haese [68] recently found a clinochlore-14A rate constant of 9.55×10^{-13} mol/m^2·s, and with an even smaller apparent activation energy than in Smith et al. [48] (16.0 kJ/mol). In all studies, except for Black and Haese [68], a reaction order with respect to protons of about 0.5 has been found. However, the low value of 0.076 found by Black and Haese [68] was attributed to the inhibiting effect of bicarbonate on the dissolution rate, largely cancelling out the catalyzing effect of protons. It is clear that it is a large range in listed rate constants and apparent activation energies, and the work by Black and Haese [68] also suggest that CO_2 and bicarbonate will significantly affect the pH dependency of the rates. It is therefore of interest to compare chlorite dissolution rates at conditions relevant for CO_2 storage. The pH of a CO_2 storage repository buffered by calcite dissolution is around 5 and quite independent of CO_2 pressure and temperature [10,17].

The temperature varies from reservoir to reservoir. At 37 °C (e.g., the Utsira CO_2 storage site), differences are only modest for the Mg-endmember, with the largest rate constants from Smith et al. [48] being approximately seven times larger than the lowest from Palandri and Kharaka [46]. The Fe-rich chlorite is at these conditions suggested to react approximately six times faster than the average found for the Mg-end-member (Table 3). At 75 °C the rate constant from Palandri and Kharaka [46], having much higher apparent activation energy, passes the value for two other studies on the Mg-chlorite. At this condition, the Palandri and Kharaka [46] rate constant is approximately 16 times that of the Black and Haese [68]. The rate constant for the Fe-chlorite, having even larger activation energy, is at 75 °C almost 100 times larger than for the average of the Mg-chlorites. This fast reactivity of the Fe-end-member fits well with studies of other Fe-rich clay minerals, such as glauconite [69], and poses a challenge in predicting the short term (<100 years) reactivity and mineral trapping potential of the Fe-chlorites, being very common in North Sea reservoirs [34,35,44,61]. The large variations in data for Mg-chlorite and the lack of data for the Fe-endmember result in a significant uncertainty in estimated dissolution rates, on top of the large uncertainties in reactive surface areas. This calls for more rate studies on chlorite dissolution, preferentially done at CO_2 storage conditions (including realistic CO_2 pressures).

5. Conclusions

The reactive surface area depends on grain size and shape, porosity and permeability, and varies according to sedimentary facies and diagenetic imprint. More accurate, or relevant ranges, of input values for reactive specific mineral surface areas as used in geochemical modelling of long term mineralization potential for CO_2 may be estimated by combining optical, physical and chemical quantification methods, and relate mineral morphology to grain size in estimations from weight %. Implementing sedimentary facies variations in reservoir models provides for volume estimations of fluid distributions in various parts of the reservoir, which may be applied for evaluating spatial variability of mineralization potentials in CO_2 storage reservoirs.

Na-plagioclase and Fe-chlorite are the main cation donors for mineral trapping of CO_2 in the Johansen Formation. Reaction rates of chamosite in reservoirs ~100 °C are likely significant on shorter time scales (100's of years), and relevant for estimation of immobilization potential and increased dissolution. The bulk reactive mineral content (feldspar and chlorite) as well as reactive surface area per weight fraction is higher in fine-grained facies. Simulations suggest that chlorites in ooids or dense aggregates may reduce the short term (<100 years) mineral trapping potential by up to 20%, compared to more reactive occurrences like grain coats. Feldspars are suggested to have the largest impact on long-term (1000–10,000 years) mineral trapping. Intact lithic fragments are less reactive, while diagenetically altered grains may be more reactive. In our geometric model, fine-grained facies have four times larger specific reactive surface areas compared to medium-grained sand, and the mineral trapping rates are correspondingly faster.

Author Contributions: A.S. and H.H. wrote, edited and reviewed the manuscript, and developed the concept, idea and methodology; H.H. performed the numerical simulations; A.S. made input for the models; A.S. handled project administration and secured funding.

Funding: This work has been partly funded by the Research Council of Norway in projects CO_2 Upslope under grant # 268512 and SUCCESS under grant # 193825/S60. SUCCESS (SUbsurface CO_2 storage—Critical Elements and Superior Strategy) is a consortium with partners from industry and science, hosted by Christian Michelsen Research, and is an Environment-friendly Energy Research (FME)-center assigned by the Research Council of Norway. The CO_2-Upslope project: Optimized CO_2 storage in sloping aquifers, is funded by the CLIMIT-programme. Gassnova SF and the Research Council of Norway collaborate on the CLIMIT-programme which finances projects within Carbon Capture and Storage (CCS).

Acknowledgments: The authors would like to thank CLIMIT and the Norwegian Research Council for funding. We acknowledge Equinor and the Norwegian Petroleum Directorate for access to sample the Johansen core, with the mineralogy presented in previously published works forming the basis for these simulations. The authors thank Equinor for sharing QemScan data from thin sections, and we appreciate discussions with C. Kruber and R. Meneguolo. We are also grateful to R. Miri, H. Dypvik, and P. Aagaard at UiO for fruitful discussions. Finally, the authors thank the two anonymous reviewers for their insightful comments.

Conflicts of Interest: The authors declare no conflict of interest. The funders had no role in the design of the study; in the collection, analyses, or interpretation of data; in the writing of the manuscript, or in the decision to publish the results.

References

1. Bachu, S. Sequestration of CO_2 in geological media: Criteria and approach for site selection in response to climate change. *Energy Convers. Manag.* **2000**, *41*, 953–970. [CrossRef]

2. Metz, B.; Davidson, O.; De Coninck, H.; Loos, M.; Meyer, L. *IPCC Special Report on Carbon Dioxide Capture and Storage*; Cambridge University Press: Cambridge, UK, 2005.

3. Michael, K.; Golab, A.; Shulakova, V.; Ennis-King, J.; Allinson, G.; Sharma, S.; Aiken, T. Geological storage of CO_2 in saline aquifers—A review of the experience from existing storage operations. *Int. J. Greenh. Gas Control* **2010**, *4*, 659–667. [CrossRef]

4. Halland, E.K.; Johansen, W.T.; Riis, F. *CO_2 Storage Atlas—Norwegian North Sea*; Norwegian Petroleum Directorate: Stavanger, Norway, 2011.

5. Bachu, S. Screening and ranking of sedimentary basins for sequestration of CO_2 in geological media in response to climate change. *Environ. Geol.* **2003**, *44*, 277–289. [CrossRef]

6. Benson, S.M.; Cole, D.R. CO_2 sequestration in deep sedimentary formations. *Elements* **2008**, *4*, 325–331. [CrossRef]

7. Xu, T.; Apps, J.A.; Pruess, K. Mineral sequestration of carbon dioxide in a sandstone–shale system. *Chem. Geol.* **2005**, *217*, 295–318. [CrossRef]

8. André, L.; Audigane, P.; Azaroual, M.; Menjoz, A. Numerical modeling of fluid-rock chemical interactions at the supercritical CO2-liquid interface during CO_2 injection into a carbonate reservoir, the Dogger aquifer (Paris Basin, France). *Energy Convers. Manag.* **2007**, *48*, 1782–1797. [CrossRef]

9. Audigane, P.; Gaus, I.; Czernichowski-Lauriol, I.; Pruess, K.; Xu, T. Two-dimensional reactive transport modeling of CO_2 injection in a saline aquifer at the Sleipner site, North Sea. *Am. J. Sci.* **2007**, *307*, 974–1008. [CrossRef]

10. Pham, V.T.H.; Lu, P.; Aagaard, P.; Zhu, C.; Hellevang, H. On the potential of CO_2–water–rock interactions for CO_2 storage using a modified kinetic model. *Int. J. Greenh. Gas Control* **2011**, *5*, 1002–1015. [CrossRef]

11. Balashov, V.N.; Guthrie, G.D.; Hakala, J.A.; Lopano, C.L.; Rimstidt, J.D.; Brantley, S.L. Predictive modeling of CO_2 sequestration in deep saline sandstone reservoirs: Impacts of geochemical kinetics. *Appl. Geochem.* **2013**, *30*, 41–56. [CrossRef]

12. Kampman, N.; Bickle, M.; Wigley, M.; Dubacq, B. Fluid flow and CO_2-fluid-mineral interactions during CO_2-storage in sedimentary basins. *Chem. Geol.* **2014**, *369*, 22–50. [CrossRef]

13. White, A.F.; Brantley, S.L. The effect of time on the weathering of silicate minerals: Why do weathering rates differ in the laboratory and field? *Chem. Geol.* **2003**, *202*, 479–506. [CrossRef]

14. White, A.F.; Peterson, M.L. Role of reactive-surface-area characterization in geochemical kinetic models. In Chemical modeling of aqueous systems II. *ACS Symp. Ser.* **1999**, *416*, 461–475.

15. Lüttge, A. Etch pit coalescence, surface area, and overall mineral dissolution rates. *Am. Miner.* **2005**, *90*, 1776–1783. [CrossRef]

16. Bolourinejad, P.; Omrani, P.S.; Herber, R. Effect of reactive surface area of minerals on mineralization and carbon dioxide trapping in a depleted gas reservoir. *Int. J. Greenh. Gas Control* **2014**, *21*, 11–22. [CrossRef]

17. Hellevang, H.; Aagaard, P. Can the long-term potential for carbonatization and safe long-term CO_2 storage in sedimentary formations be predicted? *Appl. Geochem.* **2013**, *39*, 108–118. [CrossRef]

18. Hellevang, H.; Pham, V.T.H.; Aagaard, P. Kinetic modelling of CO_2-water-rock interactions. *Int. J. Greenh. Gas Control* **2013**, *15*, 3–15. [CrossRef]

19. Gunter, W.D.; Perkins, E.H.; McCann, T.J. Aquifer disposal of CO_2-rich gases: Reaction design for added capacity. *Energy Convers. Manag.* **1993**, *34*, 941–948. [CrossRef]

20. Perkins, E.H.; Gunter, W.D. Mineral traps for carbon dioxide. In *Aquifer Disposal of Carbon Dioxide, B.*; Hitchon, Geoscience Publishing: Alberta, Canada, 1996; pp. 93–113.

21. Knauss, K.G.; Johnson, J.W.; Steefel, C.I. Evaluation of the impact of CO_2, co-contaminant gas, aqueous fluid and reservoir rock interactions on the geologic sequestration of CO_2. *Chem. Geol.* **2005**, *217*, 339–350. [CrossRef]

22. Peters, C.A. Accessibilities of reactive minerals in consolidated sedimentary rock: An imaging study of three sandstones. *Chem. Geol.* **2009**, *265*, 198–208. [CrossRef]

23. Landrot, G.; Ajo-Franklin, J.B.; Yang, L.; Cabrini, S.; Steefel, C.I. Measurement of accessible reactive surface area in a sandstone, with application to CO_2 mineralization. *Chem. Geol.* **2018**, *318*, 113–125. [CrossRef]

24. Park, J.; Baek, K.; Lee, M.; Chung, C.W.; Wang, S. The use of the surface roughness value to quantify the extent of supercritical CO_2 involved geochemical reaction at a CO_2 sequestration site. *Appl. Sci.* **2017**, *7*, 572. [CrossRef]

25. Ringrose, P.; Bentley, M. *Reservoir Model Design: A Practitioner's Guide*; Springer: Berlin/Heidelberg, Germany, 2014.

26. Northern Lights: A European CO2 Transport and Storage Project. Available online: https://www.slideshare.net/globalccs/northern-lights-a-european-CO2-transport-and-storage-project (accessed on 28 October 2019).

27. Vollset, J.; Doré, A.G. A Revised Triassic and Jurassic lithostratigraphic nomenclature for the Norwegian North Sea. *Nor. Pet. Dir. Bull.* **1984**, *3*, 1–33.

28. Sundal, A.; Miri, R.; Ravn, T.; Aagaard, P. Modelling CO_2 migration in aquifers; considering 3D seismic property data and the effect of site-typical depositional heterogeneities. *Int. J. Greenh. Gas Control* **2015**, *39*, 349–365. [CrossRef]

29. Sundal, A.; Nystuen, J.P.; Rørvik, K.L.; Dypvik, H.; Aagaard, P. The Lower Jurassic Johansen Formation, northern North Sea–Depositional model and reservoir characterization for CO_2 storage. *Mar. Pet. Geol.* **2016**, *77*, 1376–1401. [CrossRef]

30. Marjanac, T.; Steel, R.J. Dunlin Group sequence stratigraphy in the northern North sea: A model for Cook Sandstone deposition. *AAPG bulletin* **1997**, *81*, 276–292.

31. Dreyer, T.; Whitaker, M.; Dexter, J.; Flesche, H.; Larsen, E. From spit system to tide-dominated delta: Integrated reservoir model of the Upper Jurassic Sognefjord Formation on the Troll West Field. In Proceeding of the Petroleum Geology Conference series, London, UK, 1 January 2005; pp. 423–448.

32. Holgate, N.E.; Jackson, C.A.L.; Hampson, G.J.; Dreyer, T. Sedimentology and sequence stratigraphy of the middle–upper Jurassic Krossfjord and Fensfjord formations, Troll Field, northern North Sea. *Pet. Geosci.* **2013**, *19*, 237–258. [CrossRef]

33. Aagaard, P.; Jahren, J.S.; Harstad, A.O.; Nilsen, O.; Ramm, M. Formation of grain-coating chlorite in sandstones. Laboratory synthesized vs. natural occurrences. *Clay Miner.* **2000**, *35*, 261–269. [CrossRef]

34. Ehrenberg, S.N. Preservation of anomalously high porosity in deeply buried sandstones by grain-coating chlorite: Examples from the Norwegian continental shelf. *AAPG Bulletin* **1993**, *77*, 1260–1286.

35. Olivarius, M.; Sundal, A.; Weibel, R.; Gregersen, U.; Baig, I.; Thomsen, T.B.; Kristensen, L.; Hellevang, H.; Nielsen, L.H. Provenance and sediment maturity as controls on CO_2 mineral sequestration potential of the Gassum Formation in Skagerrak. *Front. Sediment. Res.* **2019**. in review.

36. Van Wagoner, J.; Posamentier, H.; Mitchum, R.; Vail, P.; Sarg, J.; Loutit, T.; Hardenbol, J. An overview of sequence stratigraphy and key definitions. In *Sea Level Changes—An Integrated Approach*; Wilgus, C., Hastings, B., Kendall, C.G.S.C., Posamentier, H., Ross, C., Van Wagoner, J., Eds.; SEPM Special Publication: Tulsa, OK, USA, 1988; pp. 39–45.

37. Van Wagoner, J.C.; Mitchum, R.; Campion, K.; Rahmanian, V. Siliciclastic sequence stratigraphy in well logs, cores, and outcrops: Concepts for high-resolution correlation of time and facies. In *Siliciclastic Sequence Stratigraphy in Well Logs, Cores, and Outcrops: Concepts for High-Resolution Correlation of Time and Facies*; American Association of Petroleum Geologists: Tulsa, OK, USA, 1990; pp. 3–55.

38. Hillier, S. Accurate quantitative analysis of clay and other minerals in sandstones by XRD: Comparison of a Rietveld and a reference intensity ratio (RIR) method and the importance of sample preparation. *Clay Miner.* **2000**, *35*, 291–302. [CrossRef]

39. Johnson, J.W.; Nitao, J.J.; Knauss, K.G. Reactive transport modeling of CO_2 storage in saline aquifers to elucidate fundamental processes, trapping mechanisms and sequestration partitioning. In *Geological Storage of Carbon Dioxide*; Bains, S.J., Worden, R.H., Eds.; Geological Society Special Publications: London, UK, 2004; pp. 107–128.

40. Deer, W.A.; Howie, R.A.; Zussman, J. *Rock Forming Minerals: Layered Silicates Excluding Micas and Clay Minerals*; Geological Society: London, UK, 2009.

41. Hillier, S. Pore-lining chlorites in siliciclastic reservoir sandstones: Electron microprobe, SEM and XRD data, and implications for their origin. *Clay Miner.* **1994**, *29*, 665–680. [CrossRef]

42. Hillier, S.; Fallick, A.; Matter, A. Origin of pore-lining chlorite in the aeolian Rotliegend of northern Germany. *Clay Miner.* **1996**, *31*, 153–171. [CrossRef]

43. Jahren, J.; Aagaard, P. Compositional variations in diagenetic chlorites and illites, and relationships with formation-water chemistry. *Clay Miner.* **1989**, *24*, 157–170. [CrossRef]

44. Maast, T.E. Reservoir Quality of Deeply Buried Sandstones—A Study of Burial Diagenesis from the North Sea. Ph.D. Thesis, University of Oslo, Oslo, Norway, 2013.

45. Jahren, J.; Aagaard, P. Diagenetic illite-chlorite assemblages in arenites. I. Chemical evolution. *Clays Clay Miner.* **1992**, *40*, 540. [CrossRef]

46. Palandri, J.L.; Kharaka, Y.K. *A Compilation of Rate Parameters of Water-Mineral Interaction Kinetics for Application to Geochemical Modeling*; OPEN-FILE-2004-1068; US Geological Survey: Menlo Park, CA, USA, 2004.

47. Critelli, T.; Marini, L.; Schott, J.; Mavromatis, V.; Apollaro, C.; Rinder, T.; De Rosa, R.; Oelkers, E.H. Dissolution rates of actinolite and chlorite from a whole-rock experimental study of metabasalt dissolution from $2 \leq pH \leq 12$ at 25 °C. *Chem. Geol.* **2014**, *390*, 100–108. [CrossRef]

48. Smith, M.M.; Wolery, T.J.; Carroll, S.A. Kinetics of chlorite dissolution at elevated temperatures and CO_2 conditions. *Chem. Geol.* **2013**, *347*, 1–8. [CrossRef]

49. Smith, M.M.; Carroll, S.A. Chlorite dissolution kinetics at pH 3–10 and temperature to 275 °C. *Chem. Geol.* **2016**, *421*, 55–64. [CrossRef]

50. Lowson, R.T.; Comarmond, M.-C.J.; Rajaratnam, G.; Brown, P.L. The kinetics of the dissolution of chlorite as a function of pH and at 25 °C. *Geochim. Cosmochim. Acta* **2005**, *69*, 1687–1699. [CrossRef]

51. Bachu, S.; Bennion, B. Effects of in-situ conditions on relative permeability characteristics of CO_2-brine systems. *Environ. Geol.* **2008**, *54*, 1707–1722. [CrossRef]

52. Morad, S. Feldspars in sedimentary rocks. In *Encyclopedia of Sediments and Sedimentary Rocks*; Middleton, G.V., Church, M.J., Coniglio, M., Hardie, L.A., Longstaffe, F.J., Eds.; Springer: Berlin/Heidelberg, Germany, 2003.

53. Aagaard, P.; Egeberg, P.K.; Saigal, G.C.; Morad, S.; Bjørlykke, K. Diagenetic albitization of detrital K-feldspars in Jurassic, Lower Cretaceous and Tertiary clastic reservoir rocks from offshore Norway; II, Formation water chemistry and kinetic considerations. *J. Sediment. Res.* **1990**, *60*, 575–581. [CrossRef]

54. Warren, E.A.; Smalley, P.C. North Sea formation waters atlas. *Oceanogr. Lit. Rev.* **1995**, *6*, 471.

55. Bachu, S.; Gunter, W.D.; Perkins, E.H. Aquifer disposal of CO_2: Hydrodynamic and mineral trapping. *Energy Convers. Manag.* **1994**, *35*, 269–279. [CrossRef]

56. Wawersik, W.R.; Rudnicki, J.W.; Dove, P.; Harris, J.; Logan, J.M.; Pyrak-Nolte, L.; Orr, F.M.; Ortoleva, P.J.; Richter, F.; Warpinski, N.R.; et al. Terrestrial sequestration of CO_2: An assessment of research needs. *Adv. Geophys.* **2001**, *43*, 97–177.

57. Parkhurst, D.L.; Appelo, C.A.J. *Description of Input and Examples for PHREEQC Version 3: A Computer Program for Speciation, Batch-Reaction, One-Dimensional Transport, and Inverse Geochemical Calculations*; U.S. Geological Survey: Reston, VA, USA, 2013; pp. 1–497.

58. Hellevang, H.; Declercq, J.; Kvamme, B.; Aagaard, P. The dissolution rates of dawsonite at pH 0.9 to 5 and temperatures of 22, 60 and 77 °C. *Appl. Geochem.* **2010**, *25*, 1575–1586. [CrossRef]

59. Hellevang, H.; Aagaard, P.; Jahren, J. Will dawsonite form during CO_2 storage? *Greenh. Gases: Sci. Technol.* **2014**, *4*, 191–199. [CrossRef]

60. Tambach, T.J.; Hellevang, H. Discussion on the paper titled "Effect of reactive surface area of minerals on mineralization and carbon dioxide trapping in a depleted gas reservoir" by Bolourinejad et al. (2014). *Int. J. Greenh. Gas Control* **2015**, *100*, 141–143. [CrossRef]

61. Sundal, A.; Hellevang, H.; Miri, R.; Dypvik, H.; Nystuen, J.P.; Aagaard, P. Variations in mineralization potential for CO_2 related to sedimentary facies and burial depth–a comparative study from the North Sea. *Energy Procedia* **2014**, *63*, 5063–5070. [CrossRef]

62. Förster, A.; Schöner, R.; Förster, H.J.; Norden, B.; Blaschke, A.W.; Luckert, J.; Beutler, G.; Gaupp, R.; Rhede, D. Reservoir characterization of a CO_2 storage aquifer: The Upper Triassic Stuttgart Formation in the Northeast German Basin. *Mar. Pet. Geol.* **2010**, *27*, 2156–2172. [CrossRef]

63. Miri, R.; Aagaard, P.; Hellevang, H. Examination of CO_2-SO_2 solubility in water by SAFT1. Implications for CO_2 transport and storage. *J. Phys. Chem.* **2014**, *118*, 10214–10223. [CrossRef]

64. Hovorka, S.D.; Doughty, C.; Benson, S.M.; Pruess, K.; Knox, P.R. The impact of geological heterogeneity on CO_2 storage in brine formations: A case study from the Texas Gulf Coast. *Geol. Soc.* **2004**, *233*, 147–163. [CrossRef]

65. Gaus, I. Role and impact of CO_2-rock interactions during CO_2 storage in sedimentary rocks. *Int. J. Greenh. Gas Control* **2010**, *4*, 73–89. [CrossRef]

66. Bachu, S. Drainage and Imbibition CO_2/Brine Relative Permeability Curves at in Situ Conditions for Sandstone Formations in Western Canada. *Energy Procedia* **2013**, *37*, 4428–4436. [CrossRef]

67. Lowson, R.T.; Brown, P.L.; Comarmond, M.C.J.; Rajaratnam, G. 2007. The kinetics of chlorite dissolution. *Geochim. Cosmochim. Acta* **2007**, *71*, 1431–1447. [CrossRef]

68. Black, J.R.; Haese, R.R. Chlorite dissolution rates under CO_2 saturated conditions from 50 to 120 °C and 120 to 200 bar CO_2. *Geochim. Cosmochim. Acta* **2014**, *125*, 225–240. [CrossRef]

69. Aagaard, P.; Oelkers, E.H.; Schott, J. Glauconite dissolution kinetics and application to CO_2 storage in the subsurface. *Geochim. Cosmochim. Acta* **2004**, *68*, A143.

Article

Injection of a CO_2-Reactive Solution for Wellbore Annulus Leakage Remediation

Laura Wasch * and Mariëlle Koenen

TNO, Princetonlaan 6, 3584 CB Utrecht, The Netherlands; marielle.koenen@tno.nl
* Correspondence: laura.wasch@tno.nl; Tel.: +31-64966157

Received: 31 July 2019; Accepted: 16 October 2019; Published: 22 October 2019

Abstract: Driven by concerns for safe storage of CO_2, substantial effort has been directed on wellbore integrity simulations over the last decade. Since large scale demonstrations of CO_2 storage are planned for the near-future, numerical tools predicting wellbore integrity at field scale are essential to capture the processes of potential leakage and assist in designing leakage mitigation measures. Following this need, we developed a field-scale wellbore model incorporating (1) a de-bonded interface between cement and rock, (2) buoyancy/pressure driven (microannulus) flow of brine and CO_2, (3) CO_2 diffusion and reactivity with cement and (4) chemical cement-rock interaction. The model is aimed at predicting leakage through the microannulus and specifically at assessing methods for CO_2 leakage remediation. The simulations show that for a low enough initial leakage rate, CO_2 leakage is self-limiting due to natural sealing of the microannulus by mineral precipitation. With a high leakage rate, CO_2 leakage results in progressive cement leaching. In case of sustained leakage, a CO_2 reactive solution can be injected in the microannulus to induce calcite precipitation and block the leak path. The simulations showed full clogging of the leak path and increased sealing with time after remediation, indicating the robustness of the leakage remediation by mineral precipitation.

Keywords: CO_2 leakage; cement; well integrity; leakage remediation; TOUGHREACT; reactive transport modelling

1. Introduction

Large scale geological storage of CO_2 can significantly reduce CO_2 emissions and limit global warming [1]. Geological reservoirs are selected for the physical containment of CO_2 which guarantees permanent storage in the subsurface. However, CO_2 injection wells (and possibly old oil and gas wells) penetrating the reservoirs and the caprocks above can compromise the integrity of the storage complex. Wells have a primary structural seal of casing and annular cement (between the casing and the geological formation) and a cement plug when abandoned. Despite these seals, many oil and gas wells leak during their operational lifetime or after abandonment through leakage pathways formed by cement shrinkage or pressure and temperature fluctuations [1]. If annular cement is placed properly, the most likely leak path for CO_2 is along the well through fractures in the cement or microannuli between the cement and the casing or adjacent rock [2–5].

CO_2 leakage through microannuli will cause dissolution of CO_2 in the pore waters which acidifies the near-wellbore environment and causes cement reactivity. Reactions of cement in contact with CO_2-rich water or brine have been extensively studied with experiments and by numerical modelling [6–13]. A typical wellbore cement mineralogy consists of mainly portlandite ($Ca(OH)_2$) and calcium silicate hydrate (CSH), with minor phases such as aluminium-, iron-, magnesium-, or calcium-containing sulphate-, carbonate-, or silicate-hydrates [14]. In general, cement-CO_2 interaction is primarily characterized by portlandite dissolution and subsequent precipitation of calcium carbonate ($CaCO_3$) because of the fast reaction kinetics. The dissolving CSH phase forms additional calcium carbonate and amorphous silica gel (SiO_2) [7]. Characteristic successive, inward moving reaction zones are observed consisting of portlandite dissolution, calcite precipitation, and subsequent calcite dissolution from the rim, leaving a porous, silica-rich rim [6,10–12]. However, cement reactivity will most-likely only lead to cement degradation under leaching/flow conditions when reaction products are quickly removed from the reaction site. At no/low flow conditions, calcite precipitation is the dominant process rather than (re-)dissolution [4,15,16]. Depending on the initial flow and chemical conditions, continuous cement leaching occurs, or cement reactivity may actually support natural sealing of the micro annulus [17–19].

In the case of sustained CO_2 leakage, a corrective measures plan must be in place and appropriate remediation measures should be taken [20,21]. Intentional clogging of the near-well area with salt has been proposed as a preventative measure against CO_2 leakage [22]. This method was based on the capacity of injected CO_2 to evaporate water and precipitate pore filling salt. The process of natural sealing of the microannulus by mineral precipitation indicates a potential for chemical clogging of the annulus leak path. Clogging with calcite or silica has already been proposed for a CO_2 leak path through the caprock [23–25]. To induce mineralization in a leak path, it was proposed to inject a silica- or calcium-rich suspension or solution into the CO_2-containing environment. The injected solution will react in the acid environment to form a solid silica (gel) or carbonate mineral. A modelling study on leakage remediation above a leaking fault through a caprock indicated a final leakage reduction of up to 95% [25]. Experimental [23] and modelling results [24,25] for caprock leakage mitigation support the feasibility of the method for reactive clogging by injecting a CO_2-reactive solution into a high permeable leak path to form solid reactants that clog the leak path, reduce permeability and stop leakage. The main objective of this study is to assess the possibility of reactive leakage mitigation for wellbore annulus leakage.

We developed a field scale reactive transport model based on the model reported by Koenen and Wasch [18] to simulate CO_2 leakage through a microannulus, resulting in either sustained flow and cement leaching or in natural sealing and reduced leakage. For the leakage cases, we study the potential of induced CO_2 mineralization in the leak path, mitigating CO_2 leakage. The numerical modelling study includes the following processes:

- Flow of supercritical CO_2 and brine along the initially water-saturated microannulus;
- Diffusion of dissolved CO_2 into the caprock;
- Diffusion of dissolved CO_2 into impermeable cement and reactions of dissolved CO_2 and cement;
- Leakage into the aquifer overlying the caprock;
- Injection of a CO_2-reactive solution in the microannulus leak path to promote clogging by calcite formation;

In this paper, we report on microannulus leakage (versus sealing), the intentional clogging process for leakage remediation, and the post-clogging phase to assess the sustainability of the clogging procedure.

2. Materials and Methods

A reactive transport model was developed in TOUGHREACT (Version 3, Lawrence Berkeley National Laboratory, Oakland, CA, USA) [26], a simulator for coupled modelling of multiphase fluid and heat flow, solute transport, and chemical reactions by introducing reactive transport into the flow simulator TOUGH2. We use the ECO2N fluid property module to include the thermodynamic and thermophysical properties of H_2O–$NaCl$–CO_2 mixtures [27].

A 2D axisymmetric field-scale model (Figure 1) was adapted from the model reported in [18]. The well consists of two cemented casings: a 9 5/8″ casing of 3590 m depth into the reservoir and a 13 3/8″ casing to a depth of 2569 m. We assume a 0.5 cm thick casing (unreactive transport barrier) and a 3 cm thick annular cement. A 500-micron thick, porous and highly permeable micro-annulus is defined between the cement and the adjacent rock. For the rock formations, three layers are defined: a reservoir (infinite volume, representing a large storage reservoir), an impermeable caprock (550 m thick) and an overlying, permeable aquifer (3040 m thick). Simulations are performed in isothermal mode, with a (fixed) temperature gradient from reservoir to surface. To reduce simulation times, the upper 2500 m of the aquifer to the surface are removed after initializing pressure and temperature. The resulting mesh contains 70 layers and 36 columns. Vertical mesh refinement is defined around the layer interfaces and at the leakage remediation interval down to 5 cm. The mesh is refined towards the well in the radial direction, down to cells of 0.5 cm width for the annular cement. The pressure is equilibrated with depth; 25 °C and 1.5 bar at the top and 90 °C and 327 bar at reservoir level. This yields a 1180 m thick model with a radius of 150 m (Figure 1). The boundary conditions are represented by the infinite volume reservoir with a 0.8 CO_2 saturation and an infinite volume upper boundary for the microannulus. The reservoir was given a 20 bar overpressure to simulate upward leakage out of the CO_2 reservoir.

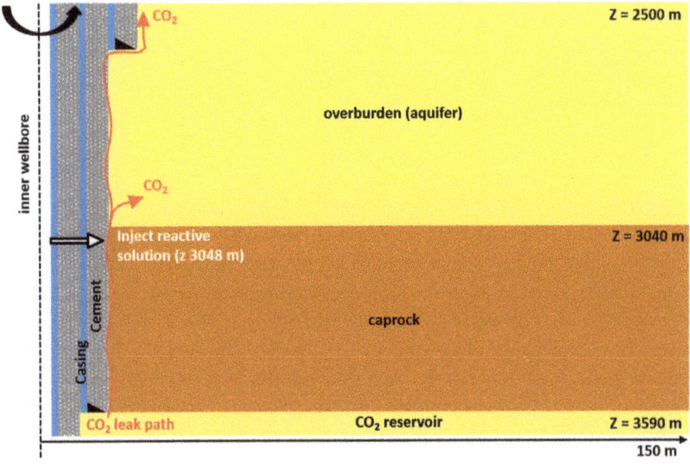

Figure 1. Schematic overview of the 2D radial symmetric TOUGHREACT model showing: the model dimensions, the different rock formations, the well and the leak path through the wellbore microannulus, and the location of leakage remediation.

The TOUGHREACT software is developed to simulate reactive transport through porous media and does not allow open space. Therefore, we defined the microannulus as cement material with a 0.9 volume fraction porosity, with the remaining 0.1 volume fraction made up by cement mineralogy. The 500-micron microannulus hence has an effective thickness of 450 micron, but the width of the microannulus may increase by 10% in case of cement mineral dissolution. The microannulus width (aperture) can be related to its permeability by the cubic law. For an aperture of 500 micron the permeability would be 4.2×10^{-9} m^2 (~4200 Darcy), however, the actual hydraulic permeability will be lower as it is affected by many factors such as fracture wall roughness. To account for this, simulations have been previously run with various initial (hydraulic) permeability values between 5×10^{-13} m^2 (~500 mDarcy) and 8.3×10^{-10} m^2, covering values reported in literature [18]. For the present study we use a permeability range between 1×10^{-13} m^2 and 1×10^{-11} m^2 to capture the uncertainty and natural variation in microannulus flow. For the base case, a permeability of 1.3×10^{-12} m^2 is selected.

The flow properties of the different materials used in the model are summarized in Table 1. For relative permeability, we use the equation from Corey's curve with a residual liquid and gas saturation of 0.02 and 0.1. The diffusion coefficient is calculated in TOUGHREACT by multiplying a standard diffusion coefficient (1×10^{-9} m^2/s) by the tortuosity, porosity and liquid saturation. Hence the effective diffusion coefficient will change over time as the porosity and saturation develop due to fluid flow and mineral reactions.

Table 1. Transport properties of the different materials used in the model. * In TOUGHREACT a permeability of 1.0×10^{-30} m^2 represents impermeable material.

Transport Properties	Unit	Reservoir	Cement	Caprock	Microannulus	Aquifer
Porosity	φ (-)	0.2	0.3	0.1	0.9	0.2
Permeability	k (m^2)	2.0×10^{-13}	1.0×10^{-30} *	1.0×10^{-30} *	1.3×10^{-12}	2.0×10^{-13}
Tortuosity	τ (-)	0.4	0.01	0.05	0.4	0.4
Capillary Curve						
Van Genuchten	λ (-)	0.2	-	-	-	0.2
	S_{lr} (-)	0.1	-	-	-	0.1
	$1/P_0$ (-)	0.002	-	-	-	0.002
	P_0 (Pa)	1.0×10^8	-	-	-	1.0×10^8
	S_{ls} (-)	1	-	-	-	1
TRUST	P_0 (Pa)	-	1.0×10^6	1.0×10^6	-	-
	S_{lr} (-)	-	0.2	0.2	-	-
	η (-)	-	0.4	0.4	-	-
	P_e (Pa)	-	4.0×10^6	4.0×10^6	-	-

The cement consists of portlandite, CSH_1.6 (CSH of a 1.6 Ca/Si ratio), monosulfoaluminate and hydrotalcite (Table 2). The secondary minerals for cement are calcite, amorphous silica, anhydrite, dolomite and gibbsite. A clastic rock was included consisting of quartz, albite, microcline, kaolinite, anhydrite, dolomite and calcite. We used the thermodynamic database Thermoddem (version 1.10, 6 Jun 2017, BRGM, France) to model chemical reactions [28]. The reaction of minerals is kinetically controlled using a rate expression of Lasaga et al. [29]. Mineral kinetics are listed in Table 3. The specific surface area is assumed 0.98 m^2/kg, except for the C-S-H phases and clays for which a value of 100 m^2/kg is assumed. The CO$_2$-reactive solution to stimulate microannulus clogging is designed by equilibrating lime with a sodium chloride brine at surface conditions. The different fluid compositions are listed in Table 4.

Table 2. Initial mineralogy of cement and aquifer/caprock and possible secondary minerals.

Mineral-Formula	Cement (Volume Fraction)	Aquifer/Caprock (Volume Fraction)
Portlandite-(Ca(OH)$_2$)	0.2	-
CSH(1.6)-(Ca$_{1.60}$SiO$_{3.6}$:2.58H$_2$O)	0.6	-
Monosulfoaluminate-(Ca$_4$Al$_2$SO$_{10}$:12H$_2$O)	0.1	-
Hydrotalcite-(Mg$_4$Al$_2$O$_7$:10H$_2$O)	0.1	-
Quartz-(SiO$_2$)	-	0.7
Calcite-(CaCO$_3$)	-	0.01
Amorphous silica-(SiO$_2$)	-	-
Anhydrite-(CaSO$_4$)	-	0.01
Dolomite-(CaMg(CO$_3$)$_2$)	-	0.01
Gibbsite-(Al(OH)$_3$)	-	-
Microcline-(K(AlSi$_3$)O$_8$)	-	0.1
Albite-(NaAlSi$_3$O$_8$)	-	0.1
Kaolinite-(Al$_2$Si$_2$O$_5$(OH)$_4$)	-	0.07

Porosity changes due to water-rock reactions are calculated in TOUGHREACT using mineral molar volumes. Porosity change can be related to permeability change by porosity-permeability relations, but they contain highly uncertain and material specific input parameters [30]. The Verma–Pruess relation (Equation (1)) is an extension of a Power Law relation considering a critical porosity at which the permeability is assumed zero. This relation is often used for mineral precipitation processes given the high impact of the critical porosity on porosities decreasing during mineral precipitation, while the relevance of critical porosity in the mathematical expression diminishes for porosities increasing during dissolution [30]. We used the porosity-permeability relationship of Verma and Pruess [31] and addressed the uncertainty of the input parameters by a sensitivity study. The base case values are a 0.80 critical porosity and a power law component of 6.

$$\frac{k}{k_i} = \left(\frac{\varphi - \varphi_c}{\varphi_i - \varphi_c}\right)^n \tag{1}$$

where k is the permeability, k_i the initial permeability, φ the porosity, φ_i the initial porosity, φ_c the critical porosity below which the permeability is assumed zero, and n is a power law exponent.

Table 3. Kinetic rate parameters (*[a] [32], *[b] [33] and *[c] [34]).

Mineral	Acid Mechanism			Neutral Mechanism		Carbonate/Base Mechanism		
	Log (k25 °C) (mol/m^2/s)	ΔH (kJ/mol)	n	Log (k25 °C) (mol/m^2/s)	ΔH (kJ/mol)	Log (k25 °C) (mol/m^2/s)	ΔH (kJ/mol)	n
Portlandite *[c]	−3.10	75	0.6	−7.66	75	-	-	-
C1.6SH *[c]	−7.23	23	−0.28	-	-	-	-	-
Monosulf. *[b]	−3.09	74.9	0.6	−11.2	15	-	-	-
Hydrotal. *[b]	−7.23	15	0.28	−17.8	15	-	-	-
Quartz *[a]	−7.52	56.9	0.5	−4.55	56.9	-	-	-
Calcite *[a]	−0.30	14.4	1	−5.81	23.5	−3.48	35.4	1
Silica(am) *[a]	-	-	-	−9.42	49.8	-	-	-
Anhydrite *[a]	-	-	-	−3.19	14.3	-	-	-
Dolomite *[a]	−3.19	36.1	0.5	−7.53	52.2	−5.11	34.8	0.5
Gibbsite *[a]	−7.65	47.5	0.99	−11.5	61.2	−16.7	80.1	−0.78
Albite *[a]	−10.2	65	0.46	−12.6	69.8	−15.6	71	−0.57

Table 4. Initial composition of the different fluids, dissolved species in mol/L.

	Cement	Aquifer/Caprock	CO$_2$ Reservoir	CO$_2$-Reactive Solution
pH (-)	10.8	6.2	3.5	13.8
Ca^{2+}	1.30×10^{-2}	3.15×10^{-2}	3.10×10^{-2}	0.68
Mg^{2+}	1.69×10^{-8}	9.32×10^{-3}	9.16×10^{-3}	-
Na$^+$	1.0	0.98	0.98	1.03
K$^+$	-	8.62×10^{-3}	8.48×10^{-3}	-
H$_4$SiO$_4$	4.10×10^{-6}	8.57×10^{-4}	8.46×10^{-4}	-
HCO$_3^-$	-	8.29×10^{-3}	1.78	-
SO$_4^{2-}$	2.07×10^{-4}	3.25×10^{-2}	3.20×10^{-2}	-
Al^{3+}	4.14×10^{-4}	7.24×10^{-8}	7.12×10^{-8}	-
Cl$^-$	1.0	1.0	0.98	1.03

3. Results of CO$_2$ Leakage and Natural Sealing

3.1. CO$_2$ Leakage and Cement Reactions

The reactive transport simulations show upward flow of CO$_2$ through the microannulus. With the applied 20 bar overpressure, the flow rate of gaseous CO$_2$ through the microannulus is 2.1×10^{-5} kg/s. During upward flow, a fraction of the CO$_2$ gas dissolves into the cement and caprock pore water and diffuses horizontally into the cement and the caprock, thereby lowering the pH of the pore waters. Cement minerals react with the carbonized water and start to dissolve to buffer the pH. Similar to e.g., Kutchko et al. [6], we observe inward progression of reaction zones in the horizontal direction, which is limited by diffusion (Figure 2). The reaction zones are characterized by: portlandite dissolution and calcite formation, CSH dissolution with amorphous silica and calcite precipitation, monosulfoaluminate and hydrotalcite dissolution with dolomite (max. 0.003 volume fraction), gibbsite (max. 0.0025 volume fraction) and anhydrite (max. 0.009 volume fraction) precipitation. The further up from the reservoir level, the less advanced the horizontal progression of these zones is, since the reactions only start as soon as the upward migration of the CO_2 reaches that level and the pH decreases. After 2 years of CO_2 leakage, approximately 1 cm of the cement just above reservoir level is affected by horizontal CO_2 diffusion and related reactions (Figure 2). The silicate minerals of the caprock do not show siginificant reactions within the simulated time. There is a minor increase of calcite in the caprock adjacent to the microannulus.

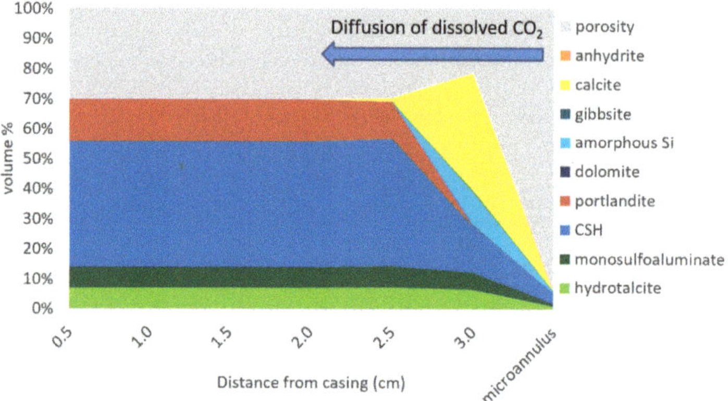

Figure 2. The cement mineralogy and porosity after two years of CO_2 leakage are plotted for a horizontal transect across the cement sheet and the microannulus, located just above the CO_2 reservoir. Cement reacts with CO_2 from the microannulus on the right. Note that, even though not well visible, small amounts of dolomite, anhydrite and gibbsite precipitate in the microannulus and cement.

After the CO_2 within the microannulus reached the top of the caprock, the pH is around 4.5 throughout the microannulus. In the cement affected by CO_2 interactions-adjacent to the microannulus-the pH is roughly 5 and in the unaltered cement nearly 11 (Figure 3). The pH decrease causes full dissolution of portlandite in the microannulus and adjacent cement cells and gradual dissolution of the other cement phases such as CSH, with decreasing amount of dissolution from the reservoir level upwards. As a result, secondary calcite is also highest at the level close to the reseroir (Figure 3). Within the microannulus calcite that precipitated is re-dissolved due to the low pH and flow conditions which allow quick removal of dissolved species. The permeability of the microannulus is predicted to increase from 1.3×10^{-12} m^2 to 1.9×10^{-12} m^2 after 2 years of CO_2 flow and corresponding cement alteration.

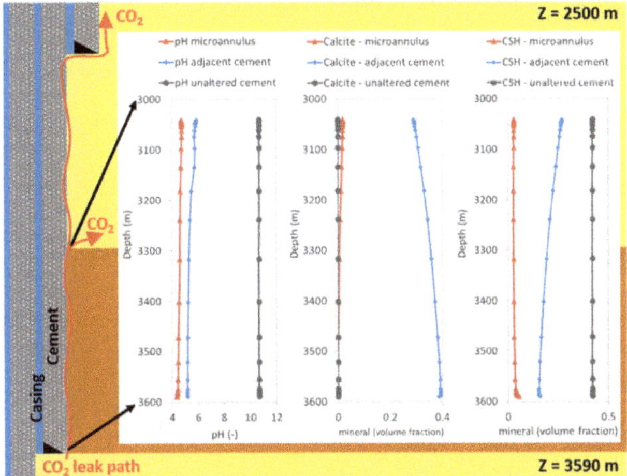

Figure 3. The background shows a schematic overview of the model with a cement sheet, the microannulus leak path and the adjacent formation rock. The pH, calcium silicate hydrate (CSH) content and calcite content after two years of CO_2 leakage are plotted for the lower part of the annulus adjacent to the caprock.

3.2. Microannulus Leakage Versus Natural Sealing

Four permeability values were selected to simulate different initial leakage rates and to assess the chemical processes within a leaking or self-sealing annulus. Results are discussed after half a year of leakage simulation. The different permeabilities yield different levels of gas saturation, with a higher gas saturation for a higher permeability (Figure 4a). For all scenarios, the flow of CO_2 and dissolution of CO_2 in the initially water saturated microannulus result in complete dissolution of portlandite within the microannulus—which was 10% cement filled—and subsequent precipitation of calcite (Figure 4b). Above the reservoir, re-dissolution of calcite can be observed. Only for the lowest permeability of 1×10^{-13} m^2, CO_2 gas does not reach the top of the caprock. This is due to natural sealing of the microannulus, with complete calcite clogging in the middle part of the microannulus. The front of calcite precipitation is characterised by a peak in calcium content (Figure 4c) and an increase in pH (Figure 4d). The permeability of the microannulus depends on the initial permeability and the dissolution and precipitation reactions. There is a high permeability near the reservoir where the calcite content is lowest (Figure 4e). The 1×10^{-11} and 5×10^{-11} m^2 scenarios show only little permeability change in the upper part of the microannulus due to portlandite dissolution and calcite precipitation. The initial permeability value 1×10^{-12} m^2 is more affected by calcite precipitation and the 1×10^{-13} m^2 scenario shows complete permeability impairment due to calcite clogging.

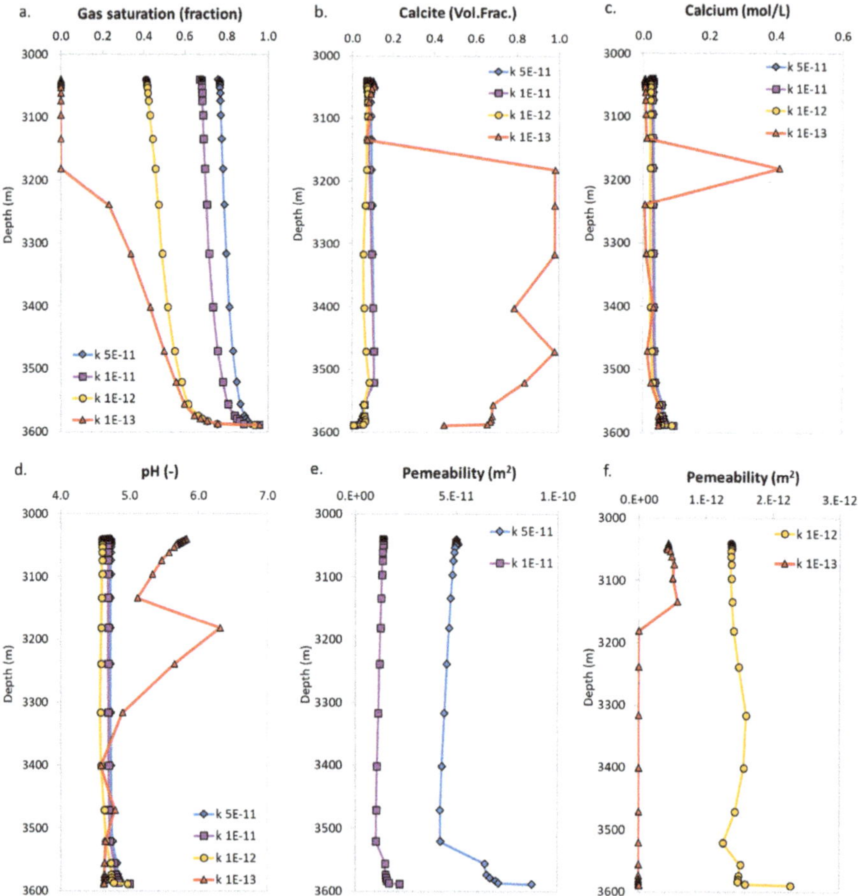

Figure 4. Simulation results along the microannulus for four different initial microannulus permeabilities showing: (**a**) gas saturation, (**b**) calcite volume fraction, (**c**) dissolved calcium, (**d**) pH, and (**e,f**) the permeability of the high and low initial permeability scenarios respectively.

4. Results of CO$_2$ Reactive Leakage Remediation

4.1. Injection of the CO$_2$-Reactive Solution with Different Injection Pressures

The 1×10^{-12} m^2 permeability scenario is selected to model initial leakage and subsequent leakage remediation. Upward CO$_2$ leakage through the microannulus was simulated for half a year before leakage remediation was applied. To remediate leakage, a CO$_2$-reactive solution (composition given in Table 4) was injected into the CO$_2$ containing microannulus. The solution is injected at a depth of 3048 m, which is 8 m below the top of the caprock. The CO$_2$-reactive, lime-saturated solution is injected in order to react with dissolved CO$_2$ to form calcite (CaO + CO$_2$- > CaCO$_3$). To inject the CO$_2$-reactive solution in the model, we used a fixed pressure cell adjacent to the microannulus to allow for pressure-controlled injection, instead of defining a fixed injection rate. The sensitivity of leakage remediation to flow rates was asessed by varying the pressure of injection with 1, 5 and 10 bar overpressure for the microannulus pressure of 324 bar.

For all three injection pressures, the injection of the CO$_2$-reactive solution leads to an initial increase followed by an interruption of the upward CO$_2$ flow in the microannulus, but flow recovers (after $t = 0$ in Figure 5, showing the results of 5 bar overpressure injection). Calcite starts to precipitate in the microannulus at the level where the CO_2-reactive solution is injected. Gradually the permeability of the microannulus reduces with a corresponding decrease of the CO_2 and water flow rate (Figure 5). The rate of solution injection reduces as well, since the permeability decreases and the pressure is fixed. Figure 6a shows the initial higher injection rate with a higher overpressure and the decrease of injection rate with time. The moment of complete flow impairment occurs at the time when the microannulus adjacent to the injection cell is nearly filled with calcite, reducing the permeability to zero (Figure 5). All overpressures yield complete calcite clogging (Figure 6b), but with differences in the development of calcite precipitation due to the differences in the balance of calcium and CO_2 supply. Within 10 days, all injection pressure scenarios predict clogging the microannulus with calcite precipitation, preventing further leakage of the CO_2.

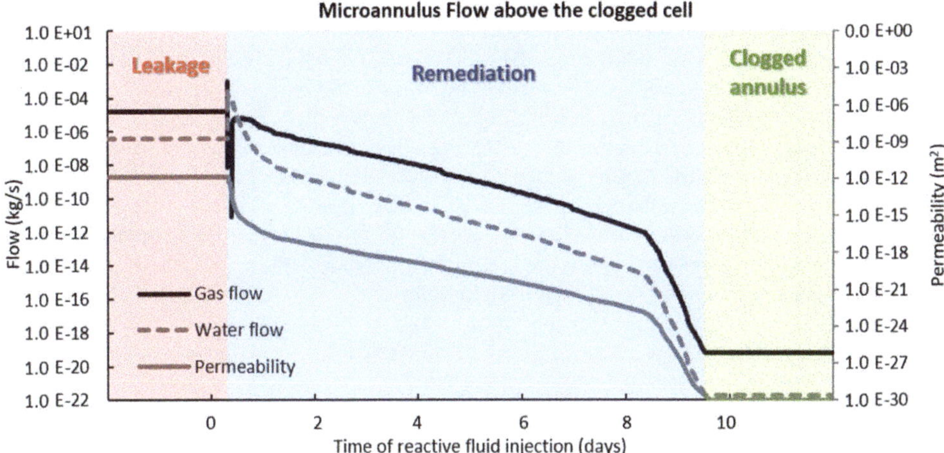

Figure 5. The development of CO_2 and water flow rate (left axis, logarithmic) and the permeability (right axis, logarithmic) in the microannulus at the level of injection of the CO_2-reactive solution. Results are from the 5 bar overpressure scenario. Data is plotted with time for the period around leakage remediation.

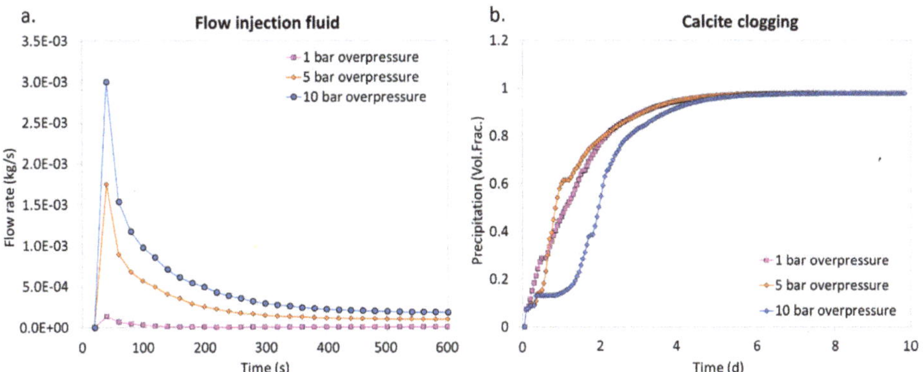

Figure 6. (**a**) The injection rate of the reactive solution for the three different injection pressures at the onset of the remediation procedure. (**b**) The evolution of calcite precipitation showing full clogging for all scenarios within the 10 days of leakage remediation.

4.2. Sensitivity to Porosity-Permeability Relation Input Parameters

The clogging process and specfically the calculation of the permeability based on the porosity development depends on highly uncertain porosity-permeability parameters [30]. A sensitivity study is performed reflecting the range in values for the critical porosity (φ_c) and power law (n) component as probed by Verma and Pruess [31] ($1 \leq n \leq 6, 0.8\varphi \leq \varphi_c \leq 0.9\varphi$) and Xu et al. [35] ($4 \leq n \leq 13, 0.88\varphi \leq \varphi_c \leq 0.94\varphi$). The initial leakage and subsequent remediation simulations (with a 1×10^{-12} m^2 permeability) were repeated for six combinations of 2 critical porosity values of 0.8 and 0.88 (80 and 88% reduction of the original porosity) with three different power law components of 2, 6 and 10.

The calcite plug formed using the different porosity-permeability relation input parameters is very similar (Figure 7). A peak of calcite precipitation is observed next to the cell from which the reactive solution is injected. Above this interval, the microannulus is only partially clogged due to calcite precipitation. The relative insensitivity of the characteristics of the calcite plug to the porosity-permeability relationship can be explained by the nature of the remediation process. The process is designed to inject the reactive solution up to full clogging, meaning that with all parameters a full permeability reduction will be simulated. There is a small difference in the calcite formed in the upper part of the plug, with the $0.88\varphi_c$–$2n$ model yielding the most precipitation and a combinition of $0.8\varphi_c$ and $10n$ the least. The used porosity-permeability parameters do have a large impact on the predicted time that is required to achieve full clogging. The time it takes to perform the remediation method ranges from 7 to 113 days (Table 5). This indicates the importance of the porosity-permeability parameters for the prediction of the duration of the remediation procedure and for the asessment of the related costs and overall feasibility.

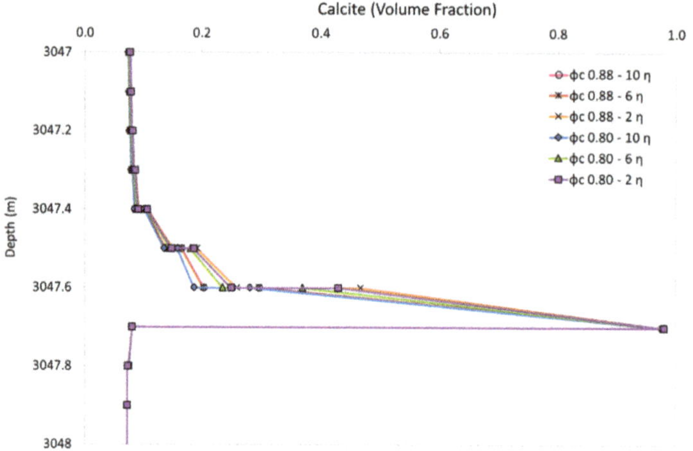

Figure 7. The calcite content for a 1 m section of the microannulus after leakage remediation, showing a peak adjacent to the level of reactive solution injection. The results are shown for six scenarios of different porosity-permeability relation input parameters.

Table 5. The predicted time of remediation up to full clogging of the microannulus.

Scenario ($\varphi_c - \eta$)	0.8–6	0.8–2	0.8–10	0.88–2	0.88–6	0.88–10
Remediation Time (d)	39	7	48	11	45	113

4.3. Stability of the Plug in Time

After the remediation procedure, the reactive transport simulation is continued for 1 year to assess the stability of the calcite plug with time. This allows for equilibration of the system and continuation of diffusion and possibly flow. To assess the stability of the plug and the chemical evolution within the microannulus, two porosity-permeability scenarios were selected representing the most $(0.88\varphi_c - 2n)$ and least calcite precipitation $(0.80\varphi_c - 10n)$.

Throughout the microannulus, the pH remains around 4.6 in the post-remediation phase, indicating the wellbore environment is still acidic and is not buffered by the cement within the year after leakage remediation that was simulated. The caprock minerals show no significant reactions in this time period. The main observed process is the increase in calcite volume fraction throughout the microannulus (Figure 8a), the thin original plug is plotted for comparison. After the remediation method stops CO_2 leakage, the process of natural sealing becomes dominant throughout the micoannulus adjacent to caprock due to the absence of flow. This results in clogging by mainly calcite precipitation and minor amorphous silica precipitation. After one year of reactions following the remediation procedure, the $0.80\varphi_c$–$10n$ scenario yields full clogging of the microannulus (Figure 8a,b). The $0.88\varphi_c$–$2n$ scenario does have two sections of partial clogging, but full clogging of the rest of the microannulus (Figure 8b). The partial clogging above the original plug is due to absence of CO_2 which has been completely consumed, yielding a zero gas saturation (Figure 8c). The gas saturation is highest just below the plug and just above the CO_2 reservoir. The zero permeability of the plug continues to block the CO_2 flow and causes the gas saturation above the clogged level to decrease with time as CO_2 migrates and is consumed by reactions. The permeability impairment forms a pressure block, with the microannulus below the plug approaching the CO_2 reservoir pressure and the microannulus above the plug retaining hydrostatic pressure.

The two porosity-permeability scenarios both yield sigificant natural sealing after clogging by injection of the remediation fluid. This indicates that reduction of leakage due to the remediation procedure enhances the natural capability of the wellbore system to seal and form a barrier against future leakage.

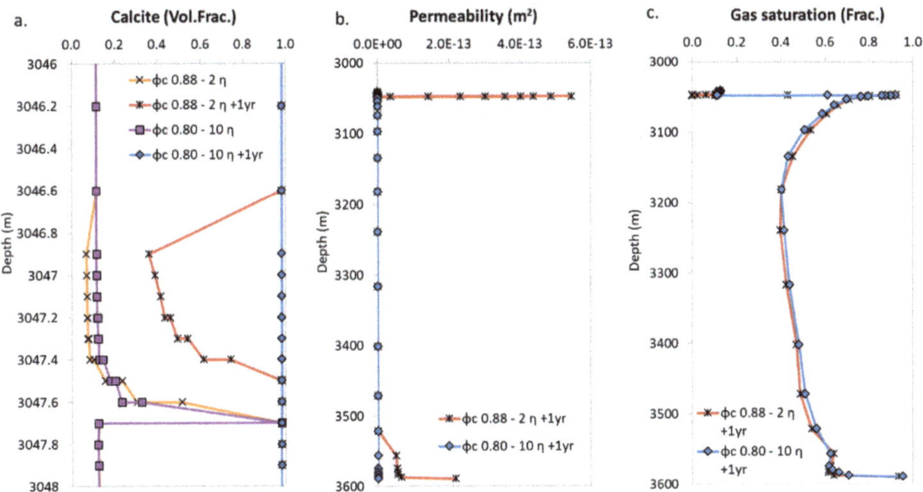

Figure 8. The development of the microannulus within 1 year after remediation. Results are shown for (**a**) calcite around the original plug, (**b**) permeability of the microannulus, and (**c**) gas saturation of the microannulus adjacent to the caprock.

5. Discussion and Conclusions

Despite the function of annular cement as a seal preventing oil, natural gas or stored CO_2 to migrate to aquifers or to the surface, wells are known to leak due to microannuli formed by processes such as cement shrinkage or pressure and temperature fluctuations [2,5]. The width of a microannulus formed tends to increase for larger temperature differences between the produced or injected fluid and the rock formation, which is especially relevant for cold CO_2 injection [36]. The higher risk of microannulus formation during CO_2 injection combined with high abandonment pressures asks for an assessment of CO_2 microannulus leakage and methods for leakage remediation. Due to the high reactivity of cement with carbonated brine, the chemical processes are key. A field-scale wellbore model was developed which successfully incorporates CO_2 migration by two-phase flow through a microannulus and diffusion of dissolved CO_2 into the adjacent caprock and cement. This enables the simulation of the complex reactive transport processes of a storage system, including a storage reservoir, wellbore cement with a continuous microannulus from reservoir to caprock, and caprock overburden. Despite the large scale of the model, it was successful in predicting the well-known small-scale reaction characteristics as found in experimental and (small scale) modelling studies [6,8–12].

Previous simulations showed an initial critical CO_2 leakage velocity of 0.1×10^{-5} m/s below which calcium released from dissolving cement minerals can diffuse towards the microannulus where calcite forms which clogs the microannulus and prevents further flow [18]. For our models, a low enough leakage rate to facilitate natural sealing was achieved with an initial microannulus permeability of 1×10^{-13} m^2. The possibility for natural microannulus clogging at low leakage rates is similar to the self-sealing potential of fractures in cement as demonstrated by e.g., Huerta et al. [16]. They discussed the critical residence time for the CO_2 fluid to be present in a cement sample for a fracture to close. However, natural sealing is not solely dependent on the flow rate as determined by the systems permeability. Nonuniformity of the microannulus geometry can lead to local changes in flow velocity affecting the sealing process [19]. The specific chemistry of the cement and rock formations has a large impact, with, for example, a high potential for calcite forming in the microannulus when the host rock is a carbonate [17]. Previous modelling showed anhydrite clogging of the microannulus related to relatively high sulphate concentration in the formation water of the caprock [18], whereas the chemical characteristics of the cement and surrounding formation water in our model led to dominant calcite clogging. The thermodynamics and kinetics of the cement phases still pose uncertainty in the chemical processes in the microannulus [17]. Our kinetic parameters for CSH and silica may be considered conservative, yielding only minor amorphous silica in the microannulus. In addition, the uncertainty and variability in the reactive surface area of cement phases may further affect the predicted leaching of cement and natural sealing process. A dedicated uncertainty assessment with varying parameters for dissolution/precipitation kinetics and mineral reactive surface areas was out of the scope of this study but would be needed to assess the sensitivity to natural sealing of the microannulus.

For high leakage rates when natural sealing is not predicted to occur, the process of microannulus clogging can be induced by adding calcium to the system [22–24]. We injected a calcium-rich brine to react with the dissolved CO_2 in the microannulus, yielding a full permeability decrease due to calcite precipitation, as graphically represented in Figure 9. There are large uncertainties in the clogging process regarding the porosity-permeability relation of mineral precipitation in a microannulus. As discussed by Ito et al. [23] and Druhan et al. [24], the porosity-permeability relation is of utmost importance for predicting effective leakage remediation. However, compared to our previous numerical modelling study [22] in which we injected the CO_2-reactive solution in an aquifer above a caprock leak, leakage remediation in the microannulus was predicted to be more successful and far less sensitive to the porosity-permeability relation. This is due to the confined nature of a microannulus and the more difficult placement of a plug above a caprock leak path. In our study, the uncertainty of the porosity-permeability relation was primarily expressed in the time it takes for remediation and not in the success of remediation. A longer remediation time is related to the larger amount of mineral precipitation and porosity reduction that is required to achieve full permeability reduction when using a more conservative porosity-permeability relationship. Hence, accurate design of the remediation procedure requires additional data on the porosity-permeability behaviour of a microannulus. The previous study [22], indicated the significance of the leakage rate on the success of leakage remediation. The design of the remediation procedure would require knowledge on the actual leakage rate or a numerical sensitivity study on the possible range in initial microannulus permeability and resulting leakage rate. With a higher initial leakage rate, injection of the reactive solution might require a higher injection pressure.

Permanent leakage remediation, considering long-term CO_2 storage, requires a chemically stable plug in the leak path. Our model results indicate that the formed calcite plug does not only remain stable, but that cessation of flow enables natural sealing in the microannulus at the level below the plug. The increase in sealing of the microannulus enhances the potential for intentional clogging as a remediation method. Future work could focus on the sensitivity of the intentional and subsequent natural clogging process to the chemistry of the CO_2 reactive solution and the cement and formation rock chemistry. The chemical nature of the plug is of less importance when subsequent natural sealing can take over the barrier function, even if a placed plug would degrade in time.

Wellbore microannulus leakage

Microannulus clogging

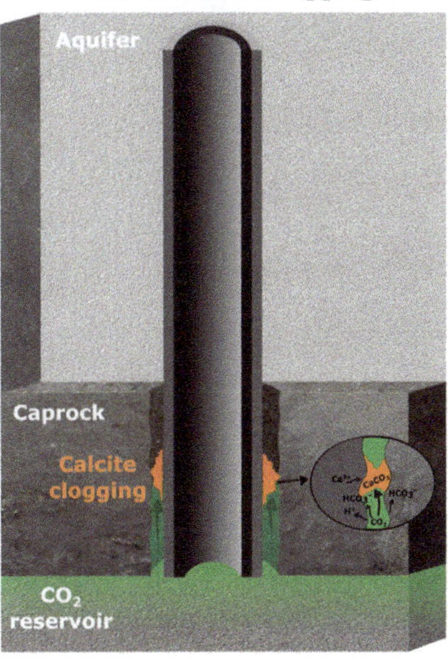

Figure 9. Graphical representation of the CO_2 leakage and CO_2-reactive remediation.

Author Contributions: Conceptualization, L.W. and M.K.; methodology, L.W. and M.K.; software, L.W. and M.K.; validation, L.W.; formal analysis, L.W.; investigation, L.W.; resources, L.W.; data curation, L.W.; writing—original draft preparation, L.W.; writing—review and editing, L.W. and M.K.; visualization, L.W.; supervision, L.W.; project administration, L.W.

Funding: This research received no external funding.

Conflicts of Interest: The authors declare no conflict of interest.

References

1. Metz, B.; Davidson, O.; Coninck, H.D.; Loos, M.; Meyer, L. (Eds.) *IPCC Special Report on Carbon Dioxide Capture and Storage*; Cambridge University Press: New York, NY, USA, 2005.
2. Dusseault, M.B.; Gray, M.N.; Nawrocki, P.A. Why oilwells leak: Cement behavior and long-term consequences. In Proceedings of the International Oil and Gas Conference and Exhibition in China, Beijing, China, 7–10 November 2000.
3. Bachu, S.; Bennion, D.B. Experimental assessment of brine and/or CO_2 leakage through well cement at reservoir conditions. *Int. J. Greenh. Gas Control* **2009**, *3*, 494–501. [CrossRef]

4. Carey, J.W.; Svec, R.; Grigg, R.; Zhang, J.; Crow, W. Experimental investigation of wellbore integrity and CO_2-brine flow along the casing-cement microannulus. *Int. J. Greenh. Gas Control* **2010**, *4*, 272–282. [CrossRef]

5. Gasda, S.E.; Bachu, S.; Celia, M.A. Spatial characterization of the location of potentially leaky wells penetrating a deep saline aquifer in a mature sedimentary basin. *Environ. Geol.* **2004**, *46*, 707–720. [CrossRef]

6. Kutchko, B.G.; Strazisar, B.R.; Dzombak, D.A.; Lowry, G.V.; Thauiow, N. Degradation of well cement by CO_2 under geologic sequestration conditions. *Environ. Sci. Technol.* **2007**, *41*, 4787–4792. [CrossRef]

7. Carey, J.W.; Wigand, M.; Chipera, S.J.; WoldeGabriel, G.; Pawar, R.; Lichtner, P.C.; Wehner, S.C.; Raines, M.A.; Guthrie, G.D., Jr. Analysis and performance of oil well cement with 30 years of CO_2 exposure from the SACROC Unit, West Texas, USA. *Int. J. Greenh. Gas Control* **2007**, *1*, 75–85. [CrossRef]

8. Huet, B.; Prevost, J.H.; Scherer, G.W. Quantitative reactive transport modeling of Portland cement in CO_2-saturated water. *Int. J. Greenh. Gas Control* **2010**, *4*, 561–574. [CrossRef]

9. Gherardi, F.; Audigane, P.; Gaucher, E.C. Predicting long-term geochemical alteration of wellbore cement in a generic geological CO_2 confinement site: Tackling a difficult reactive transport modeling challenge. *J. Hydrol.* **2012**, *420*, 340–359. [CrossRef]

10. Duguid, A.; Scherer, G.W. Degradation of oilwell cement due to exposure to carbonated brine. *Int. J. Greenh. Gas Control* **2010**, *4*, 546–560. [CrossRef]

11. Rimmelé, G.; Barlet-Gouédard, V.; Porcherie, O.; Goffé, B.; Brunet, F. Heterogeneous porosity distribution in Portland cement exposed to CO_2-rich fluids. *Cem. Concr. Res.* **2008**, *38*, 1038–1048. [CrossRef]

12. Barlet-Gouédard, V.; Rimmelé, G.; Porcherie, O.; Quisel, N.; Desroches, J. A solution against well cement degradation under CO_2 geological storage environment. *Int. J. Greenh. Gas Control* **2009**, *3*, 206–216. [CrossRef]

13. Temitope, A.; Gupta, I. A review of reactive transport modeling in wellbore integrity problems. *J. Petrol. Sci. Eng.* **2019**, *175*, 785–803.

14. Lea, F.M. The constitution of Portland cement. *Q. Rev. Chem. Soc.* **1949**, *3*, 82–93. [CrossRef]

15. Wasch, L.J.; Koenen, M.; Wollenweber, J.; Tambach, T.J. Sensitivity of chemical cement alteration—Modeling the effect of parameter uncertainty and varying subsurface conditions. *GHG S T* **2015**, *5*, 323–338. [CrossRef]

16. Huerta, N.J.; Hesse, M.A.; Bryant, S.L.; Strazisar, B.R.; Lopano, C. Reactive transport of CO_2-saturated water in a cement fracture: Application to wellbore leakage during geologic CO_2 storage. *Int. J. Greenh. Gas Control* **2016**, *44*, 276–289. [CrossRef]

17. Guthrie, G.D.; Pawar, R.J.; Carey, W.J.; Karra, S.; Harp, D.R.; Viswanathan, H.S. The mechanisms, dynamics, and implications of self-sealing and CO_2 resistance in wellbore cements. *Int. J. Greenh. Gas Control* **2018**, *75*, 162–179. [CrossRef]

18. Koenen, M.; Wasch, L.J. The Potential of CO_2 Leakage Along De-Bonded Cement-Rock Interface. In Proceedings of the 14th Greenhouse Gas Control Technologies Conference, Melbourne, Australia, 21–26 October 2018; Available online: https://ssrn.com/abstract=3365897 (accessed on 21 April 2019).

19. Wolterbeek, T.K.; Raoof, A. Meter-scale reactive transport modeling of CO_2-rich fluid flow along debonded wellbore casing-cement interfaces. *Environ. Sci. Technol.* **2018**, *52*, 3786–3795. [CrossRef]

20. Neele, F.; Grimstad, A.; Fleury, M.; Liebscher, A.; Korre, A.; Wilkinson, M. MiReCOL: Developing Corrective Measures for CO_2 Storage. *Energy Procedia* **2014**, *63*, 4658–4665. [CrossRef]

21. Pizzocolo, F.; Peters, E.; Loeve, D.; Hewson, C.W.; Wasch, L.; Brunner, L.J. Feasibility of Novel Techniques to Mitigate or Remedy CO_2 Leakage. In Proceedings of the SPE Europec featured at 79th EAGE Conference and Exhibition, Paris, France, 12–15 June 2017. [CrossRef]

22. Wasch, L.J.; Wollenweber, J.; Tambach, T.J. A novel concept for long-term CO_2 sealing by intentional salt clogging. *GHG S T* **2013**, *3*, 491–502.

23. Ito, T.; Xu, T.; Tanaka, H.; Taniuchi, Y.; Okamoto, A. Possibility to remedy CO_2 leakage from geological reservoir using CO_2 reactive grout. *Int. J. Greenh. Gas Control* **2014**, *20*, 310–323. [CrossRef]

24. Druhan, J.L.; Vialle, S.; Maher, K.; Benson, S. Numerical simulation of reactive barrier emplacement to control CO_2 migration. In *Carbon Dioxide Capture for Storage in Deep Geologic Formations—Results from the CO_2 Capture Project*; Gerdes, K.F., Ed.; CPL Press and BPCNAI: Thatcham, Berks, UK, 2015.

25. Wasch, L.J.; Wollenweber, J.; Neele, F.; Fleury, M. Mitigating CO_2 Leakage by Immobilizing CO_2 into Solid Reaction Products. *Energy Procedia* **2017**, *114*, 4214–4226. [CrossRef]

26. Xu, T.; Sonnenthal, E.; Spycher, N.; Pruess, K. TOUGHREACT—A simulation program for non-isothermal multiphase reactive geochemical transport in variably saturated geologic media: Applications to geothermal injectivity and CO_2 geological sequestration. *Computat. Geosci.* **2006**, *32*, 145–165. [CrossRef]

27. Pruess, K. *ECO2N: A TOUGH2 Fluid Property Module for Mixtures of Water, NaCl, and CO_2*; Lawrence Berkeley National Laboratory: Berkeley, CA, USA, 2005.

28. Blanc, P.H.; Lassin, A.; Piantone, P.; Azaroual, M.; Jacquement, N.; Fabbri, A.; Gaucher, E.C. Thermoddem: A geochemical database focused on low temperature water/rock interactions and waste materials. *Appl. Geochem.* **2012**, *27*, 2107–2116. [CrossRef]

29. Lasaga, A.C.; Soler, J.M.; Ganor, J.; Burch, T.E.; Nagy, K.L. Chemical weathering rate laws and global geochemical cycles. *Geochim. Cosmochim. Acta* **1994**, *58*, 2361–2386. [CrossRef]

30. Hommel, J.; Coltman, E.; Class, H. Porosity–Permeability Relations for Evolving Pore Space: A Review with a Focus on (Bio)geochemically Altered Porous Media. *Transp. Porous Med.* **2018**, *124*, 589–629. [CrossRef]

31. Verma, A.; Pruess, K. Thermohydrological conditions and silica redistribution near high-level nuclear wastes emplaced in saturated geological formations. *J. Geophys. Res.* **1988**, *93*, 1159–1173. [CrossRef]

32. Palandri, J.L.; Kharaka, Y.K. *A Compilation of Rate Parameters of Water-Mineral Interaction Kinetics for Application to Geochemical Modelling*; Open File Report; U.S. Geological Survey: Menlo Park, CA, USA, 2004.

33. Baur, I.; Keller, P.; Mavrocordatos, D.; Wehrli, B.; Johnson, C.A. Dissolution–precipitation behaviour of ettringite, monosulfate, and calcium silicate hydrate. *Cem. Concr. Res.* **2004**, *34*, 341–348. [CrossRef]

34. Marty, N.C.M.; Claret, F.; Lassin, A.; Tremosa, J.; Blanc, P.; Madé, B.; Giffaut, E.; Cochepin, B.; Tournassat, C. A database of dissolution and precipitation rates for clay-rocks minerals. *Appl. Geochem.* **2015**, *55*, 108–118. [CrossRef]

35. Xu, T.; Ontoy, Y.; Molling, P.; Spycher, N.; Parini, M.; Pruess, K. Reactive transport modeling of injection well scaling and acidizing at Tiwi field, Philippines. *Geothermics* **2004**, *33*, 477–491. [CrossRef]

36. Orlic, B.; Chitu, A.; Brunner, L.; Koenen, M.; Wollenweber, J.; Schreppers, G.-J. Numerical Investigations of Cement Interface Debonding for Assessing Well Integrity Risks. In Proceedings of the 52nd U.S. Rock Mechanics/Geomechanics Symposium, Seattle, WA, USA, 17–20 June 2018.

minerals

Article

Spontaneous Serpentine Carbonation Controlled by Underground Dynamic Microclimate at the Montecastelli Copper Mine, Italy

Chiara Boschi [1,*], Federica Bedini [1], Ilaria Baneschi [1], Andrea Rielli [1], Lukas Baumgartner [2], Natale Perchiazzi [3], Alexey Ulyanov [2], Giovanni Zanchetta [3] and Andrea Dini [1]

[1]   Istitute of Geoscience and Earth Resources, National Research Council of Italy (CNR), 56100 Pisa, Italy; kicca1000@hotmail.it (F.B.); ilaria.baneschi@igg.cnr.it (I.B.); andrea.rielli@igg.cnr.it (A.R.); andrea.dini@igg.cnr.it (A.D.)

[2]   Institute of Earth Sciences, University of Lausanne, Geopolis Building, CH-1015 Lausanne, Switzerland; Lukas.Baumgartner@unil.ch (L.B.); Alexey.Ulyanov@unil.ch (A.U.)

[3]   Earth Sciences Department, Pisa University, Via S. Maria 53, I-56126 Pisa, Italy; natale.perchiazzi@unipi.it (N.P.); zanchetta@dst.unipi.it (G.Z.)

*   Correspondence: chiara.boschi@igg.cnr.it

Received: 15 October 2019; Accepted: 5 December 2019; Published: 18 December 2019

Abstract: Understanding low temperature carbon sequestration through serpentinite–H_2O–CO_2 interaction is becoming increasingly important as it is considered a potential approach for carbon storage required to offset anthropogenic CO_2 emissions. In this study, we present new insights into spontaneous CO_2 mineral sequestration through the formation of hydromagnesite + kerolite with minor aragonite incrustations on serpentinite walls of the Montecastelli copper mine located in Southern Tuscany, Italy. On the basis of field, petrological, and geochemical observations coupled with geochemical modeling, we show that precipitation of the wall coating paragenesis is driven by a sequential evaporation and condensation process starting from meteoric waters which emerge from fractures into the mine walls and ceiling. A direct precipitation of the coating paragenesis is not compatible with the chemical composition of the mine water. Instead, geochemical modeling shows that its formation can be explained through evaporation of mine water and its progressive condensation onto the mine walls, where a layer of serpentinite powder was accumulated during the excavation of the mine adits. Condensed water produces a homogeneous film on the mine walls where it can interact with the serpentinite powder and become enriched in Mg, Si, and minor Ca, which are necessary for the precipitation of the observed coating paragenesis. The evaporation and condensation processes are driven by changes in the air flow inside the mine, which in turns are driven by seasonal changes of the outside temperature. The presence of "kerolite", a Mg-silicate, is indicative of the dissolution of Si-rich minerals, such as serpentine, through the water–powder interaction on the mine walls at low temperature (~17.0 to 18.1 °C). The spontaneous carbonation of serpentine at low temperature is a peculiar feature of this occurrence, which has only rarely been observed in ultramafic outcrops exposed on the Earth's surface, where instead hydromagnesite predominantly forms through the dissolution of brucite. The high reactivity of serpentine observed, in this study, is most likely due to the presence of fine-grained serpentine fines in the mine walls. Further study of the peculiar conditions of underground environments hosted in Mg-rich lithologies, such as that of the Montecastelli Copper mine, can lead to a better understanding of the physical and chemical conditions necessary to enhance serpentine carbonation at ambient temperature.

Keywords: CO_2 mineral sequestration; hydromagnesite; kerolite; serpentinite; Cu mine; Montecastelli; underground microclimate

---

1. Introduction

The formation of hydromagnesite ($Mg_5(CO_3)_4(OH)_2\cdot4H_2O$) is generally associated with low-temperature alteration of ultramafic rocks and has attracted an increasing number of studies because during its formation CO_2 is chemically bound within its structure, representing a natural analogue of CO_2 mineral sequestration [1–7]. The study of natural CO_2 sequestration through the formation of carbonate from ultramafic rocks can help developing more efficient engineered CO_2 storage processes involving both *in situ* and *ex situ* approaches, which are necessary to mitigate anthropogenic CO_2 emissions [8]. Southern Tuscany offers several examples of spontaneous low- and high-temperature carbonation of ultramafic rocks [1,9]. Boschi et al., (2017) showed the presence of hydromagnesite and layered double hydroxides (LDH) in association with serpentinized dunite that outcrops nearby the Montecastelli Cu mine. The authors highlighted that the efficiency of carbonation was linked to the substratum lithology on which the carbonation processes took place. The most efficient reactions were found to occur on serpentinized dunites that are rich in brucite. Brucite is preferentially dissolved at surficial conditions as compared with serpentine, producing a Mg-rich fluid necessary for the precipitation of hydromagnesite [1]. Due to the lithological selectivity of the process, only brucite-rich dunite bodies showed significant carbonation with the formation of hydromagnesite and LDH at the outcrop scale. In this study, we focus on hydromagnesite formation (+ kerolite and aragonite) in adits of the Montecastelli Cu mine. We show that the formation of hydromagnesite in an underground environment is triggered by a different process as compared with the surface, which consists of a complex interaction between meteoric water percolation, evaporation and condensation, and interaction with serpentinite fines accumulated on the adits' walls.

2. Geological Background and Topographic-Microclimatic Characteritics of the Mine

Southern Tuscany is characterized by several outcrops of ultramafic and mafic rocks belonging to the Ligurian units. These comprise serpentinized harzburgites and dunites, gabbros and basalts, with their original sedimentary cover [1,9–11]. A large ophiolite body outcrops near the village of Montecastelli Pisano. Here, the Pavone river has eroded a canyon through the ultramafic units exposing outcrops of harzburgite and dunite which locally have been strongly carbonated [1]. The Montecastelli ultramafic body has been affected by the following two main stages of deformation: (i) early ductile shearing recorded by mylonitic gabbros and (ii) late brittle deformation with the development of cataclastic zones and the recrystallization of serpentinite-gabbro-basalt assemblage. A major cataclastic zone, which runs through the entire ultramafic body WNW–ESE, hosts a reworked Cu-Fe sulphide mineralization, mainly represented by bornite, chalcopyrite, and chalcocite in veins and nodules [12–14]. The deposit was intermittently explored and mined between 1832 and 1942 through the excavation of small open pits and underground works, providing a negligible, total production of 40 tons of Cu-Fe sulphides.

The Montecastelli copper mine (43°16.4' N; 10°56.7' E) consists of about 1500 m of adits and drifts distributed over three levels from 195 m to 245 m above sea level (a.s.l.) connected by a main shaft and smaller inclined shafts (Figure 1). The three main adits have been excavated in serpentinized harzburgite and dunite reaching a NW-trending mineralized cataclastic zone from which the surface (ca. 320 m a.s.l.) plunges towards NE. The upper part of the cataclastic zone (from 320 m down to 245 m) is poorly mineralized and there are no underground works directly connecting the uppermost exploitation drifts to the surface. The entrance of the uppermost level (Santori adit) is located at 245 m a.s.l and it is connected by an inclined shaft to the intermediate level (Isabella adit, 215 m a.s.l.). Both adits are excavated through serpentinized harzburgite, with a progressive increase of serpentinized dunite and gabbro lenses towards the mineralized cataclastic zone.

A vertical internal shaft connects the Isabella adit to the lowermost level of the mine, the Ribasso adit (205 m a.s.l), which emerges 700 m further north in the Pavone valley. Most of the drifts excavated in the mineralized cataclastic zone have collapsed due to the soft and soapy character of the ore body (chlorite-, serpentine-rich); hence observation is possible only in few stopes. The Isabella adit offers the best and safest exposures of the ultramafic host rocks, crosscutting about 200 m of serpentinized harzburgite and reaching the footwall of the mineralized cataclastic zone. The intermediate part of this adit, between the entrance and the main internal shaft, displays walls and ceiling covered by white to creamy-white crusts composed mainly of carbonates (Figure 2). Something similar also occurs in the first part of the Santori adit. In particular, the whitish crusts are noticeable looking inward into the Isabella and Santori adits (Figure 2A); whereas they are not visible looking outward from the adits (Figure 2B). The Ribasso adit is flooded and observations there have been precluded. Although dripping water has been locally observed and sampled, there is not a direct spatial association between carbonate crusts and fractures discharging dripping water. Most of the walls covered by carbonates are homogeneously wet and they do not display any laminar water flow; they resemble cold surfaces covered by condensed water.

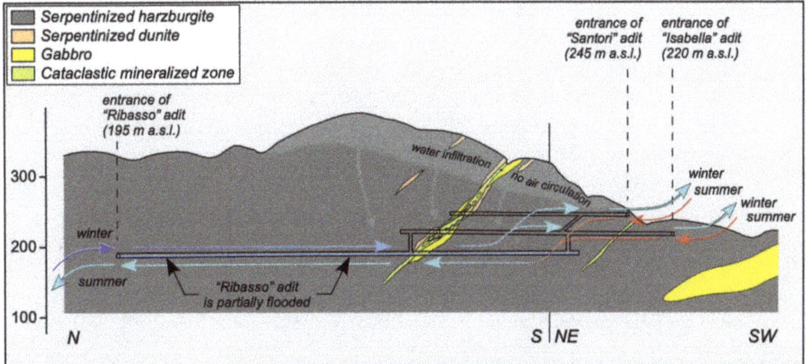

Figure 1. Cross section of the Montecastelli copper mine, showing the geological features and the overall air circulation pattern during the two main upward (winter) and downward (summer) airflow stages.

Figure 2. Isabella adit of the Montecastelli mine. (**A**) View of the adit looking inward where the white coating of hydromagnesite + kerolite and aragonite is visible and (**B**) view from the adit looking outward where the withe coating is not visible.

The hill hosting the mine is covered by a pine and oak forest and variably well-drained soils rich in organic matter. The soil allows a significant infiltration of meteoric water inside the fractured serpentinites, forming small confined aquifers which, in turn, slowly feed small perennial drippings along the underground walls, as well as small pools on the floor. The average external air temperature at Montecastelli [15] displays a typical seasonal variation with a maximum of 25 °C during August and a minimum of 6 °C during January and February (Figure 3). The maximum humidity of the external air and soil is reached during the rainy season, from October to the end of April, while the period between mid-June and late August is characterized by severe soil aridity. Air temperature and humidity inside the mine are relatively constant throughout the year (15 to 17 °C and 80% to 100%), with major fluctuations near the entrance of Isabella and Santori adits.

Owing to the configuration of the mining works, the Montecastelli mine can be classified as a dynamic underground complex [1,9–11] where ventilation is triggered by the so-called "chimney effect". The difference in elevation between the uppermost adits (Santori and Isabella) and the lowermost adit (Ribasso) triggers the air circulation, which changes seasonally. During the summer, the relatively cold air inside the mine flows out from the lowermost adit, drawing in warmer and drier air from the outside through the uppermost levels (downward airflow, Figures 1 and 3). Contrarily, during the winter, the relatively warm air inside the mine flows out from the uppermost adits (upward airflow), drawing in colder and humid air from the outside through the lowermost level (Figures 1 and 3). The direct, on-site observation of stagnant circulation during April to May and during October (weak or zero airflow), complies with the intersection between the annual cycle of external air temperature and the almost constant air temperature in the cave (Figure 3).

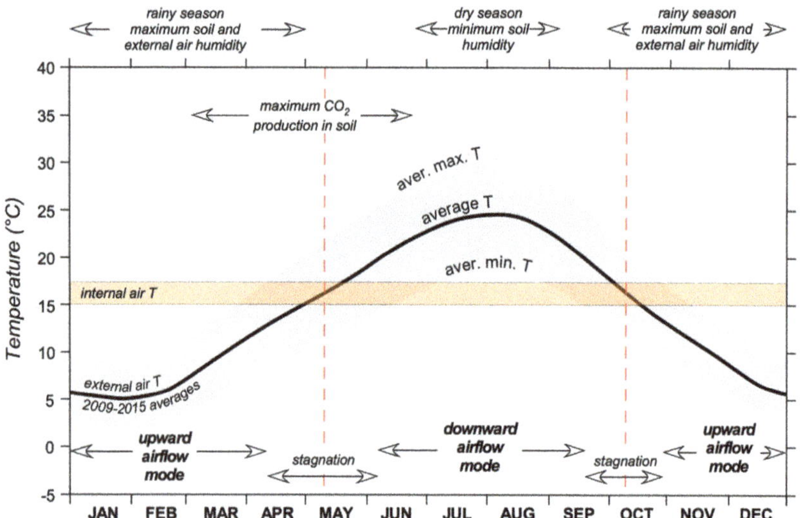

Figure 3. Diagram showing the annual cycle of external air temperature (average mean, maximum, and minimum), the air temperature in the cave, and other meteorological data with respect to the microclimatic dynamic conditions observed in the underground system. Meteorological data for Montecastelli and Castelnuovo as well as other data and information from this study are from the Tuscan Hydrologic Survey [15].

In summary, the Montecastelli underground system behaves like many multiple entrance karst undergrounds placed at midlatitudes, but with a major difference of decoupling between air and water circulation. Airflow is strictly controlled by the topography of the adits/shafts network coupled with the annual cycle of external air temperature, while the lack of large vertical fractures in the overlying rock mass (typical of carbonate-hosted karst systems) precludes the direct arrival of CO_2-enriched air during the spring and summer time. Contrarily, meteoric water interacting with soils above the mine can dissolve an aliquot of soil CO_2, bringing it into the mine by slow infiltration through a network of tiny fractures. The lack of a massive influx of CO_2-rich air during the warm season is pivotal to the understanding of mechanisms of serpentinite carbonation along the Isabella adit.

3. Methods

Morphological and structural investigations at the microscale were carried out using field emission scanning electron microscopy (FESEM-JSM-6500F) at the National Institute of Geophysics and Volcanology (INGV, Rome, Italy). Secondary electron (SE) images, backscattered electron microscopy (BSEM) images, and X-ray element distribution maps were performed to detail wall incrustations. X-ray diffraction (XRD) patterns were obtained employing different instrumentations at the Department of Earth Science (University of Pisa, Pisa, Italy) using a Philips PW 1050/1710 with a conventional Bragg-Brentano (BB) parafocusing geometry equipped with a copper tube and a secondary graphite monochromator. The data were collected in the angular range $4° \leq 2\theta \leq 60°$, with $0.02°$ 2θ scan step and counting time 2 s/step operating at 40 kV and 25 mA. Mineral chemical composition was analyzed using a JXA 8200 WD/ED electron microprobe (EMP) at the INGV (Rome, Italy). The analyses were performed using 15 kV acceleration voltage, 2.6 nA beam current, and probe diameter of 5 μm with a counting time of 10 s on the peak and 5 s on the background on both sides of the peak. Estimated precision is 0.05 wt %. The detection limit is 0.01 wt % for Cl, 0.03 wt % for Al_2O_3 and Na_2O, 0.04 wt % for SiO_2, MgO and CaO, 0.05 wt % for FeOtot, MnO, NiO, 0.06 wt % for TiO_2 and Cr_2O_3, 0.09 wt % for SO_3, 0.10 wt % for K_2O.

The concentrations of Ba, Sc, V, Cr, Ni, Cu, Zn, Rb, Sr, Y, Zr, Nb, Ce, rare earth elements (REEs), Pb, Th, and U of carbonate and serpentine were determined with a sector-field, single-collector Element 2 XR ICP-MS at the University of Lausanne in laser ablation mode (LA-ICP-MS), i.e., interfaced to a NewWave UP-193 ArF excimer ablation system (ESI). Spot size varied between 75 and 100 μm with a frequency of 20 Hz and energy of 6.0 J/cm^2, using as standard NIST SRM 612. Raw data were reduced off-line using the LAMTRACE software [16]. The analytical precision is better than 8% RSD.

Seasonal mine water sampling was performed at the following two sites: (i) mine water 1, in a deeper part of the mine; (ii) mine water 2, close to the carbonate crust along the main adit. For our model we used mine water 2, because it is associated with the carbonate crust. In the field, the temperature, pH, and electrical conductivity were measured with a portable multiparameter data logger calibrated in the laboratory; total alkalinity was determined by acidimetric titration. Accuracy is 0.5% for conductivity, 0.25% for temperature, 0.05 for pH, and 0.1 meq/L for alkalinity. After the physico-chemical measurements, water samples were collected in different modes depending on the specific chemical and isotopic analyses, performed in the laboratory. Water samples for anions ($SO_4{}^{2-}$ and Cl^-) and cations (Na^+ and K^+) analyses were collected without any pretreatment, while samples for Ca^{2+}, Mg^{2+}, trace elements, and silica were filtered in the field through 0.45 μm acetate-cellulose membrane filters and acidified in order to prevent precipitation. In addition, they were stored in bottles previously washed with diluted HNO_3. All water samples were stored at 4 °C prior to processing. The anions and cations were analyzed, at the Institute of Geoscience and Earth Resources of the CNR in Pisa (IGG-CNR; Italy), using a Dionex DX100 ion chromatograph and a Perkin-Elmer 3110 atomic absorption spectrometer. The analytical precision is 3% for both species. Silica determination was performed via spectrophotometric method. Trace elements were analyzed by ICP-MS at the IGG-CNR. The analytical precision is better than 2%.

Aqueous species speciation and mineral saturation indices (SI) were calculated using the geochemical speciation code PHREEQC [17] and the Base de Donnee Thermoddem_V1.10 database from the Bureau de Recherches Géologiques et Minières (BRGM Institute, Fontenay-aux-Roses, France; http://thermoddem.brgm.fr.) All the solutions were calculated based on Cl-charge balance.

4. Results

4.1. Serpentinite Substratum

Serpentinized harzburgites are composed mainly by lizardite (>95%), chlorite, magnetite, and only minor relicts of Mg-Al chromite (Figure 4a). Serpentine is present as mesh-textured matrix embedding bastite porphyroclasts with local chrysotile veins and can be intergrown with clinochlore. Primary Mg-Al chromites range in size from few microns up to ~1 mm. They show the typical lobate habit of spinels in mantle rocks and are characterized by rims of Fe-chromite. Along the Isabella adit, several cataclastic zones, plunging at a subvertical to medium angle and oriented like the main mineralized cataclastic zone, produced pervasive brecciation of serpentinized harzburgites (Figure 4b). Breccias are clast supported, with angular and subangular clasts ranging in size from submillimetric to pluri-decimetric. The matrix is constituted of fine-grained serpentine and andraditic garnet (identified by XRD). Small crystals of andradite (~2 to 3 μm in diameter) are observed both in matrix and clasts. They are anhedral and form interstitial aggregates (Figure 5).

The studied samples do not show evidence of pervasive Mg carbonation at the microscopic scale or carbonate precipitation. Coatings and crusts of hydrous Mg carbonates and Mg-clay have only been observed on the exposed rock surfaces in the mine adits.

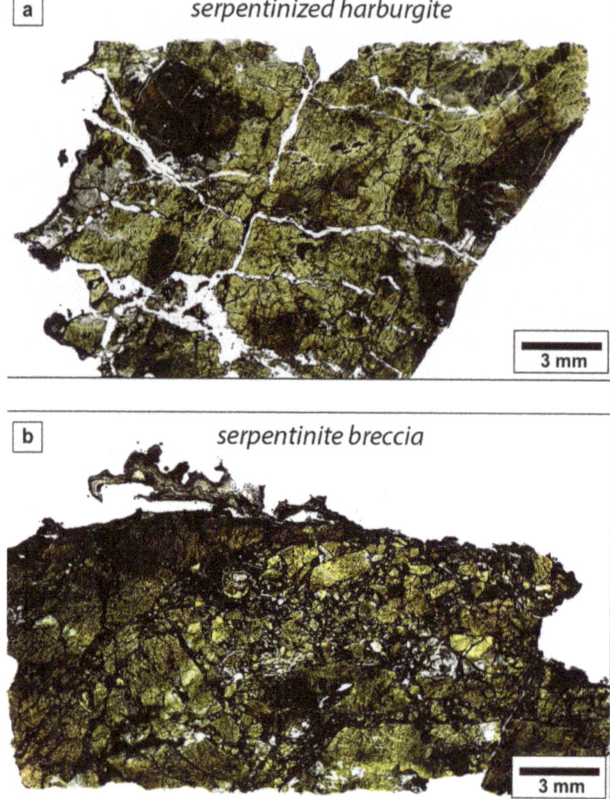

Figure 4. Thin sections of serpentinite sampled along the mine adits: (**a**) Serpentinized harzburgites and (**b**) serpentinite breccias.

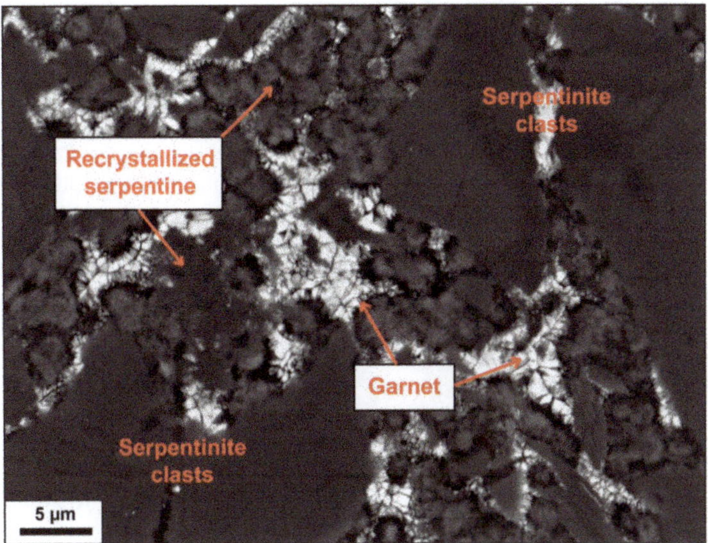

Figure 5. SEM-EDS image of interstitial andradite garnet in the matrix of serpentinite breccia.

4.2. Wall Coating

The mineral paragenesis forming the wall coating, identified by XRD and EMP analyses, is represented by hydromagnesite, with minor aragonite and a Mg-rich clay silicate (Figure S1, Table S1 and S2 in Supplementary Materials, SM). Macroscopically, hydromagnesite occurs mostly as coatings, crusts, and spherules lining the rock surfaces and tiny fractures propagating for a few millimeters into the serpentinite walls (Figure 6). Coatings, crusts, and spherules can have a large areal extension but with a limited thickness, up to few millimeters.

Figure 6. Examples of carbonate coatings: (**a**) Hydromagnesite ± kerolite coating (first precipitation), (**b**) hydromagnesite rosettes (second precipitation), (**c**) early precipitation of hydromagnesite along the rock fractures, and (**d**) preferential precipitation of hydromagnesite on the surfaces facing outwards of the mine adit.

At the microscale, two episodes of hydromagnesite precipitation are recognized (Figures 7 and 8). The first precipitation is characterized by hydromagnesite layers alternated with Mg-rich clays layers (with thicknesses ranging from 100 µm to <5 µm; Figure 8). In particular, the Mg-clays layers, generally deposited in the late stage, and showed an irregular and wavy texture with the presence of shrinkage cracks (<5 µm in width). The second episode of precipitation formed an aggregate of hydromagnesite rosettes, each made by fibrous-radiating acicular crystals (with size ranging from ~1 to >8 µm). This generation forms an external coating, distinguished from the early precipitation by a strong increase in porosity (Figure 8). Minor amounts of anhedral aragonite (with size ranging from ~500 µm to a few micrometers) have been observed in association with both hydromagnesite µm (Figure 8). We found partially dissolved serpentine fines, with a size variable from 5 to 200 microns, forming a layer between the serpentinite and the crusts (Figure 8b–f). This powdered serpentine was most likely produced during the mine excavation.

The chemical composition of Mg-clays (Table S2) can be attributed to either stevensite or a hydrated and highly disordered variety of talc-like phase, named "kerolite" $Mg_3Si_4O_{10}(OH)_2 \cdot nH_2O$ [18–20]. Kerolite-like minerals are usually associated with carbonates derived from alteration of ultramafic rocks at high pH [20], whereas stevensite is more common in low pH environments. In addition, the XRD data points to the presence of "kerolite" rather than stevensite (Figure S1). "Kerolite" has previously been found mixed with poorly crystalline serpentine, both in natural settings and in experimental reaction products [15]. The term "deweylite" has been used to describe intimate mixtures of a fine-grained, highly disordered, kerolite-like mineral and a disordered serpentine mineral, typically chrysotile, in varying proportions. This is in agreement with our variable chemical analyses, indicating an excess of Mg and a deficiency of Si, probably because of the presence of serpentine and (or) disordered kerolite (Table S2).

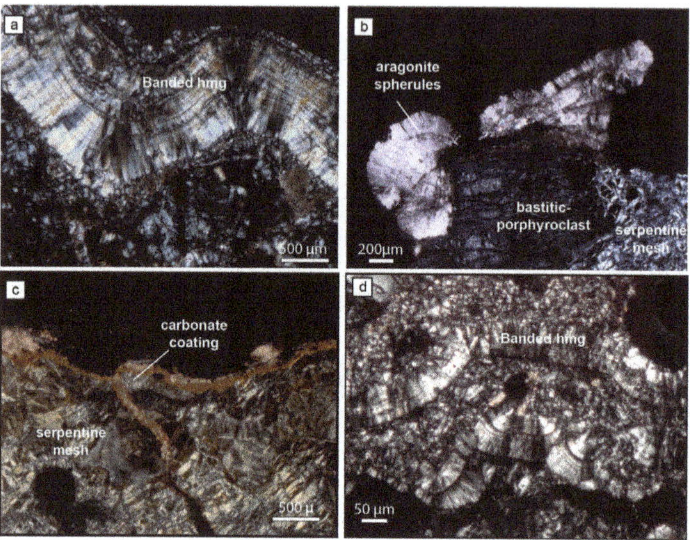

Figure 7. Microphotographs of various carbonate occurrences showing: (**a**) Banded hydromagnesite and kerolite; (**b**) aragonite spherules growth onto carbonate coating + bastite porphyroclasts; (**c**) carbonate coating composed of hydromagnesite, kerolite, and late aragonite; and (**d**) overgrowths of banded hydromagnesite.

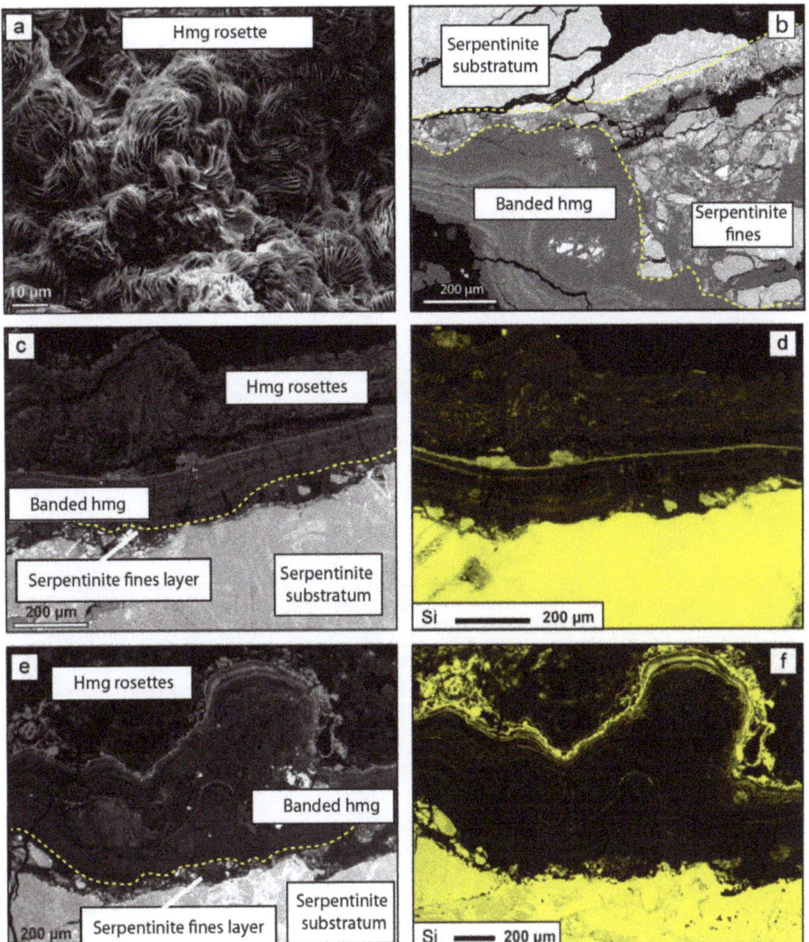

Figure 8. SEM images and X-ray element distribution maps of wall incrustations: (**a**) Secondary electron (SE) image showing intergrowth of lenticular crystals of hydromagnesite on the outer layer of the crusts; (**b**) backscattered electron microscopy (BSEM) image showing the serpentine fines embedded into hydromagnesite ± kerolite. Kerolite layers are distinguishable from hydromagnesite layers because they appear light grey in color in the BSEM images; (**c**–**f**) BSEM images (**c** and **e**) and X-ray element distribution maps for Si (**d** and **f**) showing the following: (i) serpentinite substratum, (ii) serpentinite fines at the external surface of the rock, (iii) hydromagnesite and kerolite layered crust, and (iv) late hydromagnesite rosettes. The yellow color in Figure 8d,f highlights the presence of kerolite layers and serpentinite fines on the wall surface.

4.3. Trace Element Composition of Mineral Phases

Trace element composition of serpentine and hydromagnesite analyzed by LA-ICP-MS is reported in Tables 1 and 2. Serpentine minerals have high Cr and Ni content, up to ~2600 ppm and ~1400 ppm, respectively, relatively high content of V and Cu, and detectable content of Pb, U, Th, Y, Rb, Sr, Zr, and Nb. REE serpentine patterns span over a relatively small range of contents; from ~0.1 to more ~1 times those of the chondritic values, with LREE depleted patterns and a positive Eu anomaly (Figure 9).

Table 1. Trace element composition (ppm) of serpentine determined by LA-ICP-MS.

Sample Type	MA22B	MA22D	MA22F	MA22H	MA22L	MA29B	MA29D	MA29F	MA29H	MA29L
	serp	serp	serp	serp	serp	serp	serp	serp	serp	serp
V	10	10	6	7	4	26	3	54	84	41
Cr	77	112	54	52	45	477	21	1719	3149	949
Ni	1394	1356	1363	1221	1249	1431	1391	791	719	1330
Cu	8	8	10	6	6	7	5	7	24	6
Rb	0.060	0.047	0.035	0.043	0.040	0.039	0.017	0.042	0.058	0.048
Sr	1	1	1	1	1	1	1	1	1	2
Y	0.340	1.359	0.684	0.857	0.510	1.055	0.412	0.537	1.522	1.232
Zr	0.050	0.181	0.070	0.114	0.060	0.252	0.065	0.285	0.995	0.321
Nb	0.001	0.001	0.001	0.003	0.001	0.002	0.002	0.002	0.003	0.002
La	0.020	0.028	0.028	0.042	0.020	0.034	0.025	0.023	0.063	0.042
Ce	0.060	0.111	0.097	0.143	0.060	0.113	0.078	0.067	0.154	0.144
Pr	0.010	0.017	0.016	0.023	0.010	0.015	0.010	0.010	0.022	0.017
Nd	0.060	0.116	0.094	0.133	0.063	0.088	0.053	0.062	0.136	0.100
Sm	0.020	0.049	0.031	0.042	0.022	0.031	0.017	0.023	0.050	0.036
Eu	0.020	0.036	0.032	0.039	0.021	0.036	0.022	0.021	0.037	0.042
Gd	0.030	0.094	0.050	0.067	0.040	0.059	0.030	0.042	0.109	0.072
Tb	0.004	0.019	0.009	0.013	0.007	0.011	0.005	0.008	0.022	0.014
Dy	0.030	0.170	0.073	0.103	0.054	0.099	0.043	0.065	0.196	0.126
Ho	0.010	0.044	0.019	0.026	0.013	0.027	0.010	0.016	0.052	0.034
Er	0.030	0.148	0.065	0.087	0.045	0.095	0.033	0.056	0.185	0.118
Tm	0.005	0.024	0.011	0.014	0.007	0.015	0.005	0.010	0.032	0.020
Yb	0.030	0.173	0.082	0.101	0.057	0.113	0.040	0.072	0.256	0.142
Lu	0.006	0.027	0.014	0.016	0.010	0.019	0.007	0.013	0.043	0.024
Pb	0.018	0.036	0.011	0.025	0.091	0.021	0.017	0.008	0.035	0.024
Th	0.001	0.001	0.001	0.002	0.001	0.002	0.002	0.001	0.001	0.002
U	0.003	0.005	0.003	0.003	0.004	0.004	0.004	0.003	0.003	0.006

serp = serpentine.

Table 2. Trace element composition (ppm) of hydromagnesite and aragonite determined by LA-ICP-MS.

Sample Type	MA22A	MA22C	MA22E	MA22G	MA22I	MA29A	MA29C	MA29E	MA29G	MA29A	MA29L
	Hmg	Hmg	Hmg	Hmg	Hmg	Hmg	Hmg	Hmg	Hmg	Ar	Ar
Sc	0.033	0.160	0.032	0.310	0.022	0.564	0.210	0.117	0.086	0.188	0.253
V	2	4	2	3	2	5	4	2	2	1	1.1
C	41	16	18	44	11	23	13	10	18	15	12
Ni	4	1	2	15	1	5	11	1	1	5	11
Zn	1.184	1.500	1.020	1.460	3.800	0.893	0.680	2.790	0.357	0.645	1.823
Rb	0.131	0.150	0.120	0.130	0.090	0.199	0.060	0.050	0.028	0.066	0.113
Sr	3	1	1	5	1	28	16	2	1	836	1480
Y	0.023	0.008	0.010	0.030	0.004	0.030	0.030	0.008	0.008	0.060	0.16
Zr	0.116	0.071	0.042	0.120	0.035	0.208	0.070	0.026	0.037	0.137	0.14
Nb	0.004	0.002	0.002	0.005	0.002	0.008	0.004	0.002	0.002	0.005	0.006
Ba	0.394	0.137	0.100	0.439	0.100	0.596	0.540	0.129	0.092	5.070	4.91
La	0.013	0.007	0.003	0.018	0.003	0.029	0.020	0.004	0.005	0.033	0.04
Ce	0.024	0.017	0.004	0.040	0.007	0.072	0.030	0.007	0.010	0.058	0.06
Pr	0.003	0.003	0.002	0.010	0.001	0.010	0.007	0.001	0.001	0.009	0.014
Nd	0.012	0.010	0.006	0.030	0.003	0.030	0.027	0.003	0.005	0.028	0.075
Sm	0.004	0.003	0.001	0.006	0.002	0.008	0.011	0.001	0.002	0.009	0.017
Eu	0.001	0.001	<d.l.	0.002	0.001	0.003	0.003	0.001	0.001	0.002	0.008
Gd	0.006	0.007	0.005	0.009	<d.l.	0.011	0.014	0.008	< d.l.	0.008	0.021
Tb	0.001	0.001	0.001	0.001	0.000	0.001	0.001	-	0.000	0.002	0.003
D	0.003	0.003	0.003	0.007	0.001	0.006	0.007	0.002	0.002	0.007	0.014
Ho	0.001	0.001	0.001	0.001	0.000	0.002	0.001	0.001	0.000	0.002	0.004
Er	0.003	0.002	0.002	0.006	0.001	0.005	0.004	0.001	0.001	0.004	0.011
Tm	0.000	0.001	0.000	0.001	0.000	0.001	0.001	0.000	0.000	0.001	0.001
Yb	0.003	0.003	0.003	0.007	0.001	0.005	0.005	0.003	0.002	0.004	0.004
Lu	0.000	-	0.001	0.001	-	0.001	0.001	0.000	-	-	0.001
Pb	0.345	0.276	0.199	0.365	0.277	0.538	0.603	0.332	0.248	0.322	0.385
Th	0.004	0.003	0.001	0.005	0.001	0.009	0.003	0.001	0.002	0.005	0.005
U	0.007	0.003	0.003	0.006	0.002	0.010	0.003	0.007	0.003	0.008	0.007

d.l. = detection limit; - = not determined; Hmg = Hydromagnesite; Ar = aragonite.

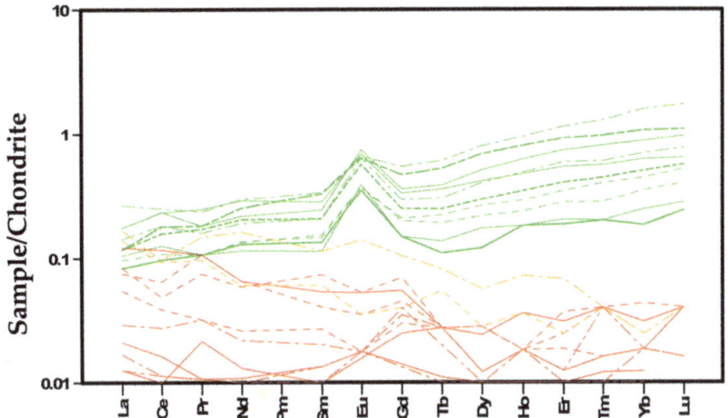

Figure 9. Trace element composition normalized to chondrite of hydromagnesite (red lines), serpentine (green lines), and aragonite (yellow lines).

Hydromagnesite shows up to 45 ppm of Cr, significant amounts of Ni, Zn, Sr, V, Ba, and Pb, and detectable REE, U, Th, Y. Hydromagnesite displays a depleted and scattered REE pattern with slightly enriched LREE ($La_N/Yb_{N(av)}$ = 1.86).

Hydromagnesite trace elements indicate that the fluid responsible for its deposition is a fluid that interacted with serpentinite host rocks. The low-temperature weathering of serpentinites produces spring waters containing a significant content of Cr, as well as Ni, Co, and Mn [1,9,21].

4.4. Chemical Composition of Mine Waters

Waters emerging from the mine walls and ceiling have a mean temperature of ~17.5 °C and a pH of ~8.4 throughout the year and can be classified as bicarbonate-magnesium type waters, because of their high Mg^{2+} and HCO_3^- concentrations, ranging from ~90 to 132 mg/L and ~7.7 to 10.0 mg/L, respectively (Table 3 and Figure 10) and show low Ca/Mg molar ratio (<0.04, Figure 10a). In the Mg-SiO_2-HCO_3 triangular diagram (Figure 10b), a mixing line between an aqueous phase in equilibrium with carbonates and Mg-rich silicates is reported. Mine waters fall to the left side of the line, indicating that such waters are in equilibrium with hydrous Mg-carbonates. Mine waters show low concentrations of trace elements, locally below detection limits, as reported also in other emerging waters in Tuscany [11]. Significant amounts of Cr (~24 ppb), Sr (~20 ppb), Ba (~85 ppb), Fe (~25 ppb) and Zn (~7 ppb) have been detected (Table 3). Other trace elements (Ni, Mn, B, V, Al, Th, and U) have very low concentration between <0.01 and up to 5 ppb. The pCO_2 and saturation indices (SI) of the relevant phases (hydromagnesite and aragonite) were calculated using the chemistry of mine waters using the PHREEQC code. Mine waters 2, collected near the carbonate crusts, show log pCO_2 varying from −3.06 to −3.04 and they are undersaturated with respect to hydromagnesite (SI_{hmg} ~−4.81) and saturated with respect to aragonite (SI_{arag} −0.03).

Table 3. Chemical composition of mine waters.

Sample		Mine Water 1	Mine Water 1	Mine Water 1	Mine Water 1	Mine Water 1	Mine Water 1	Mine Water 1	Mine Water 2	Mine Water 2
Date		08/06/2011	29/05/2013	17/07/2014	11/10/2014	03/12/2014	18/02/2015	02/05/2015	03/12/2014	18/02/2014
T	°C	18.2	17.4	17.5	17.9	17.4	17.2	17.7	17.0	17.4
pH		8.2	8.3	8.3	8.2	8.4	8.6	8.7	8.4	8.4
Alk	meq/L	9.0	8.5	10.0	9.4	9.6	9.3	9.4	4.7	4.5
Cl^-	mg/L	22.3	23.6	23.5	20.8	23.9	16.7	32.3	7.9	7.7
SO_4^{2-}	mg/L	25.5	29.2	27.1	28.8	29.4	26.1	24.7	-	-
Na^+	mg/L	10.5	10.9	11.7	11.4	12.3	10.2	15.2	-	-
K^+	mg/L	0.4	0.1	0.1	0.1	0.2	0.2	0.1	-	-
Ca^{2+}	mg/L	6.4	8.7	8.6	8.0	7.2	8.0	5.4	8.5	8.7
Mg^{2+}	mg/L	116.0	114.0	131.6	114.9	115.9	116.7	113.2	89.8	90.7
Si as SiO_2	mg/L	-	29.1	32.4	30.3	28.2	28.9	13.6	29.8	30.0
Cr	µg/L	25.0	22.7	27.0	24.0	22.7	-	22.5	-	-
Sr	µg/L	25.0	18	19	19	19	-	18	-	-
Ba	µg/L	-	-	105.0	37.0	94.0	-	103.0	-	-
B	µg/L	-	-	-	-	4.0	-	4.0	-	-
Fe	µg/L	-	-	-	-	33.3	-	17.2	-	-
Mn	µg/L	-	-	-	-	1.5	-	1.8	-	-
Ni	µg/L	-	-	-	-	5.7	-	2.2	-	-
Pb	µg/L	-	-	-	-	0.1	-	0.7	-	-
Cu	µg/L	-	-	-	-	0.4	-	1.0	-	-
Zn	µg/L	-	-	-	-	46.0	-	27.0	-	-
V	µg/L	-	-	-	-	1.3	-	1.4	-	-
Al	µg/L	-	-	-	-	<1	-	<1	-	-
Th	µg/L	-	-	-	-	<0.01	-	<0.01	-	-
U	µg/L	-	-	-	-	<0.01	-	<0.01	-	-

- = not determined.

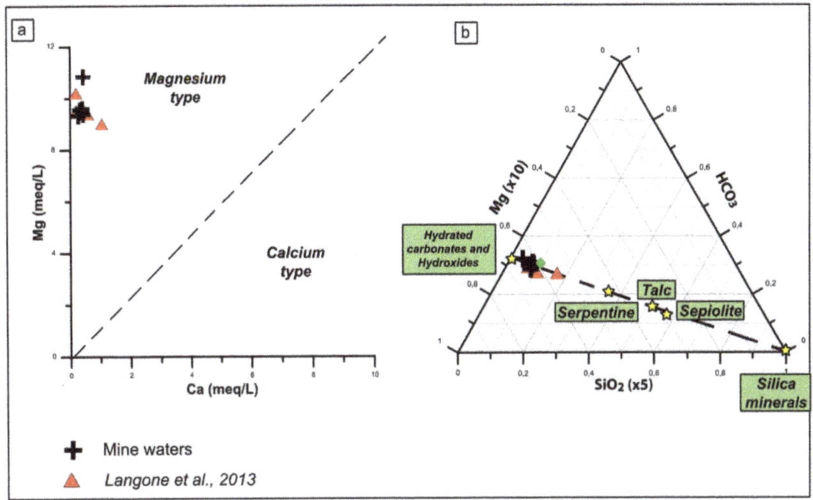

Figure 10. Chemistry of mine waters. (**a**) Mg vs. Ca plot computed from concentrations in meq/L and (**b**) Mg-SiO$_2$-HCO$_3$ triangular plot, computed from concentrations in mg/L. Red triangular points represent from another mine excavated in serpentinites from Southern Tuscany (Querceto, [11]) and stairs represent the expected compositions of aqueous phases controlled by CO$_2$-driven dissolution of Mg-bearing solid phases (i.e., sepiolite, talc, serpentine, and hydrated carbonates).

5. Discussion

Underground environments, such as mine adits and natural caves, have peculiar characteristics, such as an almost constant temperature throughout the year, high humidity, and seasonal variation of the direction of ventilation [22]. Cyclical variations of these parameters can enhance CO$_2$ degassing, as well as water evaporation and condensation that can trigger carbonate precipitation [22–24]. Even though the formation of underground carbonate concretions in natural caves has been well understood, the formation of various types of concretions in mining underground works, especially with ultramafic substratum, is still lacking a thorough investigation. The study of this latter type of concretion in an ultramafic underground environment, such as the Montecastelli Cu mine, can give the opportunity to explore the peculiar conditions at which CO$_2$-rich water spontaneously react with serpentinite rocks.

5.1. PHREEQC Models

Petrographic and XRD data shows that the wall coating paragenesis in the Montecastelli mine is composed of hydromagnesite, kerolite, and aragonite precipitated. In order to understand whether the coating precipitated directly from dripping waters we performed geochemical modeling with PHREEQC using as starting fluid waters emerging from the fractures in the mine wall close to the carbonate crusts (mine water 2). Our model indicates that the CO$_2$-rich dripping water (log pCO$_2$ = −3.05) moves towards equilibrium with the pCO$_2$ of the mine atmosphere (log pCO$_2$ = −3.40 corresponding to ~418 ppmv) measured along the mine adits (Table 4). After the pCO$_2$ of the dripping water equilibrates with that of the mine, it becomes saturated with aragonite (SI$_{arag}$ = 0.25) and undersaturated in hydromagnesite (SI$_{hmg}$ = −3.09).

Aragonite supersaturation (SI > 0.45) is reached through evaporation (more than 20%, Table 5), allowing its precipitation. Dripping water becomes supersaturated in hydromagnesite (SI = 0.46) with higher incremental evaporation (more than 55%), letting hydromagnesite precipitation after aragonite. Therefore, the observed mineral paragenesis of the wall coating, where hydromagnesite is the first phase to precipitate, cannot be generated directly from dripping waters.

Table 4. CO_2 measured along the mine adits.

Sample	Date	Site	CO_2 (ppmv)	St.dev.	log pCO_2
1	18/02/2015	adit near hysromagnesite crust	430	17	−3.37
2	18/02/2015	adit near the exit of the mine	418	11	−3.38
3	18/02/2015	adit in the upper part	399	16	−3.40
4	18/02/2015	adit in the middle part	377	1	−3.42
5	18/02/2015	Mine outside	367	2	−3.44
1	02/05/2015	adit near hysromagnesite crust	430	3	−3.37
2	02/05/2015	adit near the exit of the mine	436	1	−3.38
3	02/05/2015	adit in the upper part	415	4	−3.37
4	02/05/2015	adit in the middle part	435	4	−3.36
5	02/05/2015	Mine outside	412	3	−3.39

St.dev. = Standard deviation.

Table 5. Saturation indeces of hydromagnesite and aragonite, pH and Mg/Ca molar ratio of dripping and condensed waters computed in PHREEQC.

Evaporation %	10	20	30	40	50	55	80	85
Dripping water								
log pCO_2	−3.4	−3.4	−3.4	−3.4	−3.4	−3.4	−3.4	−3.4
pH	8.75	8.78	8.83	8.86	8.9	8.93	8.82	8.78
SI_{hmg}	−2.56	−1.98	−1.49	−0.8	−0.01	0.46	0	0
Si_{ar}	0.34	0.45	0.25	0.25	0.25	0.25	0.25	0.25
Mg/Ca (molality)	17.19	17.19	34.46	46.95	67.08	82.08	68.99	69.63
Condensed water								
log pCO_2	−3.4	−3.4	−3.4	−3.4	−3.4	−3.4	−3.4	−3.4
pH	8.42	8.45	8.52	8.58	8.65	8.69	8.96	9.05
SI_{hmg}	−8.4	−7.72	−6.95	−6.08	−5.07	−4.5	−0.46	0.81

Previously, it has been shown that the formation of Mg-carbonate wall coating in underground mines, excavated in ultramafic lithologies, can also occur through direct precipitation from dripping waters [2]. However, the dripping waters in this case [2] interacted mainly with brucite, and therefore became enriched in Mg^{2+} but not in Ca^{2+} (0.36 to 1.02 mg/L) or Si^{4+} (0.0 to 9.3 mg/L) which prevented the precipitation of aragonite and kerolite. At Montecastelli, instead, dripping waters interacted prevalently with serpentinized harzburgite (Figure 4 and Table 3), and in a minor amount with gabbro lenses which outcrop in the area, therefore, they became relatively enriched not only in Mg^{2+} but also in Ca^{2+} (5.4 to 8.7 mg/L) and Si (13.6 to 32.4 mg/L).

An alternative model has been elaborated considering the water film on the adits' walls that could have been derived from evaporation of dripping and pooled waters, as well as by ingress of humid air from outside. The chemical composition of this "condensed" water is assumed corresponding to a distilled water ($Mg^{2+} \leq 0.01$, $Si^{4+} \leq 0.01$, and $Ca^{2+} \leq 0.01$ in mg/L) in equilibrium with pCO_2 of the air in the mine. We hypothesized that the condensed water became enriched in ions after the dissolution of serpentine fines and the equilibration with pCO_2 of the air. The mineral paragenesis, dissolution reactions, and equilibrium constants (log k) at 25 °C and 1.013 bar are listed in Table S3. Our model shows that the condensed water on the mine walls after interacting with the serpentine fines would have a content of Mg^{2+} ~22 mg/L, SiO_2 ~36.2 mg/L, and HCO_3^- ~105 mg/L, with a pH of 8.38. This fluid is undersaturated in hydromagnesite (SI_{hmg} ~−9.02). Hydromagnesite supersaturation can be slowly reached by evaporation (up to 85%, SI_{hmg} ~0.81), whereas aragonite cannot precipitate due to the absence of Ca^{2+}. Therefore, this process can explain the observed wall coating paragenesis, where hydromagnesite is the first mineral to precipitate.

5.2. Interaction between Condensed Water and Mine Wall Serpentinite

Even though the dissolution of serpentine could provide the cations required for the precipitation of the wall coating paragenesis, serpentine is known to have low solubility at low-temperature conditions [25–28]. This is confirmed by our microscopic investigations showing that there is no evidence for the dissolution of serpentine in the substratum of the carbonate crusts (Figure 8). However, we have observed a discontinuous layer of serpentine fines at the interface between mine walls and carbonate crusts in all the studied samples. The presence of a layer of serpentine fines on the mine wall is most likely a consequence of the shafts and adits excavation. Underground excavation in the early half of the XIX century was done using dry jack hammers which generated a high amount of powder from the serpentine rocks which coated the adit walls and was progressively glued by condensing waters.

It is known that the dissolution rate, and reactivity, of serpentine with CO_2 is strongly enhanced by the increase of the reactive surface area [27,28]. Therefore, serpentine fines represent an efficient reagent from which CO_2-rich condensed water could have progressively stripped Mg and Si required for the formation of the wall coating paragenesis. As observed by FEG-SEM, both the serpentinite fine layer and the new hydromagnesite crusts are highly porous (Figures 7 and 8). They could play as a capillary-drying system where the condensed water infiltrates through the porosity, mostly uptaking Mg and Si along the surfaces of the serpentine particles and crystallizing new minerals (hydromagnesite or kerolite) on the outer surface of the fine layer and the previously crystallized crusts.

Such an interpretive model can be further discussed considering the boundary layer effect in the water film forming on the external surface of the crust. Here, the hydromagnesite precipitation induces a Mg-depleted boundary layer. The higher Si/Mg ratio could temporarily stabilize the precipitation of a kerolite band. The interplay between dissolution/crystallization and diffusion kinetics could generate the complex hydromagnesite-kerolite banding. Condensed water could also propagate into small fractures from the rock surface dissolving chrysotile. In this case, the chemistry of the resulting fluids would be comparable with that resulting from the interaction with the serpentine fines. However, considering the aerial extension of these veins, the overall effect contribution to the carbonation process is expected to be limited as compared with the fines layer. Therefore, we consider the spontaneous carbonation in the Montecastelli mine mainly driven by the presence of the serpentinite fines.

5.3. Seasonally Driven Deposition of the Carbonate Crusts

Geochemical modeling showed that precipitation of wall carbonates is driven by evaporation of modified condensed water on the adit walls. The most peculiar feature of the crusts is their distribution along both sides of the adit walls. Adit walls have an irregular surface with peaks and throughs of variable sizes resulting from their excavation. These irregularities produce a continuous alternation of rock surfaces, which are facing inward and outward from the adit entrance. The carbonate-kerolite crusts are present only on the outward looking faces (Figure 6d).

Considering that the mine environment has a stable temperature and humidity, such an asymmetric distribution needs to be explained by a selective depositional process. Seasonal changes of air circulation inside the mine can explain this observation because they can trigger the preferential evaporation of condensed water on the outward looking surfaces. As described before, during summer the relatively colder and heavier air inside the mine flows out from the lowermost level, drawing the hot and dry air from outside into the upper adits. Contrarily during winter, the relatively warm and light air inside the mine flows out from the upper levels, while drawing in colder and humid air from the outside through the lowermost adit. In the latter case, incoming air becomes progressively more humid flowing through the flooded lowermost adit. Therefore, the optimal condition for selective evaporation of water condensed on the outward looking surfaces of the upper adits is attained during summer because the air flow is dry (Figure 1). Instead, winter air circulation is dominated by humid air, which cannot promote evaporation of condensed water (Figure 1).

Typical carbonate-hosted caves at similar midlatitudes behave differently. Seasonal cyclicity in temperature and air density, coupled with enhanced CO_2 production in soils during the warm season, is commonly responsible for the relevant summer CO_2 buildup in cave air and the inhibition of calcite precipitation by elevated CO_2 levels in the cave waters. Maximum speleothem growth occurs during the cold season when the airflow inversion introduces CO_2-poor air in the cave and the CO_2 level in cave water is strongly reduced [22–24] The reverse behavior at the Montecastelli mine is easily understood considering the different precipitation process (evaporation-condensation-evaporation) and the peculiar topographic and geologic characters of this artificial underground system that prevent extreme CO_2 buildup during the warm season.

The interaction between condensed water and serpentine minerals on the wall surface can be summarized as by Equation (1), that is compatible with the precipitation of alternating bands of hydromagnesite and kerolite:

$$6Mg_3Si_2O_5(OH)_2 \text{ (Serpentine)} + 4HCO^-_3 + 5H_2O \rightarrow 3Mg_3Si_4O_{10}(OH)_2 \cdot H_2O \text{ (Kerolite)} + Mg_5(CO_3)_4(OH)_2 \cdot 4H_2O \text{ (hydromagnesite)} + 4Mg^{2+} + 2(OH)_2 \tag{1}$$

The presence of small amounts of aragonite can be related both to the partial dissolution of Ca-rich minerals, present in very minor amount in the serpentinites. Kerolite precipitation is thought to be promoted by low temperature, high pH, and high silica and magnesium activity [20]. Saturation of amorphous silica at high pH (>8.2) has also been proposed as a key condition for the precipitation of Mg-silicates [18–20]. Condensed waters in the Montecastelli mine has high pH (~8.9), high Mg content (>70 mg/L), and high dissolved silica activity, which are required for the precipitation of kerolite, and thus further supporting the genetic process.

6. Conclusions

In this study, we have reported a new occurrence of spontaneous CO_2 sequestration through the formation of hydromagnesite and kerolite on serpentine mine walls in the Montecastelli Cu mine located in Southern Tuscany, Italy. The formation of hydromagnesite and kerolite is triggered by the interaction between condensed water with serpentinite fines accumulated on the walls of the adits during excavation. The large surface area of the fines strongly increases the reactivity of serpentine and allows its dissolution at low temperature. This provides the required elements for the precipitation of the wall coating assemblage. The precipitation takes place due to the selective evaporation of water on upwind rock surfaces exposed to the downward circulation during summer. The peculiar features of this occurrence include the following: (i) the instauration of unusual microclimatic condition in this artificial underground system and (ii) the spontaneous carbonation of serpentine at low temperature. The latter has usually not been observed in ultramafic outcrops exposed on the Earth's surface, where, instead, hydromagnesite predominantly forms through the dissolution of the more reactive brucite.

This study shows that serpentine carbonation at low temperature is possible and it could be reproduced in ex situ applications by using fine-grained serpentine powder and a cyclical process of condensation and evaporation steps. The efficiency of this process can be tested and implemented through laboratory experiments and could have a significant impact on the applicability of CO_2 mineral sequestration because it does not require pretreatment of serpentine or high temperature during the reaction, thus, potentially reducing the costs.

Further study of the peculiar conditions of underground environments hosted in Mg-rich lithologies, such as that of the Montecastelli Cu mine, can lead to a better understanding of the physical and chemical conditions necessary to enhance serpentine carbonation at low temperature, and thus implementation of new strategies for engineered CO_2.

Supplementary Materials: The following are available online at http://www.mdpi.com/2075-163X/10/1/1/s1, Table S1: Microprobe analyses (wt %) of Montecastelli mine hydromagnesites, Table S2: Microprobe analyses (wt %) of "deweylite"—a variable mixture of kerolite and serpentine—identified in the wall coating, Table S3: Dissolution reactions and thermodynamic equilibrium constants for considered minerals at 25 °C and 1.013 bar;

and Figure S1: XRD of the coating paragenesis showing the characteristic peaks of hydromagnesite (red bars), aragonite (green bars), and clay minerals (* symbol). As it can be seen in the inserted closeup, the presence of kerolite in the specimen is demonstrated by the small angular shift and by the definitely larger FWHM of the diffraction peak at near 9.4 Å, where diffractions of kerolite and hydromagnesite overlap. In fact, the other diffraction peak at higher angle, due to the diffraction of hydromagnesite only show a definitely smaller FWHM.

Author Contributions: For research articles with several authors, a short paragraph specifying their individual contributions must be provided. The following statements should be used "conceptualization, C.B. and A.D.; methodology, C.B., I.B., L.B., N.P., A.U. and A.D.; software, F.B., I.B., L.B. and A.U.; validation, C.B., I.B., N.P., A.U. and A.D.; formal analysis, C.B., F.B., I.B., N.P., A.U.; investigation, C.B., I.B., F.B. and A.D.; resources, C.B., L.B., N.P. and G.Z.; data curation, F.B., I.B., A.R.; writing—original draft preparation, C.B., F.B., A.R. and A.D.; writing—review and editing, C.B., I.B., A.R., N.P., A.U. and A.D.; visualization, C.B., F.B., I.B., A.R., N.P. and A.D.; supervision, C.B. and A.D.; project administration, C.B.; funding acquisition, C.B., F.B. and G.Z. All authors have read and agree to the published version of the manuscript. All authors have read and agreed to the published version of the manuscript.

Funding: This research was partially funded by European Horizon 2020 GECO project (https://geco-h2020.eu), grant number 818169. The F.B. was supported by the Pegaso PhD project (Tuscan Region, Italy). C.B. was partially supported by a SNSF grant (Swiss National Science Foundation; i.e., International Short Visits, grant number IZK0Z2_158572.) and by Fondation Herbette (Universitè de Lausanne) grant for her work at the University of Lausanne.

Acknowledgments: C.B. and co-authors thank Gianluca Giorgi, the owner of the Montecastelli Mine (https://sites.google.com/site/minieradelpavone/) that recently passed away. Gianluca allowed us to study the mine geology and helped us to sample fluids, rocks, and coating. C.B. and F.B. thank Othmar Müntener for his availability and scientific and analytical support during their stay at the University of Lausanne. The authors thank two anonymous reviewers for their critical review of an earlier version of the manuscript and the Guest Editor Giovanni Ruggieri for supporting our proposal and assisting us with this special issue.

Conflicts of Interest: The authors declare no conflict of interest.

References

1. Boschi, C.; Dini, A.; Baneschi, I.; Bedini, F.; Perchiazzi, N.; Cavallo, A. Brucite-driven CO_2 uptake in serpentinized dunites (Ligurian Ophiolites, Montecastelli, Tuscany). *Lithos* **2017**, *288*, 264–281. [CrossRef]

2. Beinlich, A.; Austrheim, H. In situ sequestration of atmospheric CO_2 at low temperature and surface cracking of serpentinized peridotite in mine shafts. *Chem. Geol.* **2012**, *332*, 32–44. [CrossRef]

3. Power, I.M.; Wilson, S.A.; Thom, J.M.; Dipple, G.M.; Gabites, J.E.; Southam, G. The hydromagnesite playas of Atlin, British Columbia, Canada: A biogeochemical model for CO_2 sequestration. *Chem. Geol.* **2009**, *260*, 286–300. [CrossRef]

4. Teir, S.; Eloneva, S.; Fogelholm, C.-J.; Zevenhoven, R. Fixation of carbon dioxide by producing hydromagnesite from serpentinite. *Appl. Energy* **2009**, *86*, 214–218. [CrossRef]

5. Wang, D.; Li, Z. Gas–Liquid Reactive Crystallization Kinetics of Hydromagnesite in the $MgCl_2$–CO_2–NH_3–H_2O System: Its Potential in CO_2 Sequestration. *Ind. Eng. Chem. Res.* **2012**, *51*, 16299–16310. [CrossRef]

6. Prigiobbe, V.; Hänchen, M.; Werner, M.; Baciocchi, R.; Mazzotti, M. Mineral carbonation process for CO_2 sequestration. *Energy Procedia* **2009**, *1*, 4885–4890. [CrossRef]

7. Zhao, L.; Sang, L.; Chen, J.; Ji, J.; Teng, H.H. Aqueous carbonation of natural brucite: Relevance to CO_2 sequestration. *Environ. Sci. Technol.* **2009**, *44*, 406–411. [CrossRef]

8. IPCC; Masson-Delmotte, V.P.; Zhai, H.-O.; Pörtner, D.; Roberts, J.; Skea, P.R.; Shukla, A.; Pirani, W.; Moufouma-Okia, C.; Péan, R.; et al. Summary for Policymakers. In *Global Warming of 1.5 °C*; An IPCC Special Report on the Impacts of Global Warming of 1.5 °C above Pre-Industrial Levels and Related Global Greenhouse Gas Emission Pathways, in the Context of Strengthening the Global Response to the Threat of Climate Change, Sustainable Development, and Efforts to Eradicate Poverty; World Meteorological Organization: Geneva, Switzerland, 2018; p. 32.

9. Boschi, C.; Dini, A.; Dallai, L.; Ruggieri, G.; Gianelli, G. Enhanced CO_2-mineral sequestration by cyclic hydraulic fracturing and Si-rich fluid infiltration into serpentinites at Malentrata (Tuscany, Italy). *Chem. Geol.* **2009**, *265*, 209–226. [CrossRef]

10. Nirta, G.; Pandeli, E.; Principi, G.; Bertini, G.; Cipriani, N. The Ligurian units of southern Tuscany. *Boll. Soc. Geol. Ital.* **2005**, *3*, 29–54.

11. Langone, A.; Baneschi, I.; Boschi, C.; Dini, A.; Guidi, M.; Cavallo, A. Serpentinite-water interaction and chromium (VI) release in spring waters: Examples from Tuscan ophiolites. *Ofioliti* **2013**, *38*, 41–57.

12. Federici, F. *Relazione Geologico Mineraria sul Giacimento Cuprifero di Montecastelli Pisano*; DBGM: San Giovanni Valdarno, Italy, 1941.

13. Lotti, B. Sul Giacimento Cuprifero di Montecastelli Pisano. In *Bollettino del Reale Comitato Geologico d'Italia XVI*; Reale Comitato Geologico d'Italia: Rome, Italy, 1885.

14. Lotti, B. Rapporto Sulla Miniera Cuprifera di Montecastelli in Toscana. Rapporto interno Montecatini. Società Generale per l'Industria Mineraria e Chimica. 1924. Available online: http://www.neogeo.unisi.it/dbgmnew/ricerca.asp?act=see&id=12922 (accessed on 1 July 2019).

15. Settore Idrologico Regionale (SIR), Regione Toscana. Available online: https://www.sir.toscana.it (accessed on 1 July 2019).

16. Jackson, S.E. LAMTRACE Data Reduction Software for LAICP-MS. In *Laser Ablation ICP-MS in the Earth Sciences: Current Practices and Outstanding Issues*; Sylvester, P., Ed.; Mineralogical Association of Canada: Quebec, QC, Canada, 2008; Volume 40, pp. 305–307, Short Course Series.

17. Parkhurst, D.L.; Appelo, C. User's guide to PHREEQC (Version 2): A computer program for speciation, batch-reaction, one-dimensional transport, and inverse geochemical calculations. *Water Resour. Investig. Rep.* **1999**, *99*, 312.

18. Brindley, G.W.; Bish, D.L.; Wan, H.-M. The nature of kerolite, its relation to talc and stevensite. *Mineral. Magazine* **1977**, *41*, 443–452. [CrossRef]

19. Cathelineau, M.; Caumon, M.-C.; Massei, F.; Brie, D.; Harlaux, M. Raman spectra of Ni-Mg kerolite: Effect of Ni-Mg substitution on O–H stretching vibrations. *J. Raman Spectrosc.* **2015**, *46*, 933–940. [CrossRef]

20. Léveillé, R.J.; Longstaffe, F.J.; Fyfe, W.S. Kerolite in carbonate-rich speleothems and microbial deposits from basaltic sea caves, Kauai, Hawaii. *Clays Clay Miner.* **2002**, *50*, 514–524. [CrossRef]

21. Oze, C.; Fendorf, S.; Bird, D.K.; Coleman, R.G. Chromium Geochemistry of Serpentine Soils. *Int. Geol. Rev.* **2004**, *46*, 97–126. [CrossRef]

22. Fairchild, I.J.; Baker, A. *Speleothem Science. From Process to Past Environments*; Wiley-Blackwells: Hoboken, NJ, USA, 2012; p. 450, Blackwell Quaternary Geoscience Series Book 4.

23. Cigna, A.A. An analytical study of air circulation in caves. *Int. J. Speleol.* **1968**, *3*, 41–54. [CrossRef]

24. James, E.W.; Banner, J.L.; Hardt, B. A global model for cave ventilation and seasonal bias in speleothem paleoclimate records. *Geochem. Geophys. Geosyst.* **2015**, *16*, 1044–1051. [CrossRef]

25. Park, A.-H.A.; Fan, L.-S. CO_2 mineral sequestration: Physically activated dissolution of serpentine and pH swing process. *Chem. Eng. Sci.* **2004**, *59*, 5241–5247. [CrossRef]

26. Sanna, A.; Wang, X.L.; Lacinska, A.; Styles, M.; Paulson, T.; Maroto-Valer, M.M. Enhancing Mg extraction from lizardite-rich serpentine for CO_2 mineral sequestration. *Min. Eng.* **2013**, *49*, 135–144. [CrossRef]

27. Daval, D.; Hellmann, R.; Martinez, I.; Gangloff, S.; Guyot, F. Lizardite serpentine dissolution kinetics as a function of pH and temperature, including effects of elevated pCO_2. *Chem. Geol.* **2013**, *351*, 245–256. [CrossRef]

28. Critelli, T.; Marini, L.; Schott, J.; Mavromatis, V.; Apollaro, C.; Rinder, T.; De Rosa, R.; Oelkers, E.H. Dissolution rate of antigorite from a whole-rock experimental study of serpentinite dissolution from 2 < pH < 9 at 25 °C: Implications for carbon mitigation via enhanced serpentinite weathering. *Appl. Geochem.* **2015**, *61*, 259–271.

Article

Low Temperature Serpentinite Replacement by Carbonates during Seawater Influx in the Newfoundland Margin

Suzanne Picazo [1], Benjamin Malvoisin [1,2,*], Lukas Baumgartner [1] and Anne-Sophie Bouvier [1]

[1]   Institut des Sciences de la Terre, University of Lausanne, Quartier UNIL-Mouline Bâtiment Géopolis, CH-1015 Lausanne, Switzerland; suzanne.picazo@gmail.com (S.P.); lukas.baumgartner@unil.ch (L.B.); anne-sophie.bouvier@unil.ch (A.-S.B.)

[2]   University Grenoble Alpes, University Savoie Mont Blanc, CNRS, IRD, IFSTTAR, ISTerre, 38000 Grenoble, France

*   Correspondence: benjamin.malvoisin@univ-grenoble-alpes.fr; Tel.: +33-476-514-072

Received: 22 January 2020; Accepted: 11 February 2020; Published: 18 February 2020

Abstract: Serpentinite replacement by carbonates in the seafloor is one of the main carbonation processes in nature providing insights into the mechanisms of CO_2 sequestration; however, the onset of this process and the conditions for the reaction to occur are not yet fully understood. Preserved serpentine rim with pseudomorphs of carbonate after serpentine and lobate-shaped carbonate grains are key structural features for replacement of serpentinite by carbonates. Cathodoluminescence microscopy reveals that Ca-rich carbonate precipitation in serpentinite is associated with a sequential assimilation of Mn. Homogeneous $\delta^{18}O$ values at the µm-scale within grains and host sample indicate low formation temperature (<20 °C) from carbonation initiation, with a high fluid to rock ratio. $\delta^{13}C$ (1–3 ± 1‰) sit within the measured values for hydrothermal systems (−3–3‰), with no systematic correlation with the Mn content. $\delta^{13}C$ values reflect the inorganic carbon dominance and the seawater source of CO_2 for carbonate. Thermodynamic modeling of fluid/rock interaction during seawater transport in serpentine predicts Ca-rich carbonate production, at the expense of serpentine, only at temperatures below 50 °C during seawater influx. Mg-rich carbonates can also be produced when using a model of fluid discharge, but at significantly higher temperatures (150 °C). This has major implications for the setting of carbonation in present-day and in fossil margins.

Keywords: carbonation; CO_2 sequestration; replacement process; low temperature carbonate precipitation; Secondary Ion Mass Spectrometer; seawater influx; hydrothermal circulation; ophicalcite

1. Introduction

Carbonation of peridotites occurs along rifted margins (e.g., the Western Iberia margin, [1] and references therein) and at Mid-Ocean Ridges (MORs; e.g., [2,3]). It is also commonly observed in ophiolites in the Alpine Tethys (e.g., [4]), the Pyrenees ([5] and references therein), Norway (e.g., [6]), Oman (e.g., [7]) and Québec [8]. Carbonated peridotites include ophicarbonates commonly associated to brecciated peridotites, cemented by carbonate, and created by tectonic, hydrothermal and/or sedimentary processes [9,10]. Metasomatic processes responsible for pervasive replacement of serpentine by calcite occur at slow-spreading ridges and in magma-poor passive margins, where peridotites are exposed on the seafloor during tectonic extension [11,12] and has been described in fossil margins (Chenaillet in the Alpine Tethys; [13]). Understanding carbonation in natural systems provides constraints to develop efficient engineering strategies for atmospheric CO_2 sequestration [14,15].

The heat released during magma cooling triggers seawater circulation in the upper part of the lithosphere. During hydrothermal circulation, mantle rocks are progressively replaced by hydrous minerals including serpentine and brucite [16,17]. This serpentinization reaction is also a redox reaction in which the ferrous iron contained in peridotite is oxidized to form magnetite and ferric serpentine, whereas water is reduced to form hydrogen [18]. Recent studies on the venting temperature of hydrothermal fluids from the Lucky Strike hydrothermal system emphasize that hydrothermal fluid circulation is subdivided into major (km-scale) and minor (m-scale) convection cells; the major cells lead to focused venting of high temperatures at the outflow, whereas the minor convection cells show mixing with seawater (4 °C) at the meter-scale leading to venting of diffuse fluids at low temperatures [19,20]. The composition of the fluid formed during fluid-peridotite interaction is expected to evolve from seawater to a high pH (9–11), hydrogen- and methane-rich fluid [18]. Highly alkaline fluids are expelled at low temperature (40–75 °C, [21,22]) at the Lost City Hydrothermal Field (LCHF). Such fluids promote Ca-carbonate precipitation ± brucite in chimneys and in veins [23,24]. Similar mineralogy has been observed within the Iberian margin, where the mixing zone provides an environment favoring microbial development (Ocean Drilling Program (ODP) Leg 149 Site 897; [25]). Peridotite-hosted hydrothermal systems exist at MORs such as the Mid-Atlantic Ridge (MAR; Kane Fracture Zone [26]; 15° N [27]) and in passive margins (e.g., Western Iberia margin; [25]). Fossil systems are also found in mountain belts like the Piemonte-Liguria ophiolites (e.g., Chenaillet, [28]). Acidic (pH ~3) and high-temperature (>300 °C) fluids were also sampled at other peridotite-hosted hydrothermal systems such as Rainbow and Logatchev, suggesting a contribution of mafic bodies [18]. Ultimately, fluid-rock interactions in the presence of mafic rocks may lead to crystallization of Si-rich phases, like talc [10,29] or quartz [30].

The composition of carbonates found in ophicarbonates varies between Mg and Ca-rich end-members. Calcite, aragonite, magnesite and dolomite show variable isotopic signatures, with $\delta^{18}O_{VSMOW}$ ranging from 8.3‰ to 36.6‰ and $\delta^{13}C_{VPDB}$ from −4.5‰ to 4.5‰ (see Supplementary Table S1). As a result, temperatures of carbonate crystallization estimated using $\delta^{18}O$ range from 1 up to ~200 °C [26,27], assuming seawater isotopic compositions for the fluid. This can be used to suggest precipitation under completely different conditions [27] from seawater on the seafloor [3] to high temperature hydrothermal fluids (100–150 °C; [13]). However, details about the different conditions are not well constrained nor how it relates to the process of carbonate formation. In hydrothermal systems, carbonates may be formed during either seawater influx or hydrothermal fluid discharge, involving different temperatures, compositions and fluid flow regimes.

Here, we investigate the process of serpentinite replacement by carbonates in a sample from the Newfoundland (NF) margin. In order to constrain the timing of calcite replacement, we first establish the brecciation event succession. Then we characterize the textures of calcite growth and their in-situ oxygen isotopic signature from core to rim. We couple our micro-textural observations and O and C isotopic measurements with a thermodynamic model to provide new constraints on the conditions of carbonation in mantle rocks exposed on the seafloor. Furthermore, we compile published O–C data from present-day and fossil distal margins and slow-spreading ridge ophicarbonates to discuss the possible re-equilibration of O–C in ophicarbonates during metamorphism.

2. Geological Setting

In the Newfoundland hyper-extended margin and conjugate Iberian margin, subcontinental lithospheric mantle was exhumed on the seafloor by detachment faulting. During the mantle exposure on the seafloor, a superposition of near-seafloor processes i.e., tectonic, hydrothermal and sedimentary brecciation occurred, resulting in tectono-sedimentary breccias [31–33]. Ophicarbonates are observed from the proximal to the distal margin of both the Newfoundland and Iberian margins. Carbonation processes overlap the brecciation, leading to calcite veins, replacement of serpentine and carbonate cement. The tectono-sedimentary breccias rework exhumed serpentinized footwall and thus preserved every stage of carbonation from its onset. Previous work from the Iberian margin has estimated that the temperature of carbonation in calcite veins and cement was < 45 °C based on $\delta^{18}O_{VSMOW}$ values of 25‰ to 35‰ [1,3,34–36]. The carbonation was thus interpreted as a near-surface cold process [35].

Here, we study samples from the International Ocean Discovery Program (IODP) site 1277 (Figure 1; [37]). The drill core contains serpentinized peridotites and locally magmatic intrusions (e.g., alkaline sills of 124–112 Ma, followed by Mid-Ocean Ridge Basalt-type magmatism <112 Ma at the onset of oceanic spreading, [38,39] and references therein). The drill core is also composed of brecciated serpentinized peridotites over 20 m (Figure 1c); we focus on these serpentinite samples that show evidence of replacement by carbonates.

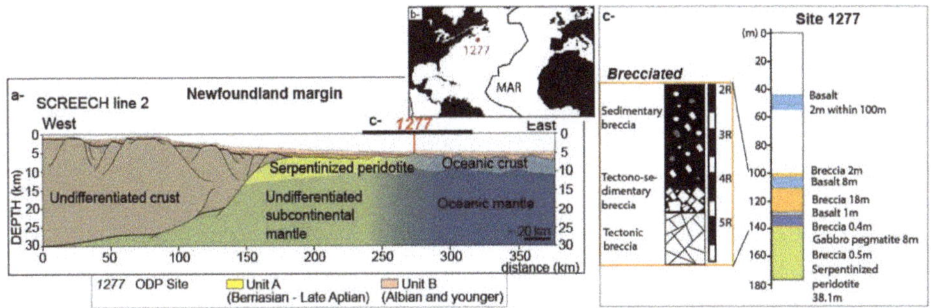

Figure 1. (**a**) Interpretative cross-section after Sutra and Manatschal [40] based on the seismic lines Lusigal 12 and of the TGS-NOPEC Geophysical Company for Iberia margin, and line 2 of the Study of Continental Rifting and Extension on the Eastern Canadian Shelf (SCREECH) project for Newfoundland. (**b**) Location of the Site 1277 in the Mid-Atlantic Ridge (MAR). (**c**) Sketch of the repartition of drilled mineralogies.

3. Methods

3.1. Microscopy

Scanning Electron Microscopy images were acquired using a Tescan Mira LMU FE-SEM (ISTe, University of Lausanne (UniL), Lausanne, Switzerland) operated at an acceleration voltage of 20 kV and a probe current of 20 nA. We used cold cathodoluminescence (CL) microscopy (model 8200 MkII (OPEA™), ISTe, UniL, Lausanne, Switzerland) to qualitatively estimate the variation in Mn content [41]. Light orange corresponds to a high Mn concentration whereas dark red is associated with low Mn concentrations. Electron microprobe analyses were conducted on a JEOL JKA-8530F (UniL, Lausanne, Switzerland), under beam of 15 nA and current of 10 nA as proposed by Lane and Dalton [42].

3.2. Secondary Ion Mass Spectrometer (SIMS) Analyses of Calcite

We measured carbonate grains in the core of the serpentine mesh textures using a Secondary Ion Mass Spectrometer Cameca 1280HR. We used a 10 kV Cs^+ primary beam, a ~1.5 nA current, resulting in a ~15 μm beam size. The electron flood gun, with normal incidence, was used to compensate charges.

For oxygen isotope measurements, ^{16}O and ^{18}O were analyzed with a mass resolving power of 2400 and collected in 2 faraday cups (10^{10} and 10^{11} Ω) in multi-collection mode. Each analysis takes ~4 min, including pre-sputtering (30 s) and automated centering of secondary electrons. This setting allowed an average reproducibility of ±0.28‰ (2 standard deviations (2 SD)) on an in house calcite reference material (UNIL_C1; $\delta^{18}O_{VSMOW}$ = 18.29 ± 0.12‰ (2 standard error (2 SE))) at the beginning of each session. A set of 4 analyses of the calcite standard were also measured every 12 analyses for monitoring the instrument stability. The variation of the calcite standard over the 24 h session was ±0.52‰ (2 SD), suggesting a magnetic drift, for which the data were corrected. The error reported for samples reflected the 2 SD bracket between the surrounding set of standard analyses.

^{12}C and ^{13}C, measured in a second SIMS session, were separated using a mass resolving power of 4600, and collected in 2 faraday cups (both 10^{11} Ω). Each analysis takes ~4–6 min, including pre-sputtering (45 s) and automated centering of secondary electrons. This setting allowed an average reproducibility of ±0.6‰ (2 SD) on a calcite in house standard (UNIL_C1; δ ^{13}C$_{VPDB}$ = 0.64 ± 0.08‰ (2 SE)) at the beginning of the session. A set of 4 analyses of a calcite standard was also measured every 10 unknowns; the variation of the calcite standard over the entire session (~3 h) was ±0.8‰ (2 SD), suggesting a slight magnetic drift. The error reported for samples reflected the 2 SD bracket between the surrounding set of standard analyses.

During both sessions, the faraday cups were calibrated at the beginning of the sessions using a calibration routine. Mass calibrations were performed at the beginning of the sessions and again after 12 h for the ^{18}O session.

We converted the measured δ^{18}O$_{VSMOW}$ values from calcite into temperatures using the standard procedure of Kim and O'Neil [43] and assumed that calcite crystallized in equilibrium with seawater with a δ^{18}O$_{VSMOW}$ = 0‰ at present-day, and a δ^{18}O$_{VSMOW}$ = −1.2‰ for an ice-free world [44].

3.3. Thermodynamic Modeling

Fluid/rock interaction was modelled at the equilibrium with the EQ3/6 software [45]. The evolution of the mineralogy and the fluid chemistry were predicted for oceanic serpentinized peridotite during hydrothermal fluid recharge or discharge. We model hydrothermal circulation in one dimension with a medium discretized in a sequence of 50 to 60 boxes. The temperature was fixed in each box and either linearly increases (recharge) or decreases (discharge) from one box to the other along the flow path (Figure 2). The simulations consisted of 3000 to 7000 iterations subdivided into two steps. First, the composition of the fluid phase in equilibrium with the solid was calculated independently in each box. This induces mass transfer between the solid and the fluid. The aim of this model was to determine the impact on mineralogical composition of fluid-mediated mass transfer from low to high temperatures (recharge) or high to low temperatures (discharge). Therefore, during the second step, the fluid of modified composition was transferred from one box to the next one in flow direction, while seawater with fixed composition was introduced into the first box (Figure 2). This latter step conserves mass. Time is not a parameter of the model since time dependent processes such as fluid flow, diffusion or reaction kinetics were not considered. To determine the extent of fluid-rock interaction, we used the cumulative mass of fluid introduced in the first box of the model ((W/R)$_d$; scaled to 1 kg of solid initially present in this box). This parameter could seem similar to the water to rock ratio commonly used in thermodynamic calculations (e.g., [46]), but it does not correspond to a single calculation at the equilibrium since over the number of iterations it integrates the amount of fluid added in the first box (it is thus a "dynamic" water to rock ratio). This water to rock ratio should ideally be of 2×10^{-2} to correspond to the porosity of ~ 5% measured in serpentinized peridotites [47]. However, with such a low amount of fluid, the computing time is too long to obtain significant mass transfer leading to carbonates formation. Therefore, at each iteration, we added into the model 38 kg (discharge) or 380 kg (recharge) of fluid (for 1 kg of solid) in the first box. Thermodynamic calculations at the equilibrium were performed with the database of Johnson et al. [48] with additional data for ferrous and ferric serpentine and ferrous brucite from Klein et al. [49] (see [50] for details). No organic components were considered. In the calculations, we suppressed the reduction of carbon by H$_2$ to account for the slow kinetics of methanogenesis reactions [51]. The composition of the reacting serpentine was initially fixed at Mg$_{2.77}$Fe^{2+}$_{0.13}$Fe^{3+}$_{0.13}$Si$_{1.94}$O$_5$(OH)$_4$ based on microprobe analyses. Serpentine composition can then evolve as a result of fluid-rock interactions. Iron oxidation state in serpentine was not determined, therefore, half of the iron was assumed to be trivalent, based on the compilation of serpentine composition by Evans [52], in agreement with the thermodynamic calculations by Klein et al. [48] and the μ-X-ray Absorption Near Edge Structure serpentine analyses on mesh textures by Andreani et al. [53]. The composition of seawater introduced in the first box of the model is given in Table S3. Simulations were run at a constant pressure of 50 MPa for simplicity.

This pressure corresponds to the lithostatic pressure expected at drill site 1277, where drilling was performed at 4600 m water-depth up to 180 m below the seafloor. As expected, tests of the respective impacts of temperature and pressure on the thermodynamic equilibrium show that temperature is the primary control of the stable assemblage whereas pressure variations are less important in the pressure and temperature ranges relevant to MORs (4 to 350 °C and 20 to 50 MPa).

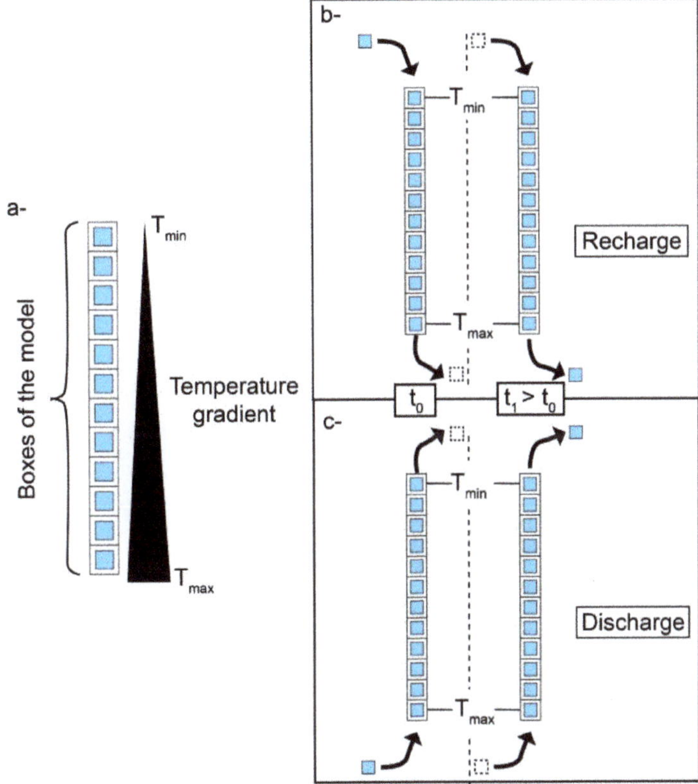

Figure 2. Sketch summarizing the modeling approach. (**a**) The model consists in calculating the equilibrium of a fluid + solid system in a series of boxes in which the fluid is transferred from one box to the next one at each time step. Two flow directions were investigated, either flow from high to low temperatures (discharge model) or from low to high temperatures (recharge model). Seawater was introduced in the first box of the model at 250 °C in the discharge model (**b**) and at 4 °C in the recharge model (**c**).

We applied equilibrium thermodynamics at low temperature in far from the equilibrium conditions. Reaction kinetics may prevent the achievement of equilibrium in these conditions. Taking into account reaction kinetics is fraught with uncertainty due to the lack of data on reaction rates in hydrothermal conditions and the need for modeling other time-dependent processes such as fluid flow. Several studies investigated the coupling between reaction kinetics and fluid flow in hydrothermal systems [54–56], but they simplified the system to a single chemical reaction and are therefore not able to model the changes in chemical composition investigated here. We therefore chose to calculate the system composition at the equilibrium towards which the system tends [57]. Such calculations were successfully applied to processes occurring at MORs (e.g., [47,48]). The use of equilibrium thermodynamics may be challenged by the production of metastable phases during reaction. The main minerals formed here are serpentine and carbonates. Metastable serpentine minerals such as chrysotile have similar composition and thermodynamic properties to stable phases (lizardite; [58]). They will thus have a limited effect on the model outputs. However, Mg-carbonates (i.e., magnesite and dolomite) have precipitation kinetics at least 6 orders of magnitude slower than calcite precipitation at ambient conditions [59]. The high activation energy of Mg-carbonate precipitation makes Mg-carbonates formation possible at high temperature. Because of the slow Mg-carbonate kinetic reaction, we run both simulations including and excluding Mg-carbonate precipitation. Finally, we assumed thermodynamic equilibrium in each box of the model. This requires that transport processes (i.e., diffusion and advection) are fast enough compared to the reaction to achieve equilibrium at a scale of several meters.

4. Results

4.1. Brecciation and Carbonate Crystallization

The 20 m-thick layer of tectono-sedimentary breccias drilled at Site 1277 (Figure 1; [37]) contains several calcite generations recognized by their crosscutting relationship and angular to rounded clasts of serpentinite (Figure 3). The calcite occurs in three contexts: it precipitates in veins crosscutting the clasts and formed during tectonic brecciation (type 1; Figure 4a,b); it constitutes the sedimentary component of the breccia cement (type 2; Figures 4b and 5b) and it crystallizes in the core of the serpentinite clasts (type 3; Figure 5a,b; [13]).

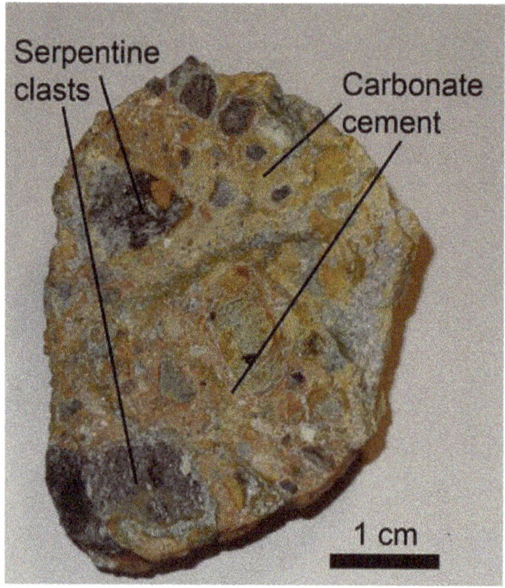

Figure 3. ODP Leg 210 Site 1277 4R-1 77–81 cm sample. Angular to rounded clasts of serpentinite are surrounded by a carbonate cement.

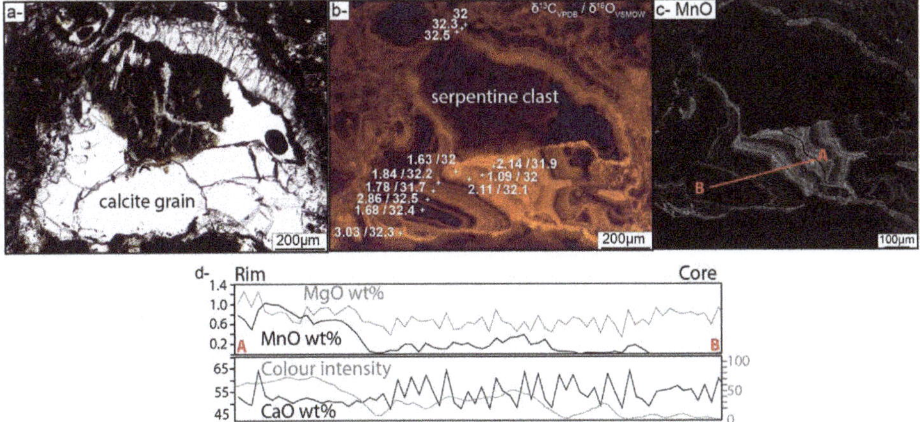

Figure 4. ODP Leg 210 Site 1277 4R-1 77–81 cm sample. (**a**) Calcite vein in serpentinite clast and (**b**) its cathodoluminescence equivalent (light orange corresponds to a high Mn concentration whereas dark red is associated with low Mn concentrations). (**c**) Microprobe map of MnO of the same area. (**d**) Microprobe profile of CaO, MgO and MnO versus the color intensity along the profile A, B shown by the red line in c. White crosses: data points of measured $\delta^{18}O_{SMOW}$ and $\delta^{13}C_{VPDB}$.

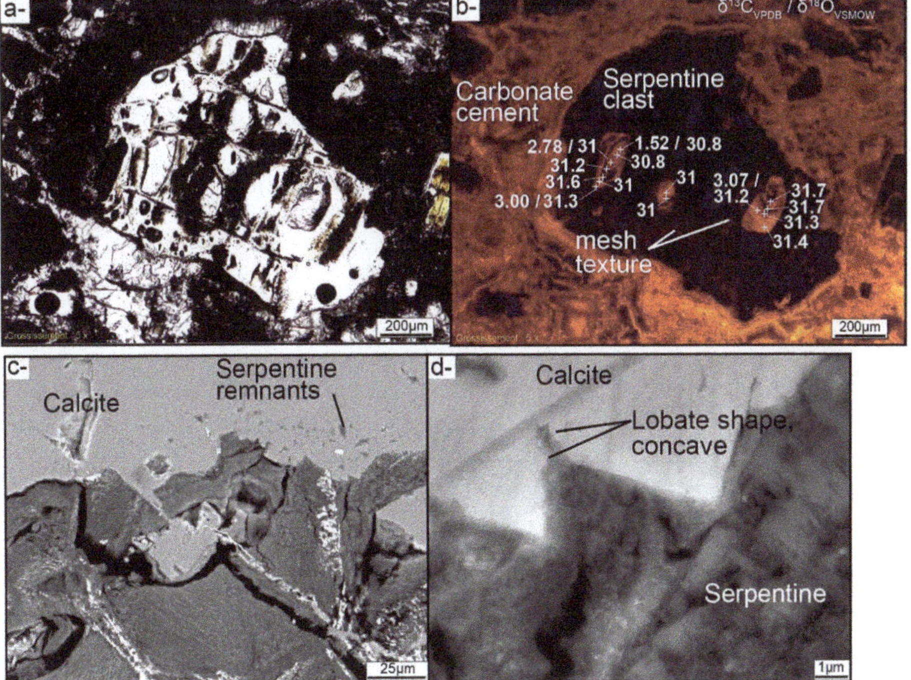

Figure 5. ODP Leg 210 Site 1277 4R-1 77–81 cm sample. (**a**) Serpentine core mesh replaced by carbonate and (**b**) its cathodoluminescence equivalent (light orange corresponds to a high Mn concentration whereas dark red is associated with low Mn concentrations). White crosses: data points of measured $\delta^{18}O_{SMOW}$ and $\delta^{13}C_{VPDB}$. (**c**,**d**) SEM images of calcite replacing serpentine.

In cathodoluminescence (CL) images, the calcite veins (type 1) display a color banding with an alternation of black and orange (Figure 4b). The banding is either parallel to fracturing, showing the opening direction of the vein (Figure 4b), or concentric, leading to scalenohedral crystals (Figure 5b). New grains appear to have nucleated in the vein at fracture edges, leading to single crystals with irregular shapes and low Mn-content (Figure 4d; dark in CL image). In addition, calcite grain growth occurs in parallel to the fracture opening direction (Figure 4b). It is common to observe in CL that the cores of the scalenohedron-shaped calcites are black (Mn-poor) whereas the last band is the brightest (Mn-rich; up to 1 wt. %, Figure 4d). The scalenohedral sparitic calcite grains found in veins vary in size from 10 to 100 μm. The banding in the veins display sharp contacts with the calcite in the cement due to brecciation (Figure 4c). The cement (type 2) is characterized by small, grain-sized calcite (<50 μm), with no CL-banding; CL colours are dark orange, with neither variation at sample or core scale (Figures 4b and 5b).

The serpentinite clasts are made of a network of mesh-like, regularly spaced fractures with two perpendicular orientations. Within the core of this regular mesh texture, 50 to 200 μm wide calcite grains are found (type 3) (Figure 5a,b). The mesh texture is preserved during replacement by carbonates (i.e., pseudomorphic replacement). In the core of the mesh, calcite grains grow concentrically from a nucleus, until reaching the rim of the mesh (Figure 5b). Depending on the nucleus location, the bands may not be continuous in the calcite grain. The CL colour of the bands is dark near the nucleus and again, bright orange in the outermost one (Figure 5b).

The banding observed in calcites of types 1 and 3 is similar (Figures 4b and 5b). At the contact between calcite and serpentine, the calcite grains are sharp and angular above the μm-scale (Figure 5c,d), and below the μm-scale the grains can display lobate shapes and are concave towards the serpentine (Figure 5d). The reaction front separates the well-crystallized calcite and fibrous serpentine and spans several micrometers. It is composed of <0.5 μm-thick calcite needles pervasively distributed in a disordered serpentine matrix (Figure 5d).

The μm-scale banding does not allow us to perform bulk isotopic analysis, as we would have obtained a mean value for all the bands. Therefore, we measured in-situ $\delta^{18}O$ in the different bands with the SIMS. We acquired profiles through calcite grains in veins (type 1) and in the serpentine mesh core (type 3). Only CL bands larger than 20 μm could be analyzed (white crosses on Figures 4b and 5b; Table S2). Measured $\delta^{18}O_{VSMOW}$ values cluster in a relative restricted range for pseudomorphous grains (30.8 ± 0.4‰ to 32.6 ± 0.4‰, n = 14), whereas calcite in veins are homogeneous (28.4 ± 0.4‰, n = 49) (Figure 6; Table S2). In all the calcites measured, no variation of $\delta^{18}O_{VSMOW}$ can be observed from core to rim. $\delta^{13}C_{VPDB}$ measurements vary from 1.09 ± 0.63‰ to 3.07 ± 0.85‰ (Table S2) but do not show systematic variation with the Mn-banding. Our dataset sits within the range of measurements available for the Iberia margin (Figure 6).

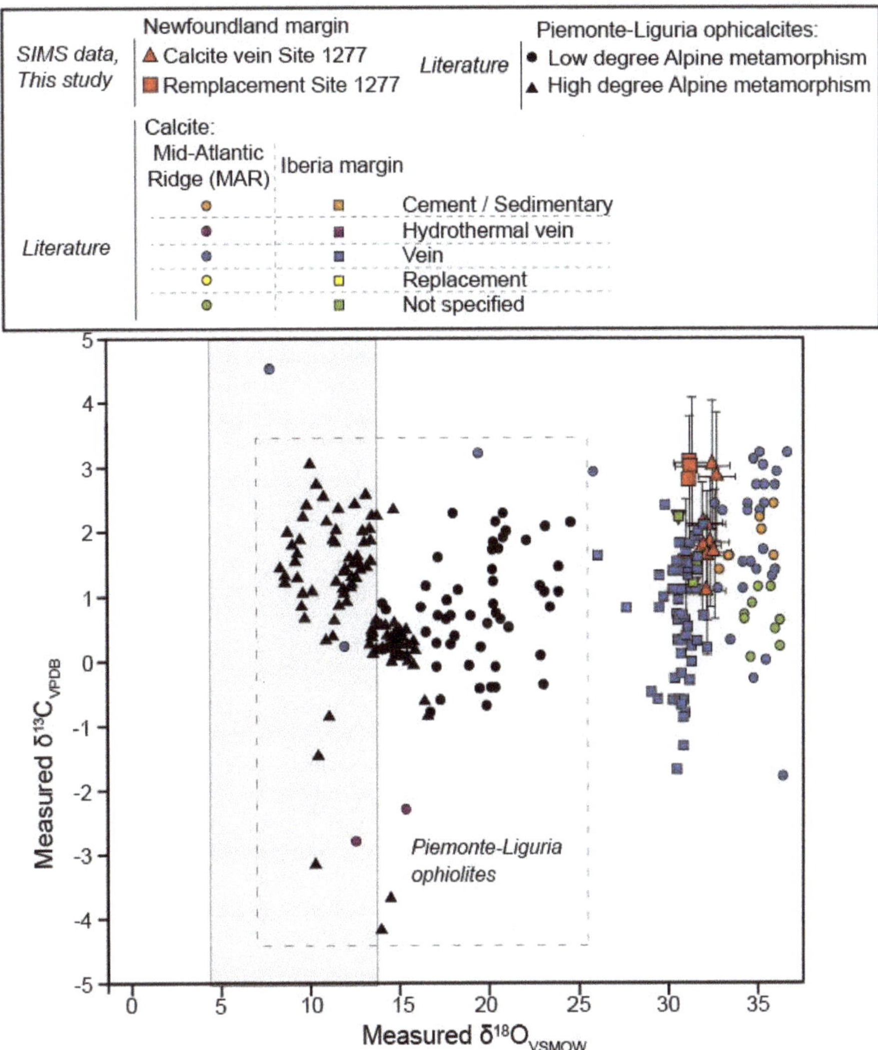

Figure 6. Diagram of $\delta^{18}O_{SMOW}$ versus $\delta^{13}C_{VPDB}$ measured with the Secondary Ion Mass Spectrometer (SIMS) facility compared to compilation of literature data of ophicarbonates from the Newfoundland-Iberia (NF-I; squares) margin, Mid-Atlantic Ridge (MAR; circles) and Alpine ophiolites, affected by low metamorphism degree (black circles) and high metamorphism degree (black triangles). See references in Table S1.

4.2. Thermodynamic Modeling of Carbonation

Opposite flow directions in the recharge and the discharge models are responsible for the formation of different mineralogical assemblages (Figures 7 and 8) and aqueous fluid compositions (Figures 9 and 10). This is due to the variations in mineral solubility with temperature. If mineral solubility increases with temperature (silicates), fluid transfer from low to high temperature (recharge) will lead to dissolution, whereas fluid transfer from high to low temperature (discharge) will lead to precipitation. Therefore, mass transfer will differ for models considering recharge or discharge. An opposite trend is expected for minerals with solubility decreasing with temperature (carbonates). Rock composition is observed to change along sharp boundaries (reaction fronts) for both recharge (Figure 7) and discharge models (Figure 8). In both models, the initial serpentine, exclusively composing the rock, is progressively replaced by talc as the fluid circulates through the rock. The talc is ultimately replaced by quartz (or clays at temperatures below ca. 40 °C) for the largest amount of fluid having circulated through the rock. The mineralogical assemblage can thus become quartz + carbonate and the modelled rock is a listvenite. Talc appears for lower $(W/R)_d$ when the precipitation of Mg-carbonates is allowed in the model, than when Mg-carbonates precipitation is excluded (Figures 7 and 8). These mineralogical evolutions require that the Mg and Si initially contained in serpentine are transported outside the rock as aqueous species in the fluid (Figures 9 and 10).

Carbonates are only formed at temperatures below 150 °C for the discharge model, and 100 °C for the recharge model (Figures 7 and 8e–h). In the discharge model, Ca concentration in the fluid slightly increases from 1.1×10^{-4} to 4.5×10^{-4} mol/kg as temperature decreases. The limitation of the carbonate stability field is thus mainly associated with pH decrease from 8.5 to 5 as temperature increases from 4 to 250 °C (Figure 10). In the recharge model, the limitation of the carbonate stability field is associated with anhydrite ($CaSO_4$) formation at temperature above 100 °C as shown, for example in Figure 9a, with the simultaneous increase in Ca and decrease in HCO_3^- concentrations at ~70 °C. Anhydrite formation only occurs in the first box of the model at high temperature in the discharge model (250 °C).

The distribution and composition of carbonates are different, even though the reactions of serpentine alteration were similar for discharge and recharge. The carbonates are first formed at the lowest temperature (4 °C) in the recharge model (Figure 7b,f). As fluid/rock interactions increase, the precipitation front progressively migrates towards higher temperatures. This front forming a serpentine + carbonate assemblage in equilibrium with a fluid of low Si concentration (Figure 9) is followed by a second front of talc + carbonate precipitation leading to increase in carbonate modal contents up to 60 mol. % (Figure 7h) and in Si concentration up to one order of magnitude (Figure 9h). Carbonates first replace serpentine at high temperatures (~150 °C) in the discharge model compared to the recharge model where this replacement starts at the lowest temperature (4 °C). The modal amounts of carbonate in the boxes are found to progressively decrease along the fluid flow path by one order of magnitude from 150 °C to 50 °C. Serpentine replacement by carbonates requires larger amount of fluids (or larger porosities for the same number of time steps in our model) to be observed in the case of recharge than in the case of discharge. This is due to the higher serpentine solubility where serpentine dissolution occurs in the discharge model at 150 °C than in the recharge at 20 °C. When Mg-carbonate precipitation is included in the model, carbonates are dolomite in the whole investigated temperature range in the recharge model and at temperatures below 20 °C in the discharge model. At temperatures above 20 °C, carbonates precipitate as magnesite in the discharge model, suggesting that Ca-rich carbonates are preferentially formed at low temperature. When Mg-carbonate precipitation is excluded, Ca-carbonates are only found to precipitate in the recharge model at temperatures below 110 °C (Figure 7).

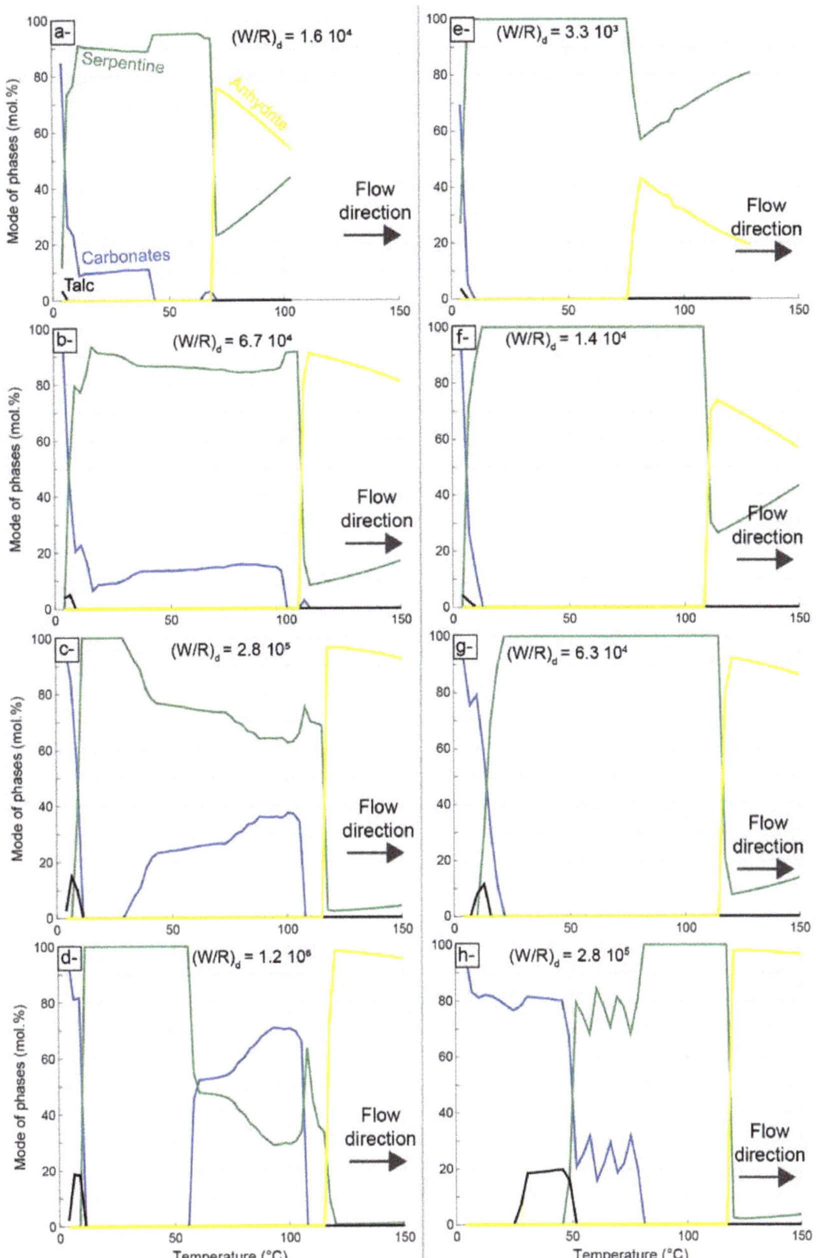

Figure 7. Serpentine (green), carbonates (blue), talc (black) and anhydrite (yellow) modes as a function of temperature in the model for recharge (flow direction is indicated). The model on the left (**a–d**) does not include Mg-carbonates whereas the model on the right does (**e–h**). For each model, the modes are displayed at various dynamic water to rock ratios $(W/R)_d$. Note that carbonates are first produced at low temperature in a reaction zone progressively extending towards higher temperatures. Carbonate formation is not predicted at temperature above 100 °C where anhydrite is the main Ca-bearing mineral to precipitate.

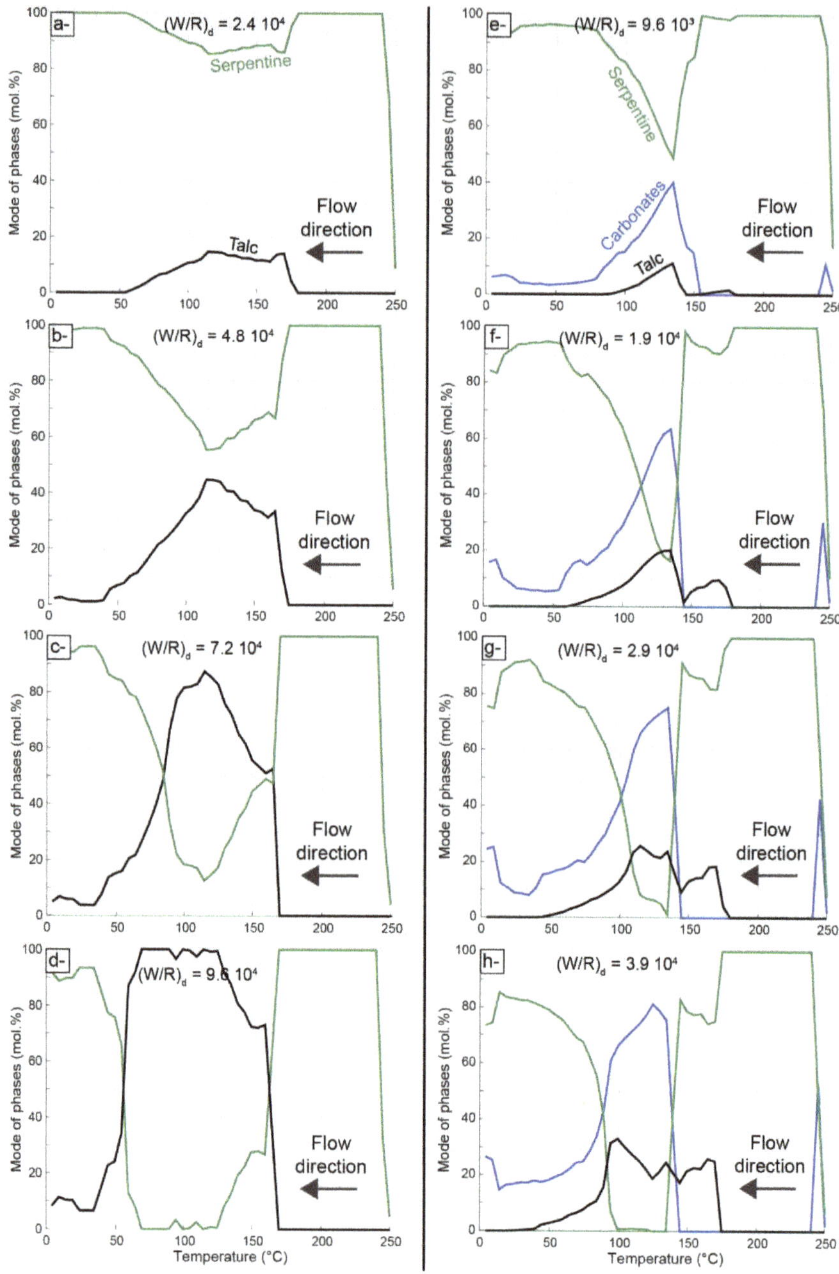

Figure 8. Serpentine (green), carbonates (blue) and talc (black) modes as a function of temperature in the model for discharge (flow direction is indicated). The model on the left (**a–d**) does not include Mg-carbonates whereas the model on the right does (**e–h**). For each model, the modes are displayed at various dynamic water to rock ratios $(W/R)_d$. Note that Mg-carbonates are mainly produced at temperatures below 150 °C in the model allowing for Mg-carbonate precipitation whereas carbonates are not produced in the model in which Mg-carbonate precipitation is not allowed.

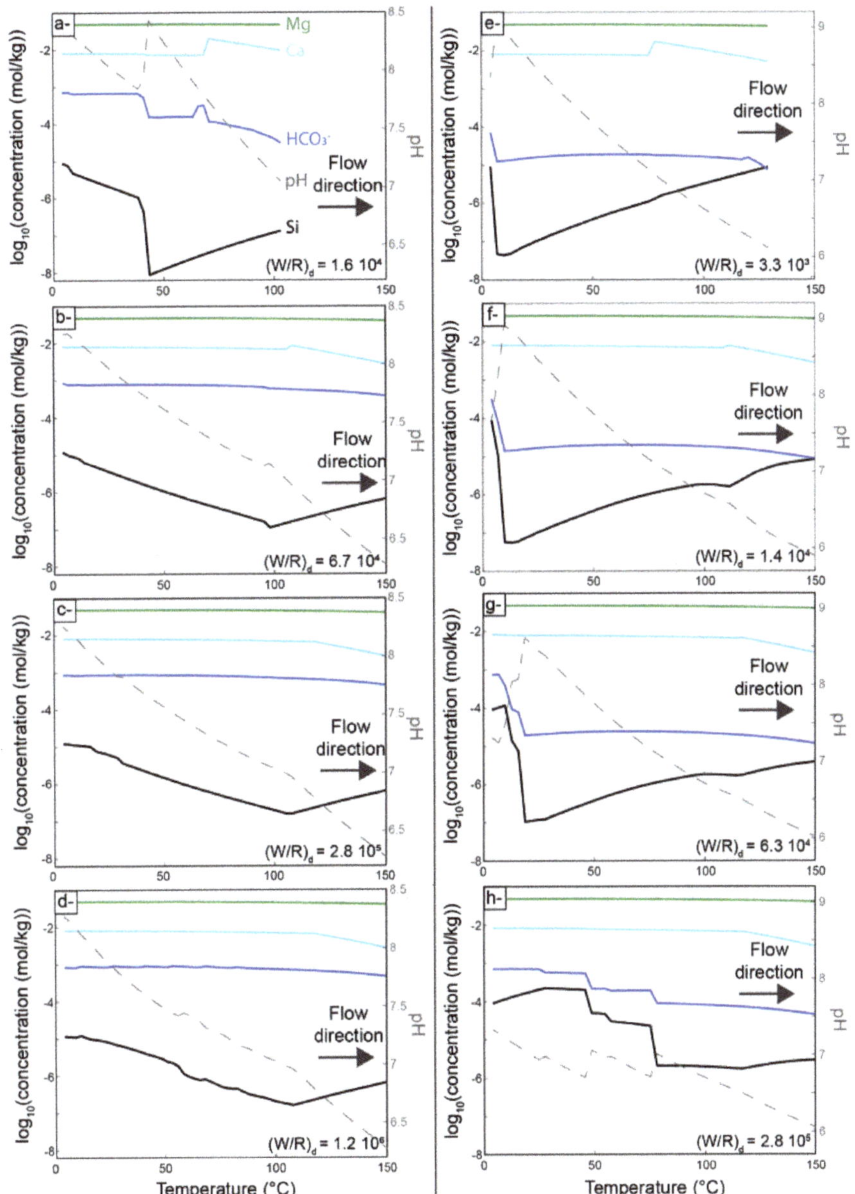

Figure 9. Si (black), Mg (green), Ca (light blue) and HCO_3^- (dark blue) concentrations in the fluid and pH (dashed line) as a function of temperature in the model for recharge (flow direction is indicated). The model on the left (**a–d**) does not include Mg-carbonates whereas the model on the right does (**e–h**). For each model, concentrations and pH are displayed at various dynamic water to rock ratios $(W/R)_d$.

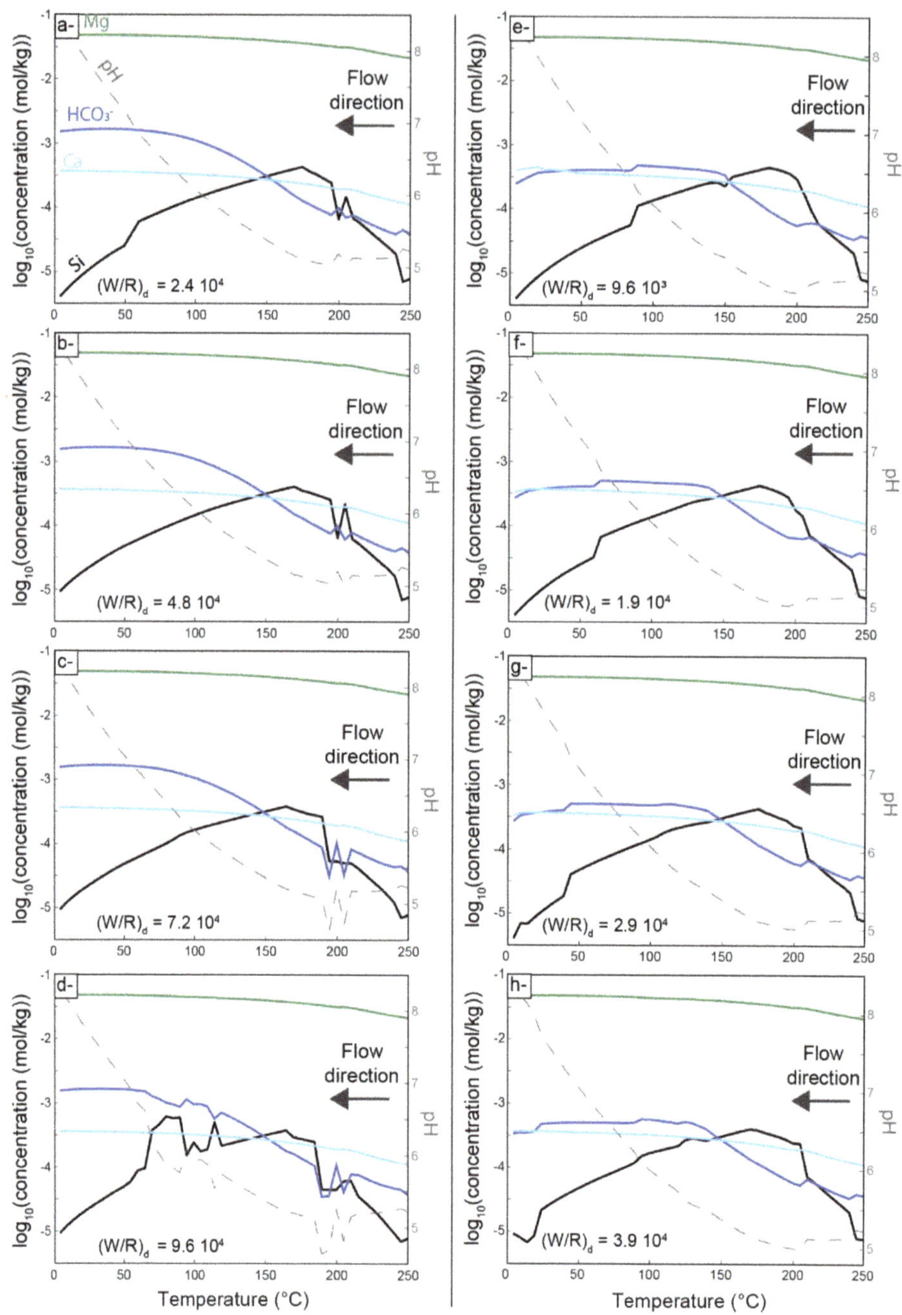

Figure 10. Si (black), Mg (green), Ca (light blue) and HCO_3^- (dark blue) concentrations in the fluid and pH (dashed line) as a function of temperature in the model for discharge (flow direction is indicated). The model on the left (**a–d**) does not include Mg-carbonates whereas the model on the right does (**e–h**). For each model, concentrations and pH are displayed at various dynamic water to rock ratios $(W/R)_d$.

5. Discussion

5.1. Serpentine Replacement by Carbonate During Seawater Influx in the NF Margin

We identify three mechanisms of carbonation in the Newfoundland serpentinites: veining, cementation and pseudomorphic replacement. Tectonic deformation during mantle exhumation along passive rifted margins leads to cataclasis in the footwall of detachment faults [32] and to fluid circulation. Both mechanisms result in calcite vein formation (e.g., [5,60]). They follow the pre-existing fracture network of the serpentinized peridotite. Near the seafloor, sedimentation is active, leading to calcite cementation into serpentinite breccia. All along this tectono-sedimentary sequence, serpentine is partially replaced by carbonate in the core of the mesh texture. To determine when replacive carbonation occurs during the mantle exhumation, we measured the temperature record of calcite in the inferred oldest carbonate.

During mantle exhumation, the anisotropic thermal contraction of peridotite together with tectonic stresses generates a primary microfracturing responsible for the onset of serpentinization [61]. This fracture network is then re-used for reaction induced fracturing [56,62,63]. The serpentine growing along the microfractures is organized in lizardite pseudocolumns constituting the rim of the mesh texture [61,63,64]. For higher reaction degree, preserved olivine grains can be altered into isotropic serpentine or proto-serpentine [65,66]. The core of meshes can be composed of either preserved olivine or proto-serpentine, depending on the extent of the serpentinization reaction. Residual olivine grains are extremely rare in the Iberia and Newfoundland exhumed serpentine basement, where the degree of serpentinization is >90% complete [67]. Therefore, the different habits of serpentine crystals and their coherence is probably responsible for the preferential carbonate growth in the mesh core of the studied sample. The lobate shape of calcite grains suggests serpentine replacement. Our interpretation is reinforced by the fact that calcite needles at the reaction front start to grow from isotropic serpentine. In the veins, calcite grains grow with a similar CL banding, but the absence of lobate shape suggests that these grains are not replacive.

Hydrothermal systems are an important source of manganese as described in several present-day active hydrothermal vents, e.g., East Pacific Rise, [68] or in the acidic hydrothermal systems (pH < 5) from the MAR where Mn concentration varies from 59 to 2250 μM [18]. Ca^{2+} from calcite can be substituted by Mn^{2+} to form a solid solution of $MnCO_3$ and $CaCO_3$ [69]. CL of calcite is activated by Mn^{2+}, whose emission colour is orange [70]. Temperature is not strongly influencing the Mn uptake in calcite grains [71]. Therefore, the Eh-dependent solubility of Mn into the fluid allows using CL colour variation as a proxy for oxygen fugacity variations in the fluid [72].

The CL-banding sequence is comparable between calcite grains growing in veins and in the serpentine mesh core, indicating a synchronous growth. The low content of Mn in calcite grains core reflects oxidizing conditions in both cases. In contrast, the last band in calcite is enriched in Mn testifying for more reduced fluids [72]. The fluid responsible for carbonation is oxidizing from the beginning of the reaction, even though fluids formed during serpentinization are reduced due to the coupled oxidation of the ferrous iron from olivine and pyroxenes to form magnetite [73]. This indicates that carbonation, from its onset, occurs after serpentinization near the seafloor, through interaction with oxidizing seawater.

The measured $\delta^{18}O$ are homogeneous in calcite precipitating in the veins ($\delta^{18}O_{VSMOW} \approx 31‰$) and vary by less than 2‰ in the calcite formed in the cores of the serpentine meshes. There is no systematic variation in the $\delta^{18}O$ with respect to the Mn content variation or the location in the grain. The narrow ranges of the in-situ isotopic measurements at the tens of micrometer scale indicate limited variations of the composition of infiltrating fluids, both spatially, as suggested by the bulk data at MORs, and temporally, as indicated by the absence of zoning. Assuming seawater isotopic composition, the $\delta^{18}O$ measurements give carbonation temperatures between T_{min} −1 to 11 °C and T_{max} 4 to 18 °C (Table S2). This low temperature of carbonate formation is consistent with experiments of inorganic carbonate precipitation at 25 °C [74–76]. It also suggests that serpentinization and carbonation can be two temporally decoupled processes. Indeed, the highest serpentinization rates measured in experiments are found in the range of 250–300 °C [77,78].

The temporal decoupling between serpentinization and carbonation can also be shown by subtracting $\delta^{18}O_{calcite}$ from $\delta^{18}O_{serpentine}$ [79]. For this calculation, we used a mean value calculated from a compilation of $\delta^{18}O_{serpentine}$ from the Iberia margin (Leg 103, [34]; Leg 149, [3]) and from MAR [80]. The mean value for $\delta^{18}O_{serpentine}$ used here is 10.65‰. We also used a mean value for $\delta^{18}O_{calcite}$ from the NF margin i.e., 31‰. If calcite was formed in equilibrium with serpentine, $\delta^{18}O_{calcite}-\delta^{18}O_{serpentine} \approx 10‰$. Here, the difference exceeds 20‰ (Figure 6 and Table S2), therefore, calcite and serpentine are not at equilibrium. Fast advective fluid transport (e.g., in a fracture network) during fluid recharge or discharge would lead to limited interaction between the fluid and the serpentine, and to the preservation of the fluid isotopic signature. However, pervasive fluid flow in a low permeable rock favors the fluid–serpentine interactions and can thus modify the initial isotopic signature of the fluid and of the precipitating carbonate [81]. In addition to the fluid–serpentine disequilibrium, the extensive carbonation as well as the absence of $\delta^{18}O$ zoning in calcite indicate that carbonation is fluid-buffered. As the fluid has a seawater signature, carbonation probably occurs near the surface during the onset of recharge into the serpentinized peridotites. The reproducibility of data in the MAR and Iberia-Newfoundland margins suggests that the recharge is a common process.

$\delta^{13}C_{VPDB}$ measured in NF margin are in agreement with previous data from the conjugated Iberia margin (i.e., between −3‰ and 3‰) and are typical for marine hydrothermal systems (e.g., [1]). This suggests that C is dominantly inorganic derived from seawater as suggested for the Iberian margin [1]. Thermodynamic modeling indicates that such a precipitation of seawater-derived inorganic carbon is possible and requires the circulation of more than 10^3 kg of seawater per kg of rock (Figures 7 and 8). The need for high dynamic water to rock ratios is associated with the low concentration of inorganic carbon in seawater (~2 mM). There are no systematic correlations between Mn-banding and $\delta^{13}C$ variation.

5.2. Carbonation in Passive Margins and Slow-Spreading MORs

We compiled calcite $\delta^{18}O$ and $\delta^{13}C$ bulk measurements from the Iberia and Newfoundland margins and from the Central MAR ophicarbonates (Figure 6). Based on their $\delta^{18}O$ record and calculated temperatures of formation, we infer that the ophicarbonates can be separated into two distinct groups, (1) calcite formed during seawater influx in an evolving carbonation front; or (2) during hydrothermal fluid discharge, in the case that during recharge, fluid-peridotite interaction was small.

As calcites from the Iberia margin and MAR, formed by veining, cementation and pseudomorphic replacement, record high $\delta^{18}O$, we infer this calcite to be formed during seawater influx. This has already been proposed in calcite veins from Iberia with O isotopic compositions varying from 27.6‰ to 31.2‰ [1] and in sedimentary carbonates (e.g., botryoidal calcite) at the Iberia margin ($\delta^{18}O$ from 30‰ to 32‰; [2,35]) and in the MAR ($\delta^{18}O$ from 33‰ to 35‰; [23]).

Calcite formed during hydrothermal fluid discharge is created by veining or hydrothermal deposits (e.g., [21,23,24]). In the MAR, carbonates found in veins and deposits are calcite and aragonite [21]. Mg content in calcite can be as high as 28 wt. % [24]. These high temperature calcite veins are found at the vicinity of hydrothermal vents, and record $\delta^{18}O$ between 26‰ and 8‰ (Mid-Atlantic Ridge, Kane Fracture Zone [26]; ODP Leg 209, [27]; Lost City, [21,23]). However, those samples represent a minority of the collected samples (see references in Table S1). Such low $\delta^{18}O_{calcite}$ indicates carbonation temperatures higher than 100 °C [23]. No pseudomorphic replacement is observed in environments in close relationship with an active hydrothermal vent. Based on O and C isotopic signature of replacive calcites, we interpret carbonate replacement in the samples from Newfoundland, the Iberian margin and the MAR as related to near-surface alteration during seawater influx.

We demonstrate that carbonation on passive margins is a low temperature, near surface process. This is true, even if magmatism is present as demonstrated at Site 1277, where syn- and post-rifting magmatism was identified (e.g., [39]). Indeed, calcite recorded temperatures below 20 °C. In the case of the Iberia-Newfoundland margins, magmatism has also no effect on the isotopic record, and hence does not influence carbonate precipitation.

5.3. Insights from Numerical Modeling

Carbonates are formed at two distinct temperature ranges at MORs, either below 10 °C or locally at the Lost City above 100 °C in veins feeding hydrothermal vents. Interestingly, this difference in carbonation temperature is also the main difference between the two simulations performed here with recharge and discharge models. Our simulations reproduced carbonate formation during recharge at near-surface conditions. They also predict carbonate formation at approximately 150 °C during discharge. Moreover, carbonate production required a higher fluid amount $((W/R)_d)$ for recharge than for discharge models, suggesting more open system conditions. Talc formation is decoupled from carbonate precipitation in the recharge models only, leading to rocks exclusively composed of serpentine and carbonates as observed in the natural samples. In all the simulations, Ca-rich carbonates were only formed at low temperature, whereas carbonation at temperature above 150 °C only produced Mg-rich carbonates. Based on these results, we interpret carbonate replacement in the samples from Newfoundland, the Iberian margin and the MAR as related to near-surface alteration during recharge. Carbonation at high temperature (>100 °C; [23]) is also expected to occur at MORs during discharge and, in particular, near hydrothermal vents such as the Lost City. The simulations predict the observation of calcite in the natural samples studied here, but they do also predict Ca-rich carbonates formation at temperatures above 100 °C, as is observed at the Lost City Hydrothermal Field. Due to missing compatible thermodynamic data, we do not include phases such as layered double hydroxides (coalingite-pyroaugite, LDHs), and hydrous Mg-carbonates (hydromagnesite and nesquehonite) (e.g., [82]).

5.4. Implications for Ophicarbonates Formation Preserved in Ophiolites

Ophicarbonates have been described first in remnants of the central Jurassic Tethys, in the Klosters, Totalp and Arosa ophiolites, in the Central-Eastern Alps [4] and in the Chenaillet ophiolite in the Western Alps [9]. Over the last 30 years, multiple studies have been conducted with a renewed interest, since they contain a key to the understanding of mantle exhumation to the seafloor, and because of their implication as an analogue for present-day passive margins. In addition, carbonation of serpentinite and other mantle rocks is a potential engineering solution to CO_2 sequestration [83]. Alpine ophicarbonates are interpreted to be the result of oceanic alteration of serpentinized peridotites during mantle exhumation associated with the Jurassic hyperextension phase [13], already proposed by Weissert and Bernoulli [4] and discussed by Früh-Green et al. [84].

It is important to note that carbonates are mostly calcites, and minor aragonite [2], never dolomite. Mineralogy of carbonates encountered in the Alpine Tethys ophicarbonates is mostly calcite, and minor dolomite (e.g., in Val Ventina, Alps, [82]; and in the Chenaillet, [13] and references therein). We compiled $\delta^{18}O$ and $\delta^{13}C$ measured in ophicarbonates from the literature in several locations in the Alpine Tethys. The data set contains bulk measurements on carbonate veins [84], disseminated carbonates [13], as well as sedimentary ophicarbonates [4,85,86].

Regional metamorphism experienced by ophicarbonates in the Platta nappes did not exceed 200 to 250 °C ([87] and references therein). The $\delta^{18}O$ values of carbonates range from 21‰ to 14‰ [85] in ophicarbonates towards the thrust plane, which separates them from serpentinites in the Platta nappe. Metamorphism and fluid flow during metamorphism seems to play in important role in the $\delta^{18}O$ record in ophicarbonates, and hence a careful re-evaluation of these values should be attempted to ascertain the initial condition of formation of these carbonates prior to metamorphism.

This explanation is supported by the low $\delta^{18}O$ values also recorded in sedimentary ophicarbonates (from 8.6‰ to 24.7‰ in the Western and Central Alps [4,13,86,88]). These low $\delta^{18}O$ values correspond to temperatures from 40 to 70 °C (Table S1). As in the Iberian margin, near-surface sedimentary carbonation precipitation also occurs in a range of temperatures below 20 °C, we infer that these temperatures are too high for true sedimentary deposition.

A way to optically verify if carbonate re-equilibration by metamorphism has started is to observe the Mn-banding in carbonates by cathodoluminescence. Diffusion of Mn is approximately 10 orders of magnitude faster (e.g., 1 mm at 300 °C in 25 Ma) than diffusion of O in calcite (30 μm to 10^{-6} μm for the same setting [89–91]). Many carbonates precipitating in shallow hydrothermal settings will obtain rhythmic layers of CL-active elements, as is shown in this study. Since diffusion of these elements (like Mn) is fast, they will disappear upon heating, and are obliterated by recrystallization. This means that if a carbonate preserves its banding structure acquired during carbonation on the seafloor, it should also preserve its $\delta^{18}O$ signature. Therefore, a test for recognizing re-equilibration in ophicarbonates from ophiolites would be the presence of thin CL-banding. This criterion can only be applied if diffusion is the only process acting during re-equilibration since calcite dissolution-reprecipitation may also modify the composition of the carbonates at low temperature. Nevertheless, this will also eliminate potentially the isotopic signature.

To follow up, we estimate the $\delta^{18}O$ expected in calcite that is supposed to be equilibrated with the surrounding serpentine during metamorphism. $\delta^{18}O_{serpentine}$ in Alpine ophiolites ranges from 4.3‰ to 13.4‰ [84]. If calcite and the surrounding serpentine are in equilibrium, according to the temperature of metamorphism reached, $\delta^{18}O_{calcite}$-$\delta^{18}O_{serpentine}$ should be $\Delta^{18}O_{cc-serp} \approx 5.27$‰ to 5.98‰ for a metamorphism at 200–250 °C (e.g., Platta); and $\Delta^{18}O_{cc-serp} \approx 2.82$‰ to 3.41‰ for a metamorphism at 450–550 °C (e.g., Val Ventina; [79]). Our compilation shows that samples displaying the least degree of metamorphism have $\Delta^{18}O_{cc-serp}$ higher than the estimated equilibrium values; whereas samples displaying the highest degree of metamorphism have $\Delta^{18}O_{cc-serp}$ corresponding to the estimated calcite–serpentine fractionation.

An alternative to the onset of differential re-equilibration during Alpine metamorphism could be that instead of precipitating during seawater influx, the ophicarbonates were formed during hydrothermal fluid discharge. We want to emphasize that this explanation would explain only the high temperatures recorded by the $\delta^{18}O$ in carbonates, as the textures and brecciation processes are similar to the present-day passive margin samples and are easily explained by the seawater influx model.

In an ophiolite where hydrothermal vents are clearly observed (e.g., Chenaillet), it should be easier to establish a relationship between carbonation and the distance between the hydrothermal fluid pathways. Magmatism may also play an important role in the high temperature recorded by ophicarbonates in distal parts of the Ocean Continent Transitions such as the Chenaillet, as late syn-rift to early post-rift magmatism with intrusions crosscutting the detachment faults are observed [28].

In the case of mostly discharge-driven carbonation, as the Alpine Tethys ophicarbonates are a good analogue for the present-day Iberia passive margin, near-surface ophicarbonates with high $\delta^{18}O$ (25–35‰) should also be reported in the remnant ophiolites, which is still not the case.

5.5. Implications for CO$_2$ Mineral Sequestration

The storage of CO_2 through carbonate precipitation appears to be one of the solutions to compensate for the anthropogenic greenhouse gas emissions [92]. Among the Earth's crust rocks, ultramafic rocks have one of the highest potentials for carbon sequestration due to their high magnesium content [6,93–99]. The results of this study provide constraints for the process of CO_2 sequestration in ultramafic rocks. We show here that low-temperature carbonation on the seafloor (T < 50 °C) produces Ca-carbonates. The calcium source is the seawater, and high water to rock ratios are needed to induce carbonate precipitation. The high MgO content of the ultramafic rock is therefore not an asset for CO_2 storage in the conditions prevailing on the seafloor. However, thermodynamic modeling indicates that high temperature carbonation (T > 100 °C) requires smaller water to rock ratios since magnesium can be incorporated into the carbonates. Mimicking the high temperature process thus appears to be more relevant to develop an efficient CO_2 storage solution. Such a solution would also need to circumvent the potential issues associated with the modification of the porous network by carbonation. Serpentine reaction occurring at high temperature to form talc and then quartz indeed requires significant magnesium and silica transport.

6. Conclusions

We provide microtextural observations of pervasive serpentine replacement by calcite and synchronous calcite growth in veins in exhumed mantle from the Newfoundland passive margin. We interpret those calcite grain cores as the first carbonate to grow in exhumed mantle. This is shown by the crosscutting relationship between brecciation and calcite grains textures revealed by CL images. Replacive calcite maintains the serpentine mesh texture and grows as scalenohedral crystals with a characteristic Mn-compositional banding. Preciseness of SIMS allowed us to measure O and C isotopic composition of each band in a single calcite grain. Measured O isotopes highlight no systematic variation, C isotope measurements display seawater range values and O isotope thermometry reveals that carbonation is cold (<20 °C) since the onset of the reaction.

Our thermodynamic modeling predicts Ca-rich carbonate crystallization near-surface during seawater influx and hydrothermal fluid discharge. Si-rich phases appear in the system with carbonation front evolution through time and space (e.g., talc).

In the discharge model, this stability field limitation is associated with carbonate dissociation into $HCO_3^-{}_{,aq}$ as the pH decreases with temperature. In the recharge model, the limitation of the carbonate stability field is associated with anhydrite ($CaSO_4$) formation at temperatures above 100 °C. Anhydrite formation only occurs in the first box of the model at high temperature in the discharge model (250 °C). We summarize the petrological, geochemical and numerical modeling results of this study in Figure 11, where the differences between recharge and discharge regarding carbonate precipitation are highlighted.

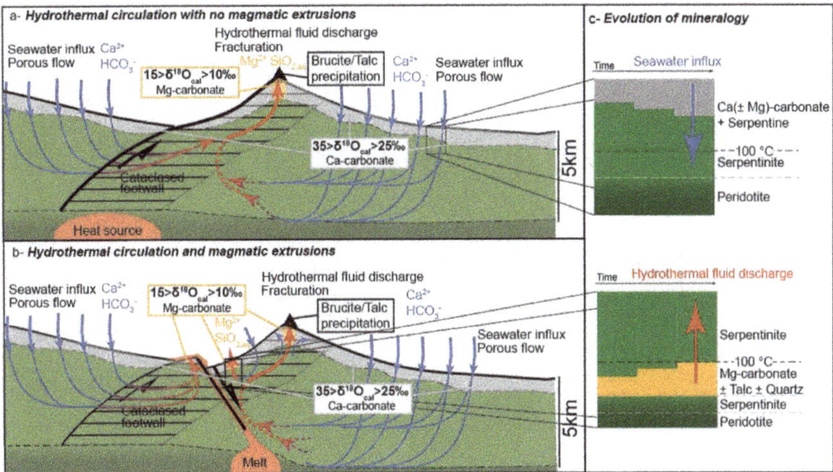

Figure 11. Conceptual model for the relationship between carbonation, hydrothermal circulation and magmatism. Two different models are proposed without (**a**) and with (**b**) magmatic extrusions. They both include recharge and discharge zones. Low-temperature carbonation (<50 °C) occurs in the recharge zones leading to Ca(±Mg)-carbonate formation (**c**). High-temperature carbonation starts at 150 °C in the discharge zones where Mg-carbonates only precipitate (**c**). Carbonation induces more magnesium and silica transport in the discharge zones leading to serpentine replacement by talc and then quartz as the water to rock ratio increases. These elements may precipitate at the surface in hydrothermal chimney as it is observed at the Lost City Hydrothermal Field [23]. The discharge model requires large scale fluid transport in permeable zones whereas carbonation in the recharge model is widespread and occurs through pervasive fluid transport.

According to our petrological and modeling results, we wonder why there are no cold ophicarbonates sampled yet in the Alpine ophiolites, and we propose that, in the case of absence of post-rift melting, O and C isotopic composition might have been re-equilibrated during Alpine deformation.

We show that carbonation of ultramafic rocks is more efficient (lower water to rock ratio) at high temperature (>100 °C) since Mg-bearing carbonates can be formed. This result may guide the development of future engineering solutions for CO_2 sequestration.

Supplementary Materials: The following are available online at http://www.mdpi.com/2075-163X/10/2/184/s1; Table S1: Compilation of carbon and oxygen isotope composition on calcite in serpentinized peridotites collected on the seafloor and in metamorphic environments. Table S2: Carbon and oxygen isotope composition on calcite from IODP Site 1277. Table S3: Electron microprobe analyses of serpentine and carbonate, and seawater composition used in thermodynamic modeling.

Author Contributions: S.P. and B.M. contributed to the data acquirement, interpretation of the data and wrote the article; L.B. contributed to the interpretation; A.-S.B. contributed to the setting of the SIMS. All authors have read and agreed to the published version of the manuscript.

Funding: Suzanne Picazo acknowledges Center of Excellence in Basin Analysis Grant from ExxonMobil for financial support. Benjamin Malvoisin acknowledges support from the Swiss National Science Foundation (Ambizione grant n°PZ00P2_168083). Lukas Baumgartner obtained funding from Swiss National Science Foundation and KIP6 PCI CASA.

Acknowledgments: The authors are grateful to Tiffany Barry for language corrections. We thank Susanne Seitz and Guillaume Siron for SIMS technical support.

Conflicts of Interest: The authors declare no conflict of interest.

References

1. Schwarzenbach, E.M.; Früh-Green, G.L.; Bernasconi, S.M.; Alt, J.C.; Plas, A. Serpentinization and carbon sequestration: A study of two ancient peridotite-hosted hydrothermal systems. *Chem. Geol.* **2013**, *351*, 115–133. [CrossRef]

2. Morgan, J.K.; Milliken, K.L. Petrography of calcite veins in serpentinized peridotite basement rocks from the Iberia Abyssal Plain, Sites 897 and 899: Kinematic and environmental implications. In *Proceedings of the Ocean Drilling Program, Scientific Results*; Whitmarsh, R.B., Sawyer, D.S., Klaus, A., Eds.; National Science Foundation: College Station, TX, USA, 1996; Volume 149.

3. Agrinier, P.; Cornen, G.; Beslier, M.O. Mineralogical and oxygen isotopic features of serpentinites recovered from the ocean/continent transition in the Iberia Abyssal Plain. In *Proceedings-Ocean Drilling Program Scientific Results*; National Science Foundation: College Station, TX, USA, 1996; pp. 541–552.

4. Weissert, H.; Bernoulli, D. Oxygen isotope composition of calcite in Alpine ophicarbonates: A hydrothermal of Alpine metamorphic signal? *Eclogae Geol. Helv.* **1985**, *77*, 29–43.

5. Clerc, C.; Boulvais, P.; Lagabrielle, Y.; de Saint Blanquat, M. Ophicalcites from the northern Pyrenean belt: A field, petrographic and stable isotopic study. *Int. J. Earth Sci.* **2014**, *103*, 141–163. [CrossRef]

6. Beinlich, A.; Plümper, O.; Hövelmann, J.; Austrheim, H.; Jamtveit, B. Massive serpentinite carbonation at Linnajavri, N-Norway. *Terra Nova* **2012**, *24*, 446–455. [CrossRef]

7. Surour, A.A.; Arafa, E.H. Ophicarbonates: Calichified serpentinites from Gebel Mohagara, Wadi Ghadir area, eastern desert, Egypt. *J. Afr. Earth Sci.* **1997**, *24*, 315–324. [CrossRef]

8. Lavoie, D.; Cousineau, P.A. Ordovician ophicalcites of southern Quebec Appalachians: A proposed early seafloor tectonosedimentary and hydrothermal origin. *J. Sedimentol. Res.* **1995**, *65*, 337–347.

9. Lemoine, M.; Tricart, P.; Boillot, G. Ultramafic and gabbroic ocean floor of the Ligurian Tethys (Alps, Corsica, Apennines): In search of a genetic model. *Geology* **1987**, *15*, 622–625. [CrossRef]

10. Picazo, S.; Cannat, M.; Delacour, A.; Escartín, J.; Rouméjon, S.; Silantyev, S. Deformation associated with the denudation of mantle-derived rocks at the Mid-Atlantic Ridge 13°–15° N: The role of magmatic injections and hydrothermal alteration. *Geochem. Geophys. Geosyst.* **2012**, *13*. [CrossRef]

11. Cannat, M. Emplacement of mantle rocks in the seafloor at mid-ocean ridges. *J. Geophys. Res. Solid Earth* **1993**, *98*, 4163–4172. [CrossRef]

12. Manatschal, G.; Bernoulli, D. Architecture and tectonic evolution of nonvolcanic margins: Present-day Galicia and ancient Adria. *Tectonics* **1999**, *18*, 1099–1119. [CrossRef]

13. Lafay, R.; Baumgartner, L.P.; Schwartz, S.; Picazo, S.; Montes-Hernandez, G.; Vennemann, T. Petrologic and stable isotopic studies of a fossil hydrothermal system in ultramafic environment (Chenaillet ophicalcites, eastern Alps, France): Processes of carbonate cementation. *Lithos* **2017**, *294*, 319–338. [CrossRef]

14. Kelemen, P.B.; Matter, J. In situ carbonation of peridotite for CO_2 storage. *Proc. Natl. Acad. Sci. USA* **2008**, *105*, 17295–17300. [CrossRef]

15. Matter, J.; Kelemen, P. Permanent storage of carbon dioxide in geological reservoirs by mineral carbonation. *Nat. Geosci.* **2009**, *2*, 837–841. [CrossRef]

16. Johannes, W. Experimental investigation of the reaction forsterite + H2O ⇌ serpentine + brucite. *Contrib. Mineral. Petrol.* **1968**, *19*, 309–315. [CrossRef]

17. Mével, C. Serpentinization of abyssal peridotites at mid-ocean ridges. *Comptes Rendus Geosci.* **2003**, *335*, 825–852. [CrossRef]

18. Charlou, J.L.; Donval, J.P.; Fouquet, Y.; Jean-Baptiste, P.; Holm, N. Geochemistry of high H_2 and CH_4 vent fluids issuing from ultramafic rocks at the Rainbow hydrothermal field (36 14′ N.; MAR). *Chem. Geol.* **2002**, *191*, 345–359. [CrossRef]

19. Barreyre, T.; Escartin, J.; Sohn, R.A.; Cannat, M.; Ballu, V.; Crawford, W. Temporal variability and tidal modulation of hydrothermal exit-fluid temperatures at the Lucky Strike deep-sea vent-field, Mid-Atlantic Ridge. *J. Geophys. Res.* **2014**, *119*, 2543–2566. [CrossRef]

20. Cooper, M.J.; Elderfield, H.; Schultz, A. Diffuse hydrothermal fluids from Lucky Strike hydrothermal vent field: Evidence for a shallow conductively heated system. *J. Geophys. Res.* **2000**, *105*, 19369–19375. [CrossRef]

21. Kelley, D.S.; Karson, J.A.; Früh-Green, G.L.; Yoerger, D.R.; Shank, T.M.; Butterfield, D.A.; Hayes, J.M.; Schrenk, M.O.; Olson, E.J.; Proskurowski, G.; et al. A serpentinite-hosted ecosystem: The Lost City hydrothermal field. *Science* **2005**, *307*, 1428–1434. [CrossRef]

22. Lang, S.Q.; Butterfield, D.A.; Schulte, M.; Kelley, D.S.; Lilley, M.D. Elevated concentrations of formate, acetate and dissolved organic carbon found at the Lost City hydrothermal field. *Geochim. Cosmochim. Acta* **2010**, *74*, 941–952. [CrossRef]

23. Früh-Green, G.L.; Kelley, D.S.; Bernasconi, S.M.; Karson, J.A.; Ludwig, K.A.; Butterfield, D.A.; Boschi, C.; Proskurowski, G. 30,000 years of hydrothermal activity at the Lost City vent field. *Science* **2003**, *301*, 495–498. [CrossRef] [PubMed]

24. Ludwig, K.A.; Kelley, D.S.; Butterfield, D.A.; Nelson, B.K.; Früh-Green, G. Formation and evolution of carbonate chimneys at the Lost City Hydrothermal Field. *Geochim. Cosmochim. Acta* **2006**, *70*, 3625–3645. [CrossRef]

25. Klein, F.; Humphris, S.E.; Guo, W.; Schubotz, F.; Schwarzenbach, E.M.; Orsi, W.D. Fluid mixing and the deep biosphere of a fossil Lost City-type hydrothermal system at the Iberia Margin. *Proc. Natl. Acad. Sci. USA* **2015**, *112*, 12036–12041. [CrossRef] [PubMed]

26. Alt, J.C.; Shanks, W.C. Serpentinization of abyssal peridotites from the MARK area, Mid-Atlantic Ridge: Sulfur geochemistry and reaction modeling. *Geochim. Cosmochim. Acta* **2003**, *67*, 641–653. [CrossRef]

27. Bach, W.; Rosner, M.; Jöns, N.; Rausch, S.; Robinson, L.F.; Paulick, H.; Erzinger, J. Carbonate veins trace seawater circulation during exhumation and uplift of mantle rock: Results from ODP Leg 209. *Earth Planet. Sci. Lett.* **2011**, *311*, 242–252. [CrossRef]

28. Manatschal, G.; Sauter, D.; Karpoff, A.M.; Masini, E.; Mohn, G.; Lagabrielle, Y. The Chenaillet Ophiolite in the French/Italian Alps: An ancient analogue for an oceanic core complex? *Lithos* **2011**, *124*, 169–184. [CrossRef]

29. Schroeder, T.; John, B.E. Strain localization on an oceanic detachment fault system, Atlantis Massif, 30 N.; Mid-Atlantic Ridge. *Geochem. Geophys. Geosyst.* **2004**, *5*. [CrossRef]

30. Bonnemains, D.; Escartín, J.; Mével, C.; Andreani, M.; Verlaguet, A. Pervasive silicification and hanging wall overplating along the 13°20′ N oceanic detachment fault (Mid-Atlantic Ridge). *Geochem. Geophys. Geosyst.* **2017**, *18*, 2028–2053. [CrossRef]

31. Sibson, R.H. Fault rocks and fault mechanisms. *J. Geol. Soc.* **1977**, *133*, 191–213. [CrossRef]

32. Manatschal, G.; Froitzheim, N.; Rubenach, M.; Turrin, B.D. The role of detachment faulting in the formation of an ocean-continent transition: Insights from the Iberia Abyssal Plain: From Wilson, R.L.C.; Whitmarsh, R.B.; Taylor, B.; Froitzheim, N. Non-volcanic rifting of continental margins: A comparaison of evidence from land and sea. *Geol. Soc. Lond. Spec. Publ.* **2001**, *187*, 405–428.

33. Woodcock, N.H.; Mort, K. Classification of fault breccias and related fault rocks. *Geol. Mag.* **2008**, *145*, 435–440. [CrossRef]

34. Agrinier, P.; Mével, C.; Girardeau, J. Hydrothermal alteration of the peridotites cored at the ocean/continent boundary of the Iberian margin: Petrologic and stable isotope evidence. In *Proceedings-Ocean Drilling Program Scientific Results*; Boillot, G., Winterer, E.L., Eds.; National Science Foundation: College Station, TX, USA, 1988; Volume 103, pp. 225–234.

35. Milliken, K.L.; Morgan, J.K. Chemical evidence for near seafloor precipitation of calcite in serpentinites (Site 897) and serpentinite breccias (Site 899), Iberia Abyssal Plain. In *Proceedings of the Ocean Drilling Program, Scientific Results*; Whitmarsh, R.B., Sawyer, D.S., Eds.; National Science Foundation: College Station, TX, USA, 1996; Volume 149.

36. Skelton, A.D.; Valley, J.W. The relative timing of serpentinisation and mantle exhumation at the ocean–continent transition, Iberia: Constraints from oxygen isotopes. *Earth Planet. Sci. Lett.* **2000**, *178*, 327–338. [CrossRef]

37. Whitmarsh, R.B.; Beslier, M.O.; Wallace, P.J. *Proceedings ODP, Initial Reports*; Ocean Drilling Program: College Station, TX, USA, 1998; Volume 173. [CrossRef]

38. Jagoutz, O.; Muntener, O.; Manatschal, G.; Rubatto, D.; Péron-Pinvidic, G.; Turrin, B.D.; Villa, I.M. The rift-to-drift transition in the North Atlantic: A stuttering start of the MORB machine? *Geology* **2007**, *35*, 1087–1090. [CrossRef]

39. Eddy, M.P.; Jagoutz, O.; Ibañez-Mejia, M. Timing of initial seafloor spreading in the Newfoundland-Iberia rift. *Geology* **2017**, *45*, 527–530. [CrossRef]

40. Sutra, E.; Manatschal, G. How does the continental crust thin in a hyperextended rifted margin? Insights from the Iberia margin. *Geology* **2012**, *40*, 139–142. [CrossRef]

41. Machel, H.G.; Mason, R.A.; Mariano, A.N.; Mucci, A. Causes and emission of luminescence in calcite and dolomite. In *Luminescence Microscopy and Spectroscopy—Qualitative and Quantitative Applications. SEPM (Society for Sedimentary Geology) Short Course*; Barker, C.E., Kopp, O.C., Eds.; SEPM: Tulsa, OK, USA, 1991; Volume 25, pp. 9–25.

42. Lane, S.J.; Dalton, J.A. Electron microprobe analysis of geological carbonates. *Am. Mineral.* **1994**, *79*, 745–749.

43. Kim, S.T.; O'Neil, J.R. Equilibrium and nonequilibrium oxygen isotope effects in synthetic carbonates. *Geochim. Cosmochim. Acta* **1997**, *61*, 3461–3475. [CrossRef]

44. Zachos, J.; Pagani, M.; Sloan, L.; Thomas, E.; Billups, K. Trends, rhythms, and aberrations in global climate 65 Ma to present. *Science* **2001**, *292*, 686–693. [CrossRef]

45. Wolery, T.J. *EQ3/6: A Software Package for Geochemical Modeling of Aqueous Systems: Package Overview and Installation Guide (Version 7.0)*; Lawrence Livermore National Laboratory: Livermore, CA, USA, 1992; p. 41.

46. Klein, F.; Bach, W.; Jöns, N.; McCollom, T.; Moskowitz, B.; Berquó, T. Iron partitioning and hydrogen generation during serpentinization of abyssal peridotites from 15° N on the Mid-Atlantic Ridge. *Geochim. Cosmochim. Acta* **2009**, *73*, 6868–6893. [CrossRef]

47. Tutolo, B.M.; Mildner, D.F.; Gagnon, C.V.; Saar, M.O.; Seyfried, W.E., Jr. Nanoscale constraints on porosity generation and fluid flow during serpentinization. *Geology* **2016**, *44*, 103–106. [CrossRef]

48. Johnson, J.W.; Oelkers, E.H.; Helgeson, H.C. SUPCRT92: A software package for calculating the standard molal thermodynamic properties of minerals, gases, aqueous species, and reactions from 1 to 5000 bar and 0 to 1000 C. *Comput. Geosci.* **1992**, *18*, 899–947. [CrossRef]

49. Klein, F.; Bach, W.; McCollom, T.M. Compositional controls on hydrogen generation during serpentinization of ultramafic rocks. *Lithos* **2013**, *178*, 55–69. [CrossRef]

50. Malvoisin, B. Mass transfer in the oceanic lithosphere: Serpentinization is not isochemical. *Earth Planet. Sci. Lett.* **2015**, *430*, 75–85. [CrossRef]

51. Milési, V.; Guyot, F.; Brunet, F.; Richard, L.; Recham, N.; Benedetti, M.; Dairou, J.; Prinzhofer, A. Formation of CO_2, H_2 and condensed carbon from siderite dissolution in the 200–300 °C range and at 50 MPa. *Geochim. Cosmochim. Acta* **2015**, *154*, 201–211. [CrossRef]

52. Evans, B.W. Control of the products of serpentinization by the $Fe^{2+} Mg^{-1}$ exchange potential of olivine and orthopyroxene. *J. Petrol.* **2008**, *49*, 1873–1887. [CrossRef]

53. Andreani, M.; Munoz, M.; Marcaillou, C.; Delacour, A. µXANES study of iron redox state in serpentine during oceanic serpentinization. *Lithos* **2013**, *178*, 70–83. [CrossRef]

54. Emmanuel, S.; Berkowitz, B. Suppression and stimulation of seafloor hydrothermal convection by exothermic mineral hydration. *Earth Planet. Sci. Lett.* **2006**, *243*, 657–668. [CrossRef]

55. Rudge, J.F.; Kelemen, P.B.; Spiegelman, M. A simple model of reaction-induced cracking applied to serpentinization and carbonation of peridotite. *Earth Planet. Sci. Lett.* **2010**, *291*, 215–227. [CrossRef]

56. Malvoisin, B.; Brantut, N.; Kaczmarek, M.A. Control of serpentinisation rate by reaction-induced cracking. *Earth Planet. Sci. Lett.* **2017**, *476*, 143–152. [CrossRef]
57. Lasaga, A.C. *Kinetic Theory in the Earth Sciences*; Princeton University Press: Chichester, UK, 2014; Volume 402.
58. Frost, B.R.; Beard, J.S. On silica activity and serpentinization. *J. Petrol.* **2007**, *48*, 1351–1368. [CrossRef]
59. Saldi, G.D.; Jordan, G.; Schott, J.; Oelkers, E.H. Magnesite growth rates as a function of temperature and saturation state. *Geochim. Cosmochim. Acta* **2009**, *73*, 5646–5657. [CrossRef]
60. Artemyev, D.A.; Zaykov, V.V. The types and genesis of ophicalcites in Lower Devonian olistostromes at cobalt-bearing massive sulfide deposits in the West Magnitogorsk paleoisland arc (South Urals). *Russ. Geol. Geophys.* **2010**, *51*, 750–763. [CrossRef]
61. Boudier, F.; Baronnet, A.; Mainprice, D. Serpentine mineral replacements of natural olivine and their seismic implications: Oceanic lizardite versus subduction-related antigorite. *J. Petrol.* **2010**, *51*, 495–512. [CrossRef]
62. Kelemen, P.B.; Hirth, G. Reaction-driven cracking during retrograde metamorphism: Olivine hydration and carbonation. *Earth Planet. Sci. Lett.* **2012**, *345*, 81–89. [CrossRef]
63. Rouméjon, S.; Cannat, M. Serpentinization of mantle-derived peridotites at mid-ocean ridges: Mesh texture development in the context of tectonic exhumation. *Geochem. Geophys. Geosyst.* **2014**, *15*, 2354–2379. [CrossRef]
64. Rumori, C.; Mellini, M.; Viti, C. Oriented, non-topotactic olivine—Serpentine replacement in mesh-textured, serpentinized peridotites. *Eur. J. Mineral.* **2004**, *16*, 731–741. [CrossRef]
65. Viti, C.; Mellini, M. Mesh textures and bastites in the Elba retrograde serpentinites. *Eur. J. Mineral.* **1998**, *10*, 1341–1359. [CrossRef]
66. Andreani, M.; Grauby, O.; Baronnet, A.; Muñoz, M. Occurrence, composition and growth of polyhedral serpentine. *Eur. J. Mineral.* **2008**, *20*, 159–171. [CrossRef]
67. Müntener, O.; Manatschal, G. High degrees of melt extraction recorded by spinel harzburgite of the Newfoundland margin: The role of inheritance and consequences for the evolution of the southern North Atlantic. *Earth Planet. Sci. Lett.* **2006**, *252*, 437–452. [CrossRef]
68. Klinkhammer, G.; Bender, M.; Weiss, R.F. Hydrothermal manganese in the Galapagos Rift. *Nature* **1977**, *269*, 319–320. [CrossRef]
69. Pingitore, N.E.; Eastman, M.P.; Sandidge, M.; Oden, K.; Freiha, B. The coprecipitation of manganese (II) with calcite: An experimental study. *Mar. Chem.* **1988**, *25*, 107–120. [CrossRef]
70. Haberman, D.; Neuser, R.D.; Richter, D.K. Low limit of Mn^{2+}—Activated cathodoluminescence of calcite: State of the art. *Sediment. Geol.* **1996**, *116*, 13–24. [CrossRef]
71. Dromgoole, E.L.; Walter, L.M. Iron and manganese incorporation into calcite: Effects of growth kinetics, temperature and solution chemistry. *Chem. Geol.* **1990**, *81*, 311–336. [CrossRef]
72. Barnaby, R.J.; Rimstidt, J.D. Redox conditions of calcite cementation interpreted from Mn and Fe contents of authigenic calcites. *Geol. Soc. Am. Bull.* **1989**, *101*, 795–804. [CrossRef]
73. Barnes, I.; O'Neil, J.R. The relationship between fluids in some fresh alpine-type ultramafics and possible modern serpentinization, western United States. *Geol. Soc. Am. Bull.* **1969**, *80*, 1947–1960. [CrossRef]
74. Kitano, Y.; Hood, D.W. Calcium carbonate crystal forms formed from sea water by inorganic processes. *J. Oceanogr. Soc. Jpn.* **1962**, *18*, 35–39. [CrossRef]
75. Müller, G.; Irion, G.; Förstner, U. Formation and diagenesis of inorganic Ca-Mg carbonates in the lacustrine environment. *Naturwissenschaften* **1972**, *59*, 158–164. [CrossRef]
76. Romanek, C.S.; Jiménez-Lopez, C.; Rodriguez Navarro, A.; Sanchez-Roman, M.; Sahai, N.; Coleman, M. Inorganic synthesis of Fe-Ca-Mg carbonates at low temperature. *Geochim. Cosmochim. Acta* **2009**, *73*, 5361–5376. [CrossRef]
77. Martin, B.; Fyfe, W.S. Some experimental and theoretical observations on the kinetics of hydration reactions with particular reference to serpentinization. *Chem. Geol.* **1970**, *6*, 185–202. [CrossRef]
78. Malvoisin, B.; Brunet, F.; Carlut, J.; Rouméjon, S.; Cannat, M. Serpentinization of oceanic peridotites: 2. Kinetics and processes of San Carlos olivine hydrothermal alteration. *J. Geophys. Res. Solid Earth* **2012**, *117*. [CrossRef]
79. Zheng, Y.F. Calculation of oxygen isotope fractionation in hydroxyl-bearing silicates. *Earth Planet. Sci. Lett.* **1993**, *120*, 247–263. [CrossRef]
80. Wenner, D.B.; Taylor, H.P. Temperatures of serpentinization of ultramafic rocks based on O^{18}/O^{16} fractionation between coexisting serpentine and magnetite. *Contrib. Mineral. Petrol.* **1971**, *32*, 165–185. [CrossRef]

81. Baumgartner, L.P.; Valley, J.W. Stable isotope transport and contact metamorphic fluid flow. *Rev. Mineral. Geochem.* **2001**, *43*, 415–467. [CrossRef]

82. Boschi, C.; Dini, A.; Baneschi, I.; Bedini, F.; Perchiazzi, N.; Cavallo, A. Brucite-driven CO_2 uptake in serpentinized dunites (Ligurian Ophiolites, Montecastelli, Tuscany). *Lithos* **2017**, *288*, 264–281. [CrossRef]

83. Kelemen, P.B.; Matter, J.; Streit, E.E.; Rudge, J.F.; Curry, W.B.; Blusztajn, J. Rates and mechanisms of mineral carbonation in peridotite: Natural processes and recipes for enhanced, in situ CO_2 capture and storage. *Annu. Rev. Earth Planet. Sci.* **2011**, *39*, 545–576. [CrossRef]

84. Früh-Green, G.L.; Weissert, H.; Bernoulli, D. A multiple fluid history recorded in Alpine ophiolites. *J. Geol. Soc.* **1990**, *147*, 959–970. [CrossRef]

85. Pozzorini, D.; Früh-Green, G.L. Stable isotope systematics of the Ventina Ophicarbonate Zone, Bergell contact aureole. *Schweiz. Mineral. und Petrol. Mitt.* **1996**, *76*, 549–564.

86. Abart, R.; Pozzorini, D. Implications of kinetically controlled mineral-fluid exchange on the geometry of stable-isotope fronts. *Eur. J. Miner.* **2000**, *12*, 1069–1082. [CrossRef]

87. Manatschal, G.; Nievergelt, P. A continent-ocean transition recorded in the Err and Platta nappes (Eastern Switzerland). *Eclogae Geol. Helv.* **1997**, *90*, 3–28.

88. Driesner, T. Aspects of petrographical, structural and stable isotope geochemical evolution of ophicarbonates breccias from ocean floor to subduction and uplift: An example from Chatillon, Middle Aosta Valley, Italian Alps. *Schweiz. Mineral. und Petrol. Mitt.* **1993**, *73*, 69–84.

89. Farver, J.R. Oxygen self-diffusion in calcite: Dependence on temperature and water fugacity. *Earth Planet. Sci. Lett.* **1994**, *121*, 575–587. [CrossRef]

90. Labotka, T.C.; Cole, D.R.; Riciputi, L.R. Diffusion of C and O in calcite at 100 MPa. *Am. Mineral.* **2000**, *85*, 488–494. [CrossRef]

91. Rosenbaum, J.M. Stable isotope fractionation between carbon dioxide and calcite at 900 °C. *Geochim. Cosmochim. Acta* **1994**, *58*, 3747–3753. [CrossRef]

92. Benson, S.M.; Cole, D.R. CO_2 sequestration in deep sedimentary formations. *Elements* **2008**, *4*, 325–331. [CrossRef]

93. Xu, T.; Apps, J.A.; Pruess, K. Numerical simulation of CO_2 disposal by mineral trapping in deep aquifers. *Appl. Geochem.* **2004**, *19*, 917–936. [CrossRef]

94. Andreani, M.; Luquot, L.; Gouze, P.; Godard, M.; Hoise, E.; Gibert, B. Experimental study of carbon sequestration reactions controlled by the percolation of CO_2-rich brine through peridotites. *Environ. Sci. Technol.* **2009**, *43*, 1226–1231. [CrossRef]

95. Boschi, C.; Dini, A.; Dallai, L.; Ruggieri, G.; Gianelli, G. Enhanced CO_2-mineral sequestration by cyclic hydraulic fracturing and Si-rich fluid infiltration into serpentinites at Malentrata (Tuscany, Italy). *Chem. Geol.* **2009**, *265*, 209–226. [CrossRef]

96. King, H.E.; Plümper, O.; Putnis, A. Effect of secondary phase formation on the carbonation of olivine. *Environ. Sci. Technol.* **2010**, *44*, 6503–6509. [CrossRef]

97. Paukert, A.N.; Matter, J.M.; Kelemen, P.B.; Shock, E.L.; Havig, J.R. Reaction path modeling of enhanced in situ CO_2 mineralization for carbon sequestration in the peridotite of the Samail Ophiolite, Sultanate of Oman. *Chem. Geol.* **2012**, *330*, 86–100. [CrossRef]

98. Klein, F.; McCollom, T.M. From serpentinization to carbonation: New insights from a CO_2 injection experiment. *Earth Planet. Sci. Lett.* **2013**, *379*, 137–145. [CrossRef]

99. Power, I.M.; Wilson, S.A.; Dipple, G.M. Serpentinite carbonation for CO_2 sequestration. *Elements* **2013**, *9*, 115–121. [CrossRef]

Article

Mineralogical Transformations of Heated Serpentine and Their Impact on Dissolution during Aqueous-Phase Mineral Carbonation Reaction in Flue Gas Conditions

Clémence Du Breuil [1], Louis César-Pasquier [1,*], Gregory Dipple [2], Jean-François Blais [1], Maria Cornelia Iliuta [3] and Guy Mercier [1]

[1] Institut National de la Recherche Scientifique (Centre Eau, Terre et Environnement), University of Quebec, 490 rue de la Couronne, Quebec, QC G1K 9A9, Canada; clemence.jouveau_du_breuil@inrs.ca (C.D.B.); jean-francois.blais@ete.inrs.ca (J.-F.B.); guy.mercier@ete.inrs.ca (G.M.)

[2] Department of Earth, Ocean and Atmospheric Sciences, University of British Columbia, 2020–2207 Main Mall, Vancouver, BC V6T 1Z4, Canada; gdipple@eoas.ubc.ca

[3] Chemical Engineering Department, Laval University, Quebec, QC G1V 0A6, Canada; maria-cornelia.iliuta@gch.ulaval.ca

* Correspondence: louis-cesar.pasquier@ete.inrs.ca; Tel.: +1-418-654-2606; Fax: +1-418-654-2633

Received: 16 August 2019; Accepted: 23 October 2019; Published: 3 November 2019

Abstract: Mineral carbonation is known to be among the most efficient ways to reduce the anthropogenic emissions of carbon dioxide. Serpentine minerals ($Mg_3Si_2O_5(OH)_4$), have shown great potential for carbonation. A way to improve yield is to thermally activate serpentine minerals prior to the carbonation reaction. This step is of great importance as it controls Mg^{2+} leaching, one of the carbonation reaction limiting factors. Previous studies have focused on the optimization of the thermal activation by determining the ideal activation temperature. However, to date, none of these studies have considered the impacts of the thermal activation on the efficiency of the aqueous-phase mineral carbonation at ambient temperature and moderate pressure in flue gas conditions. Several residence times and temperatures of activation have been tested to evaluate their impact on serpentine dissolution in conditions similar to mineral carbonation. The mineralogical composition of the treated solids has been studied using X-ray diffraction coupled with a quantification using the Rietveld refinement method. A novel approach in order to quantify the meta-serpentine formed during dehydroxylation is introduced. The most suitable mineral assemblage for carbonation is found to be a mixture of the different amorphous phases identified. This study highlights the importance of the mineralogical assemblage obtained during the dehydroxylation process and its impact on the magnesium availability during dissolution in the carbonation reaction.

Keywords: serpentinite; X-ray diffraction; rietveld refinement; magnesium leaching; thermal activation; meta-serpentine; carbonation; heat activation optimization

---

1. Introduction

The increasing greenhouse gas emissions and particularly anthropogenic carbon dioxide (CO_2) in the atmosphere are known to play a major role in climate change [1]. Mitigation solutions are needed more than ever. Among the methodologies proposed for mitigation, mineral carbonation appears to be one of the most sustainable [2,3]. This natural and spontaneous phenomenon involves the reaction between CO_2 (aqueous or gas) and divalent cations bearing minerals in order to form the associate carbonates [3]:

Equation (1) Carbonation reaction [4]

$$(Mg, Ca)_x Si_y O_{x+2y} + xCO_2 => x(Mg,Ca)CO_3 + yCO_2 \tag{1}$$

The reaction products are stable and inert solids where CO_2 is sequestered. The composition of the resulting carbonates depends on the major cations present in the reactant mineral [5]. Carbonation reaction can be divided in three main steps: (i) the CO_2 dissolution in water (ii) the material dissolution and (iii) the precipitation of carbonates as final products. The process is essentially controlled by the first two steps [6]. Serpentine minerals, due to their high amount of Mg^{2+} [7] are considered for carbonation [8]. Thermal treatment acts on serpentine dissolution by enhancing Mg^{2+} availability, making it a key step for the process [9]. Serpentine dissolution first results in a rapid exchange of surfacing Mg^{2+} with protons (H^+) before being extracted from the structure into the solution, during a much slower phase [10,11]. The dissociation of CO_2 added to the solution will generate protons and HCO_3^- ions, therefore enhancing Mg^{2+} availability (Pasquier et al., 2014b).

Lizardite, antigorite, and chrysotile are the main minerals of the serpentine group ($Mg_3Si_2O_5(OH)_4$), belonging to the phyllosilicate class [7,12–14]. Serpentine structure is made of stacked layers composed of two sheets: the tetrahedral layer composed of silicon tetrahedral (SiO_4), linked to the lateral Mg of the octahedral layer by its apical oxygen atoms, forming a covalent bond [14,15]. Outer hydroxyl groups contribute to Van der Waals interactions between the two layers, whereas inner hydroxyl groups contribute to intrafoliar Van der Waals interactions [15–17].

Under high temperatures, hydroxyl groups, linked to Mg atoms, escape the structure. During this dehydroxylation process, serpentine transformed into amorphous phases (between 550 and 750 °C—Equation (2)), and then recrystallized into forsterite ($Mg_2SiO_4 > 750$ °C), associated with enstatite ($MgSiO_3 > 800$ °C) as the temperature increased (Equation (3)) [18–20]. Two types of amorphous phases have been described [21]: pseudo-amorphous phases, named α-meta-serpentine, appearing at 50% of the total dehydroxylation reaction, and amorphous meta-serpentine, appearing at 90% of the total dehydroxylation. The formation of αmeta-serpentine component can be observed at a temperature close to 580 °C visualized on a diffractogram by a feature in the lower angle domain ($2\theta = \pm 6°$) [21].

Equations (2) and (3): Serpentine dihydroxylation

$$Mg_3Si_2O_5(OH)_{4\,(s)} \rightarrow Mg_3Si_2O_{7\,(s)} + 2H_2O_{\,(g)} \tag{2}$$

$$2Mg_3Si_2O_{7\,(s)} + SiO_{2\,(s)} \rightarrow 3Mg_2SiO_{4\,(s)} + MgSiO_{3\,(s)} + SiO_{2\,(s)}. \tag{3}$$

It has been observed that amorphous meta serpentine tends to promote Mg^{2+} leaching and thus carbonation [21–23]. Therefore, optimized conditions for carbonations have been prescribed to be between 630 °C and 650 °C for 30 to 120 min [22,24,25]. However, in the previous studies, carbonation reactions have essentially been performed using pure CO_2 gas at high temperature and high pressure [21,25–27], strong acids or salts to promote dissolution [22,28]. To date no studies have been conducted on optimizing thermal activation from the mineralogical point of view, especially for direct aqueous mineral carbonation using diluted gas. In these conditions, serpentine dissolution is only promoted by carbonic acid at room temperature and low/mild CO_2 partial pressure and a good activation is more than ever critical for reaction.

This study is part of the follow-up work on direct flue gas carbonation process initiated by Mercier et al. at INRS, Québec [29]. Using mining residues available in the Province of Québec, the process uses a simulated cement plant flue gas to perform direct flue gas aqueous carbonation [30]. Carbonation reaction parameters have been optimized by Pasquier [31], optimized conditions for the precipitation of carbonates have been determined by Moreno [32] whereas a technical and economical evaluation of the process have shown its feasibility and sustainability in the Province of Québec [33]. However, a pilot scale test revealed that thermal treatment conditions needed to be optimized for the INRS process as well [34,35].

In the present paper, only the proportion of magnesium prior to precipitation will be studied and considered as an intermediate product of the carbonation, as thermal activation can only acts on enhancing serpentine dissolution. Therefore, post-carbonation solids were not considered in the present study for the given reasons. Furthermore, it serves to give a novel approach of evaluating the influence of amorphous phases on serpentine dissolution and thus Mg^{2+} leaching during direct flue gas aqueous mineral carbonation by introducing a new quantifying method of those phases. Those new mineralogical data will provide a further understanding of the relation between thermal activation and serpentine dissolution and therefore, improve this step in the INRS carbonation process.

2. Materials and Methods

2.1. Sample Preparation, Characterization and Analytical Methods

Serpentinite residues were sampled on stockpiles from Jeffrey Mine, near the town of Asbestos, southern of the Province of Québec. Lizardite is identified to be the major serpentine polymorph [36]. Fibres, representing 20 wt % of the residue, were removed by gravimetric separation based on their buoyancy. Iron oxides were also removed by wet gravimetric separation using the Wifley table, due to their potential commercial value for the process.

The sample was ground using a ring mill (Retsch RS200, Dusseldorf, Germany). The grain size distribution is given in Table 1 Values were obtained and measured using a particle size distribution analyzer (LA-950V2 Horiba, Kyoto, Japan).

Table 1. Size distribution of the sample.

Mean Size (μm)	d_{90}	d_{50}	d_{10}
16.00	1.99	8.16	43.28

The chemical composition of the starting material is given in Table 2. The chemical composition of liquid and solid samples was obtained using inductively coupled plasma-atomic emission spectrometry (ICP-AES) analysis (Varian, Palo Alto, CA, USA). Solid samples were first fused using the Claisse Method [37]. Loss on ignition (LOI) was obtained from mass difference after placing the sample into a ceramic crucible inside a muffle furnace for 6 h at 1025 °C.

Table 2. Composition of the raw solid feedstock [1].

Elements	Values (wt %)
CaO	0.7
Cr_2O_3	0.2
Fe_2O_3	6.8
K_2O	0.2
MgO	41.0
MnO	0.1
NiO	0.3
SiO_2	39.9
LOI	10.8

[1] Major compounds only.

Phases were identified using XRPD analysis (Bruker AXS, 2004, Karlsruhe, Germany), performed at the University of British Columbia. To prepare the sample, 1.6 g were mixed with 0.4 g of pure corundum (Al_2O_3) [38], used as an internal standard, representing a 20.0 wt % spike. Samples were ground in ethanol using agate grinding pellets for seven minutes, in a McCrone micronizing mill to ensure homogenization. Scans were acquired for 30 min with 2θ ranging from 3° to 80° with scanning step size of $2\theta = 0.3°$ with a counting time of 7 s per step, on a Siemens D5000 Bragg-Brentano θ-2θ

diffractometer (Bruker AXS, 2004, Karlsruhe, Germany) with radiation CuKα (40kV, 40mA). Matches were obtained using Bruker identification software DIFFRACplus EVA and the ICDD PDF-2 database.

Quantification of phases was performed using the Rietveld method ([39–42]) based on a calibration factor obtained from the mass and volume of each phase's unit cell. However, this method requires that all of the phases show high degrees of crystallinity with well-defined crystal structures [42]. Serpentine minerals are known to show discrepancies from their ideal crystal structures [38,43]. Therefore, when the crystalline structure of a phase is unknown or partially known, it can be quantified through the use of the Partial Or No Known Crystalline Structure method (PONKCS), combined with the Rietveld method [44]. A standard sample of pure chrysotile (90.0 wt %) and fluorite (10.0 wt %), provided by The University of British Columbia (Vancouver, British Columbia, Canada)., whose composition is well known [38] was used in order to calibrate the PONKCS model. A calibrated mass value for the unit cell of both phases was acquired by Rietveld refinements and the chrysotile peaks were fitted using the Le Bail method [45]. The unit cell parameters and the space group were extracted from Falini [46]. The generated PONKCS model was then used in the Rietveld refinements as a crystallographic information files in the software TOPAS (Bruker AXS) [44,47].

2.2. Experimental Apparatus and Conditions

2.2.1. Thermal Activation

Heat treatments were performed in a muffle furnace, Furnatrol I33 (Thermolyne Subron Corporation). Twenty grams of samples were placed in a cast-iron skillet in a thin surface, then introduced into the furnace. Cast iron was chosen for its resistance to drastic temperature changes. No major iron contaminations were observed after thermal treatments according to chemical analysis. Tests were performed under isothermal conditions, meaning that the furnace was set to the targeted temperature beforehand to the test. At the end of the treatment, the furnace was open to help the skillet cool down, for five minutes. The skillet and the sample were weighted and the difference between initial and final masses corresponded to the mass lost induced by heat treatment. A series of seven tests were conducted at different residence times and temperatures, as shown in Table 3.

Table 3. Treatment conditions.

Residence Time (min)	Temperature (°C)		
	550	650	750
15	A	C	F
30	-	D	G
60	B	E	H

2.2.2. Dissolution in Aqueous Carbonation Conditions

Dissolution reactions were performed using carbonation conditions in order to evaluate thermal treatment effects on the developing process. Carbonation reactions were conducted in a 300 mL stirred reactor, model 4561 of Parr Instrument Company (Moline, IL, USA) [29,31]). The tests were performed using a certified composition gas of 4.0 vol% O_2 and 18.2 vol% CO_2, balanced with N_2, simulating cement plant flue gas. The pulp density was set to 15% (150 g·L^{-1}), with a volume of 75 mL of water, 11.25 g of solid and a gas phase volume of 225 mL, as optimized by Mercier et al. [29]. A batch of 10.2 bar of gas was introduced into the stirred reactor and was allowed to react with the pulp for 15 min at room temperature 22 °C $\pm$ 3 °C with the agitation sets to 600 RPM (Figure 1). As the reaction happens, the pressure decreased between 0.1 and 3.5 bar, depending on the amount of gas initially introduced. At the end of each batch, the pulp was filtered to obtain the liquid for chemical composition analyses. The reactivity of the eight thermally treated samples were tested along with an untreated sample (U). They were only observed throughout the proportion of magnesium leached from the solid during the reaction, using Equation (4) [Mg]$_{liq}$ corresponds to the measured concentration of Mg^{2+} at the end of the reaction, V and m are the volume of the solution and the mass of solid, respectively, and C_{Mg} is the measured concentration of Mg in the post-thermal treatment solid. Carbonates were not precipitated from the solution as thermal activation impacts on the Mg leaching. Mg analysis was performed on liquid sample after reaction. The liquid fraction was obtained after filtration of the resulting pulp. Consistency in the procedure was validated by performing mass balance to highlight any precipitation occurring during manipulation or during the pressure release of the vessel.

Equation (4): Proportion of Mg^{2+} leached

$$\%Mg = \frac{([Mg]_{liq} \times V) \times 100}{(C_{Mg} \times m)}. \tag{4}$$

In a successive batches test, the solid was used for 12 batches of gas. Every two batches, the solid was filtered and reused with fresh liquid in the subsequent batches. After six batches, the solid was filtered, dried at 60 °C, and ground for 1 min at 700 RPM in a ring mill, to partially remove the silica layer formed around the grains and then re used for another series of 6 batches as described by Figure 2. The liquid phase was sampled and renewed every two batches to prevent saturation. Long term reactivity of samples D and F treated at 650 °C for 30 min and 750 °C for 15 min, respectively, were tested in a successive batches experiment.

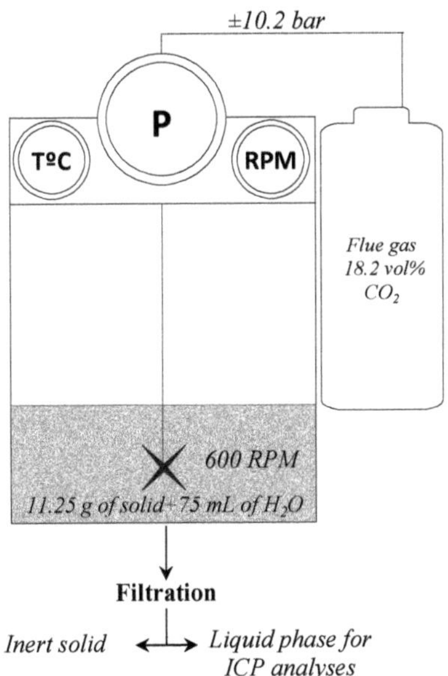

Figure 1. Parr reactor experimental set up.

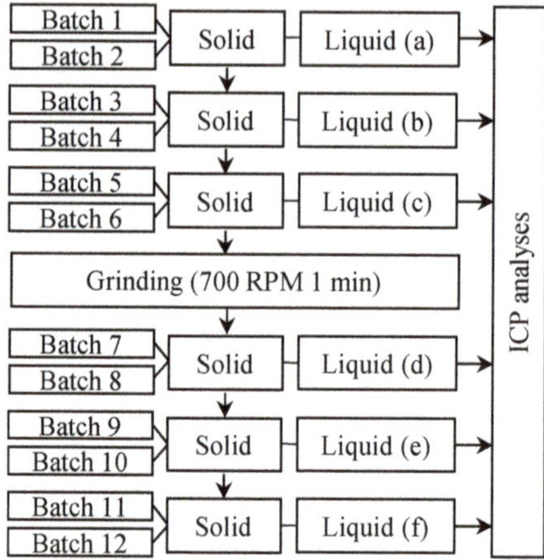

Figure 2. Batches dissolution experiments.

3. Results and Discussion

3.1. Mass Loss

The proportion of mass lost by each sample during thermal treatments has been registered and is presented in Table 4. As expected, the proportion of mass lost during treatment increased with the temperature. It reached a peak at 14.5% for sample F and was treated at 750 °C for 15 min. This value is in agreement with the expected one, between 12.0% and 14.0% [48,49].

Table 4. Mass lost during each thermal treatment, expressed as percent of initial mass of sample.

Temperature (°C)	550		650			750			
Residence Time (min)	15	60	15	30	60	15	30	60	LOI
Sample Name	A	B	C	D	E	F	G	H	
% Mass Lost	5.5	5.5	5.0	9.9	12.5	14.2	13.1	11.7	10.8

3.2. Mineralogical Transformations Along with Activation Temperatures

The evolution in the mineral composition at different temperatures and residence times has been studied using XRPD. Serpentine shows high crystallinity in samples U, A, and B, respectively untreated, treated at 550 °C for 15 min, and treated for 60 min. It then decreases in samples C, D, E, and F, respectively treated at 650 °C for 15 min, 30 min, and 60 min, and at 750 °C for 15 min. Crystalline features disappear in samples F and G, respectively, at 750 °C for 15 and 30 min. Amorphous contents can be identified in all of the treated samples as the crystallinity decreases. Forsterite is observed in samples E, G, and H, shown by highly crystalline peaks.

The remaining magnetite (the small proportion not removed during gravimetric separation) shows peaks in all samples, whereas the hematite (Fe_2O_3) which appears during the duration and temperature of the test increases. Due to the tests being performed in atmospheric conditions, the iron in the ferrous form (Fe^{2+}) contained in the serpentine structure is oxidized into ferric iron (Fe^{3+}) [50]. As iron rich olivine (fayalite—Fe_2SiO_4) can essentially incorporate Fe^{2+} in its structure [51], hematite (Fe_2O_3) is preferentially formed.

Table 5 presents phases quantification as measured using the Rietveld refinement. Three issues are faced: (i) these values do not consider the mass loss occurring during thermal treatment (ii) amorphous components are identified in the untreated sample, due to the stacking disorder of serpentine, making the identification of thermally induced amorphous components difficult, and finally (iii) a small peak is observed in the low angle that can be attributed to illite thus undermining the observation of the formation of meta-serpentine as described by [21]. Wilson et al. [38] determine that absolute quantification errors (wt %) for serpentine (chrysotile) and non-serpentine phases, regardless of their abundance in a sample, to be under 5.0 wt %. Consequently, illite is not considered in the Rietveld refinement as their peaks are too low and would fall under the estimation limit.

Table 5. Mineral composition using Rietveld refinements on XRD patterns, given in wt %.

		Amorphous	Serpentine	Forsterite	Magnetite	Hematite
Untreated	**U**	39.6	55.8	0.0	4.4	0.0
550 °C	A (15 min)	40.4	54.9	0.0	4.1	0.4
	B (60 min)	39.2	56.3	0.0	3.5	0.8
650 °C	C (15 min)	44.5	50.8	0.0	3.6	0.9
	D (30 min)	43.5	51.6	0.0	3.4	1.3
	E (60 min)	46.9	46.3	2.4	2.4	1.8
750 °C	F (15 min)	70.1	19.8	0.0	6.0	4.1
	G (30 min)	67.9	0.0	28.5	0.0	3.6
	H (60 min)	61.8	0.0	34.6	0.0	3.7

In an attempt to overcome these issues, a mass factor (MF in Equation (5)) is computed based on the mass loss of each sample (Table 4). Using this factor, the abundance of each phase can be expressed as grams per 100 g of starting material as given in Equation (5).

Equation (5): Proportion of phases expressed in mass

$$m_{phase} = \%phase \times \left(\frac{100}{100 + \%mass\ loss}\right) = \%phase \times MF. \tag{5}$$

As dehydroxylation is considered to be the loss of H_2O from the structure, the mass of H_2O lost per gram of serpentine is computed in order to obtain the proportion of dehydroxylated serpentine (Equation (6)). The value used as maximum mass loss "%mass loss ^{max}" was obtained experimentally and found to be 14.2% for this material.

Equation (6): Proportion of dehydroxylated serpentine

$$\% \text{ dehydroxylated serpentine} = \frac{\%mass\ loss / \sum (m_{amorphous} + m_{serpentine})}{\% mass\ loss^{max}} = \frac{m_{H2O\ lost}}{\%mass\ loss^{max}}. \tag{6}$$

The initial remaining material is decomposed into a non-reacted serpentine (serpentine$_{(i)}$) associated with a non-reacted amorphous phase (amorphous$_{(i)}$) induced by the layered structure of the serpentine. Their masses are calculated according to Equation (7), assuming that amorphous phase and crystalline initial serpentine both dehydroxylated in the same proportion.

Equation (7): Mass of initial phases

$$m_{phase_{(i)}} = (m_{phase}) - (m_{phase} \times \% \text{ dehydroxylated serpentine}). \tag{7}$$

The amount of dehydroxylated serpentine and amorphous phase corresponding to the first amorphous observed, (respectively named serpentine$_{(d)}$ and amorphous$_{(d)}$) are given by Equation (8).

Equation (8): Mass of intermediate amorphous phases

$$m_{phase(d)} = m_{phase} - m_{phase(i)}. \tag{8}$$

Further dehydroxylation leads to the formation of meta-serpentine, whose mass is obtained by Equation (9) This formation is marked by the total loss of the hydroxyls groups at close to 10 wt % of the starting material mass.

Equation (9): Mass of meta-serpentine

$$m_{\text{meta-serpentine}} = \left(m_{\text{amorphous}} \times m_{\text{H2O lost}}\right) - m_{\text{amorphous}_{(i)}} \cdot \tag{9}$$

As a result, three phases emerge from this calculation: first an initial serpentine, resulting from the sum of amorphous$_{(i)}$ and serpentine$_{(i)}$, then an intermediate amorphous components which is the sum of amorphous$_{(d)}$ and serpentine$_{(d)}$ corresponding to the first stage of amorphization, and finally meta-serpentine. Forsterite and iron oxides (magnetite and hematite) remain unaltered by the calculation.

As shown in Table 6, Serpentine is gradually replaced by intermediate amorphous phases in samples treated at temperatures lower than 650 °C and peaks for 60 min treatment at 70.3 g/100 g of starting material. Meta-serpentine is first found in samples treated at 650 °C for 15 min. Its proportion increases with the temperature and peaks at 27.2 g/100 g of starting material in the sample treated at 750 °C for 15 min. The increase of meta-serpentine is combined with a decrease of intermediate amorphous components contents. As seen previously (Table 5), forsterite is observed in samples E, G and H, respectively treated at 650 °C for 60 min and at 750 °C for 30 and 60 min. A treatment at 750 °C for 15 min produced a sample with no initial serpentine and no forsterite but only amorphous phases, associated with iron oxides. These observations are in agreement with previous studies which observed the formation of an intermediate amorphous component, α meta-serpentine, progressively replacing serpentine below 580 °C. It is then followed by the appearance of an amorphous meta-serpentine material by 650 °C prevailing by 750 °C [21].

Table 6. Mineralogical compositions based on Rietveld refinements, expressed in grams per 100 g of starting material) at given temperature and residence times. In. Serp: Initial serpentine, Inter. Am.: Intermediate amorphous components, Meta-serp.: Meta-serpentine, For.: forsterite, Mag: magnetite, Hem: hematite and ML: Mass loss.

		In. Serp.	Inter. Am.	Meta-Serp.	For.	Mag.	Hem.	ML
Untreated	U	95.6	0.0	0.0	0.0	4.4	0.0	0.0
550 °C	A (15 min)	52.8	36.5	0.0	0.0	3.9	0.4	5.5
	B (60 min)	53.5	37.0	0.0	0.0	3.3	0.8	5.5
650 °C	C (15 min)	54.0	31.9	4.8	0.0	3.4	0.9	5.0
	D (30 min)	21.6	60.3	4.0	0.0	3.0	1.2	9.8
	E (60 min)	4.2	70.3	7.2	2.1	2.1	1.6	12.5
750 °C	F (15 min)	0.0	49.9	27.2	0.0	5.2	3.5	14.2
	G (30 min)	0.0	33.6	25.5	24.8	0.0	3.1	13.1
	H (60 min)	0.0	34.3	20.3	30.5	0.0	3.2	11.7

3.3. Impact of Mineralogy on Dissolution

3.3.1. Two Batches Dissolution

It is known that the amount of Mg^{2+} available for leaching will directly control, along with the amount of CO_2 treated, the quantity of carbonates being precipitated from the liquid phase after carbonation [30,31]. This study focuses on the proportion of Mg^{2+} leached from thermally treated serpentine samples. Figure 3 shows the mass of intermediate amorphous components and of meta-serpentine added up and plotted against the proportion of Mg^{2+} leached into the liquid phase during the carbonation reaction. As the amount of amorphous components increases from none (U-untreated sample) to 77 g/ 100 g of starting material (F—750 °C 15 min), the proportion of Mg^{2+}

leached during two batches of gas increases too, respectively from 3.3 wt % to 13.5 wt % of initial Mg^{2+} concentration in solid. Samples D, G and H show similar proportions of Mg^{2+} leached and a close amount of amorphous components. However, initial serpentine constitutes a third of the former composition, whereas forsterite is formed in the two latter. As observed in previous studies at 650 °C, the solubility of Mg^{2+} ions is first increased by thermal treatment until it is reduced with the decreasing content of amorphous phases and the formation of forsterite. The amount of Mg^{2+} leached from the heat activated serpentine appears to be linearly dependent on the proportion of amorphous phases.

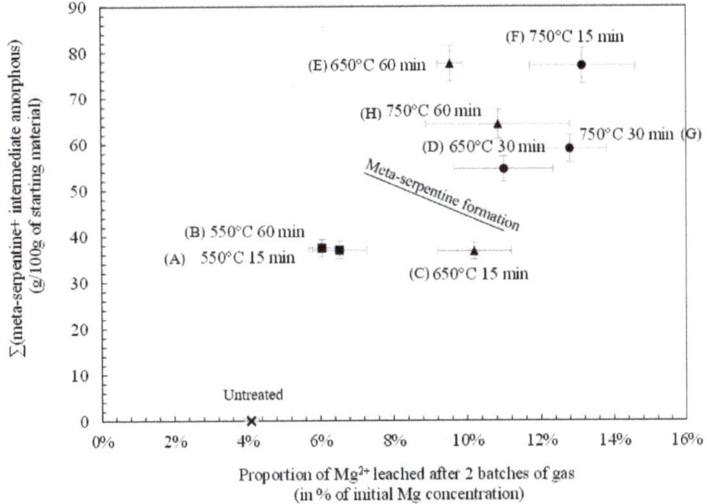

Figure 3. Extracted plotted against the quantity of amorphous phases, being the sum of the intermediate amorphous components and meta-serpentine. (Squares, triangle, and circles stands for test temperature of 550 °C, 650 °C, and 750 °C, respectively).

3.3.2. Successive Batches Dissolution

Thermal treatment conditions of samples D and F (650 °C for 30 min and 750 °C for 15 min) are chosen to be tested on successive batches as they respectively are the recommended conditions in literature [22] and the conditions giving the highest proportion of Mg^{2+} leached after two batches of gas in our conditions. Figure 4 shows the cumulative proportion of Mg^{2+} leached after twelve batches of gas. After two batches of gas, the proportion of Mg^{2+} leached demonstrates a significant discrepancy from the previous results and the present one. Indeed, the sample treated at 650 °C for 30 min shows a similar proportion of Mg^{2+} leached to the one treated at 750 °C for 15 min. After 4 batches, the sample treated at 750 °C for 15 min is catching up with a proportion of Mg^{2+} leached higher by 5 wt % compared to the other sample. At the end of the 12 batches, 44.6 wt % of Mg^{2+} has been leached from the sample treated at 750 °C for 15 min against 32.4 wt % for the one treated at 650 °C for 30 min. For the sample treated at 650 °C for 30 min, the proportions of Mg^{2+} leached reached a plateau close to 0.5 wt % during the tenth batch, suggesting that almost all of the Mg^{2+} available in the present dissolution conditions might have been leached. This occurred with at 750 °C after 15 min, which indicates that the plateau has not been reached yet, suggesting that more batches of CO_2 could allow a higher proportion of Mg^{2+} leached. As the solution is refreshed for every two batches of gas, the limiting factor is the availability of the Mg^{2+} and not the saturation of the solution.

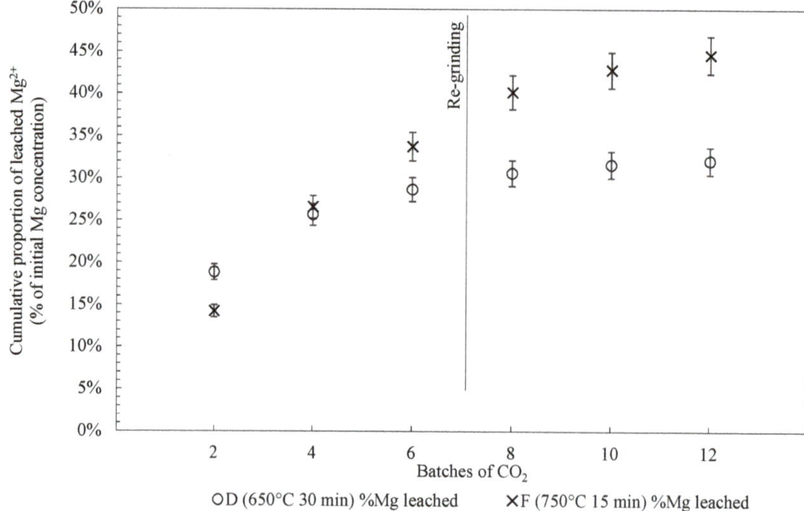

Figure 4. The Mg^{2+} leached (expressed in percent of initial Mg^{2+} concentration in the solid) for samples treated at 650 °C for 30 min (blue) and at 750 °C for 15 min (green).

The increase in the slope of the curves between the batches 6 and 8 demonstrate the slight effect of the grinding on the material after the batch 6. Pasquier et al. [30] demonstrated that the effect of the passivation silica layer, formed during dissolution, can be reduced by grinding and so revive the leaching of Mg^{2+}. Nevertheless, studies from the Carmex project [52,53]) show that a continuous mechanical exfoliation of the passivation layer as it forms on the grains would be a promising way to avoid the need for regrinding after six batches of gas.

3.4. Mineralogical Assemblage and Carbonation

As the hydroxyl groups escape the serpentine octahedral sheets, the remaining atoms of Mg and Si are reorganized through the amorphous components. As the temperature increases, serpentine is transformed into amorphous phases, then recrystallizes into forsterite.

The best mineralogical assemblage is shown to be a mixture of amorphous phases, as observed in sample F with the highest proportion of Mg^{2+} leached after both two batches and successive batches carbonation tests. McKelvy et al. [21] observed significantly higher carbonation reaction rates in the presence of amorphous meta-serpentine, formed above 610 °C, than in the presence of α-meta-serpentine formed below 580 °C and identified as intermediate amorphous components here. When structurally stable Mg-bearing phases are present in the assemblage, such as serpentine or forsterite, material reactivity decreases. In serpentine, Mg atoms are bonded to the hydroxyls groups whereas in forsterite, they are ionically bonded to silica tetrahedron [54]. The differentiation of the two amorphous phases based on the proposed calculation in this paper were revealed to be accurate because the sample showing the highest value of metaserpentine appeared to be the one demonstrating the highest proportion of Mg^{2+} leached, in accordance with observations proposed in the literature.

Moreover, reactivity appears to be affected more by mineralogy than by surface area. Larachi et al. [15] provide surface area measurements on calcined samples, from 300 °C to 1200 °C, showing that it tends to decrease with increasing temperature as dehydroxylation occurs.

Furthermore, observations made here are in accordance to those made by other authors regarding the mineralogical transformations occurring during thermal activation and dehydroxylation [21,22]. Nevertheless, a shift of ideal temperature is observed, as observations made by others suggest that reactions at 650 °C were more likely to occur, rather than at 750 °C as occurred in the present study. Such a change can be attributed to numerous factors such as the initial material mineralogy, the experimental set up, and the methodology used to evaluate activation efficiency. For instance, Li et al., [22] used hydrochloric acid to perform lixiviation tests, which is far from neutral pH and weak acid conditions used in the present study. On the other hand, McKelvy et al.'s [21] work set the basis of serpentine dehydroxilation understanding using TGA/DTA and XRD. Conversely, their carbonation conditions used high temperature and supercritical CO_2, which is again far from the conditions tested here. Based on their results, past ideal activation conditions were shown to be effective, but not necessarily optimal. Therefore, the present results highlight the importance of considering the mineralogical assemblage alongside the thermal treatment parameters (temperature and residence time). Such an approach will allow us to take into account the effect of the initial material composition and potential specificity of the activation conditions/technique. Indeed, conditions in a rotary kiln will be very different from a furnace or a fluidized bed. As results, a study of the mineralogical assemblage can lead to an accurate optimization of heat activation operating conditions in accordance with the material activated and the equipment used.

4. Conclusions

In this study, a novel approach of amorphous phase quantification, resulting from serpentine thermal activation, is introduced. It enables a better understanding of their implications in serpentine dissolution using carbonic acid as a lixiviant, in similar conditions to those used in the direct flue gas mineral carbonation process developed at INRS. The following conclusions can be made from this study:

(i)   It is possible to differentiate and quantify intermediate amorphous phases and metaserpentine formed during dehydroxylation of serpentine and correlate these values to the efficiency of carbonation reaction. In a static furnace, treatment at 750 °C for 15 min leads to the formation of 27.2 g/100 g of starting material of meta-serpentine.

(ii)   Thermally produced amorphous phases enhance Mg^{2+} solubility during carbonation reaction. Furthermore, the formation of meta-serpentine, resulting in a complete dehydroxylation, significantly upgrades Mg^{2+} leaching yield.

(iii)   The crystallization of forsterite decreases the sample dissolution potential by limiting the amount of Mg^{2+} accessible for leaching in the present dissolution conditions.

(iv)   Adjusting thermal activations parameters (temperature and residence time) led to an increase of 39% of Mg^{2+} leached during the carbonation reaction.

Author Contributions: Methodology, C.D.B., L.C.-P., G.D., J.-F.B., M.C.I., G.M.; formal analysis, C.D.B.; writing—original draft preparation, C.D.B., L.C.-P., G.D., J.-F.B., M.C.I. and G.M.; writing—review and editing, C.D.B., L.C.-P.; supervision, L.C.-P., G.D., J.-F.B., M.C.I. and G.M.; project administration, G.M.; funding acquisition, L.C.-P., G.D., J.-F.B., M.C.I. and G.M.

Funding: This research was funded by Fond de Recherche Quebecois en Nature et Technology, projet de recherche en équipe 2015–2016.

Acknowledgments: This research was funded by 'projet de recherche en équipe' grant from FRQNT. The authors would like to thank Matti Raudsepp. Kate Carroll. Ian Power from the University of British Columbia (Vancouver. Canada) and Connor Turvey from Monash University (Melbourne. Australia) for their advice and help on the XRD and Rietveld refinement application to the serpentine minerals.

Conflicts of Interest: The authors declare no conflict of interest.

References

1.   IPCC. *Climate Change 2014: Synthesis Report. Contribution of Working Groups I, II and III to the Fifth Assessment Report of the Intergovernmental Panel on Climate Change*; Core Writing Team, Pachauri, R.K., Meyer, L.A., Eds.; IPCC: Geneva, Switzerland, 2014; p. 151.

2.   Lackner, K.S.; Wendt, C.H.; Butt, D.P.; Joyce, B.L.; Sharp, D.H. Carbon dioxide disposal in carbonate minerals. *Energy* **1995**, *20*, 1153–1170. [CrossRef]

3. Seifritz, W. CO_2 disposal by means of silicates. *Nature* **1990**, *345*, 486. [CrossRef]

4. Maroto-Valer, M.M.; Fauth, D.J.; Kuchta, M.E.; Zhang, Y.; Andrésen, J.M. Activation of magnesium rich minerals as carbonation feedstock materials for CO_2 sequestration. *Fuel Process. Technol.* **2005**, *86*, 1627–1645. [CrossRef]

5. Hänchen, M.; Prigiobbe, V.; Baciocchi, R.; Mazzotti, M. Precipitation in the Mg-carbonate system: Effects of temperature and CO_2 pressure. *Chem. Eng. Sci.* **2008**, *63*, 1012–1028. [CrossRef]

6. Harrison, A.L.; Power, I.M.; Dipple, G.M. Accelerated carbonation of brucite in mine tailings for carbon sequestration. *Environ. Sci. Technol.* **2012**, *47*, 126–134. [CrossRef]

7. Page, N.J. Chemical differences among the serpentine polymorphs. *Am. Mineral.* **1968**, *53*, 201–215.

8. Goff, F.; Guthrie, G.D.; Lipin, B.; Fite, M.; Chipera, S.; Counce, D.A.; Kluk, E.; Ziock, H. *Evaluation of Ultramafic Deposits in the Eastern United States and Puerto Rico as Sources of Magnesium for Carbon Sequestration*; Los Alamos National Laboratory: Los Alamos, NM, USA, 2000; p. 30.

9. Nagamori, M.; Plumpton, A.J.; Le Houillier, R. Activation of magnesia in serpentine by calcination and the chemical utilization of asbestos tailings, A review. *C. Bull.* **1980**, *73*, 144–156.

10. Luce, R.W.; Bartlett, R.W.; Parks, G.A. Dissolution kinetics of magnesium silicates. *Geochim. Cosmochim. Acta* **1972**, *36*, 35–50. [CrossRef]

11. Stumm, W.; Morgan, J.J. *Aquatic Chemistry: An Introduction Emphasizing Chemical Equilibria in Natural Waters*; John Wiley and Sons: Hoboken, NJ, USA, 1981.

12. Aruja, E. An X-ray study of crystal-structure of antigorite. *Mineral. Mag.* **1994**, *27*, 11.

13. Nagy, B.; Faust, G. Serpentines: Natural mixtures of Chrysotile and antigorite. *Am. Mineral.* **1956**, *41*, 817–838.

14. Wicks, F.J.; Whittaker, E.J.W. Serpentine textures and serpentinization. *Can. Mineral.* **1977**, *15*, 459–488.

15. Larachi, F.; Daldoul, I.; Beaudoin, G. Fixation of CO_2 by chrysotile in low-pressure dry and moist carbonation: Ex-situ and in-situ characterizations. *Geochim. Cosmochim. Acta* **2010**, *74*, 3051–3075. [CrossRef]

16. Mellini, M.; Zanazzi, P.F. Crystal structures of lizardite-1T and lizardite-2H1 from Coli, Italy. *Am. Mineral.* **1987**, *72*, 943–948.

17. Turci, F.; Tomatis, M.; Mantegna, S.; Cravotto, G.; Fubini, B. A new approach to the decontamination of asbestos-polluted waters by treatment with oxalic acid under power ultrasound. *Ultrason. Sonochem.* **2008**, *15*, 420–427. [CrossRef] [PubMed]

18. Ball, M.C.; Taylor, H.F.W. The dehydration of chrysotile in air and under hydrothermal conditions. *Mineral. Mag.* **1963**, *33*, 467–482. [CrossRef]

19. Brindley, G.W.; Hayami, R. Mechanism of formation of forsterite and enstatite from serpentine. *Mineral. Mag.* **1965**, *35*, 189–195. [CrossRef]

20. Brindley, G.W.; Hayami, R. Kinetics and mecanisms of dehydration and recrystallization of serpentine. *Clays Clay Miner.* **1965**, *12*, 35–37. [CrossRef]

21. McKelvy, M.J.; Chizmeshya, A.V.; Diefenbacher, J.; Béarat, H.; Wolf, G. Exploration of the role of heat activation in enhancing serpentine carbon sequestration reactions. *Environ. Sci. Technol.* **2004**, *38*, 6897–6903. [CrossRef]

22. Li, W.; Li, W.; Li, B.; Bai, Z. Electrolysis and heat pretreatment methods to promote CO_2 sequestration by mineral carbonation. *Chem. Eng. Res. Des.* **2009**, *87*, 210–215. [CrossRef]

23. Farhang, F.; Rayson, M.; Brent, G.; Hodgins, T.; Stockenhuber, M.; Kennedy, E. Insights into the dissolution kinetics of thermally activated serpentine for CO_2 sequestration. *Chem. Eng. J.* **2017**, *330*, 1174–1186. [CrossRef]

24. Gerdemann, S.J.; O'Connor, W.K.; Dahlin, D.C.; Penner, L.R.; Rush, H. Ex situ aqueous mineral carbonation. *Environ. Sci. Technol.* **2007**, *41*, 2587–2593. [CrossRef] [PubMed]

25. O'Connor, W.K.; Dahlin, D.C.; Rush, G.E.; Gedermann, S.J.; Penner, L.R.; Nilsen, D.N. Aqueous Mineral Carbonation; Final Report DOE/ARC-TR-04-002. 2005, p. 462. Available online: https://www.researchgate.net/profile/William_Oconnor8/publication/315844800_Aqueous_Mineral_Carbonation_Mineral_Availability_Pretreatment_Reaction_Parametrics_and_Process_Studies/links/58ebb247a6fdcc965767765f/Aqueous-Mineral-Carbonation-Mineral-Availability-Pretreatment-Reaction-Parametrics-and-Process-Studies.pdf (accessed on 3 November 2019).

26. Gerdemann, S.J.; Dahlin, D.C.; O'Connor, W.K.; Penner, L.R. Carbon dioxide sequestration by aqueous mineral carbonation of magnesium silicate minerals. In *Greenhouse gas Technologies*; Albany Research Center (ARC): Albany, OR, USA, 2003; p. 8.

27. Mann, J. Serpentine Activation for CO_2 Sequestration. Ph.D. Thesis, University of Sydney, Sydney, Australia, 2014; p. 274.

28. Sanna, A.; Wang, X.; Lacinska, A.; Styles, M.; Paulson, T.; Maroto-Valer, M.M. Enhancing Mg extraction from lizardite-rich serpentine for CO_2 mineral sequestration. *Miner. Eng.* **2013**, *49*, 135–144. [CrossRef]

29. Mercier, G.; Blais, J.-F.; Cecchi, E.; Veetil, S.P.; Pasquier, L.-C.; Kentish, S. Carbon Dioxide Chemical Sequestration from Industrial Emissions by Carbonation. U.S. Patent 9.440,189 B2, 13 September 2016.

30. Pasquier, L.-C.; Mercier, G.; Blais, J.F.; Cecchi, E.; Kentish, S. Reaction mechanism for the aqueous-phase mineral carbonation of heat-activated serpentine at low temperatures and pressures in flue gas conditions. *Environ. Sci. Technol.* **2014**, *48*, 5163–5170. [CrossRef] [PubMed]

31. Pasquier, L.-C.; Mercier, G.; Blais, J.F.; Cecchi, E.; Kentish, S. Parameters optimization for direct flue gas CO_2 capture and sequestration by aqueous mineral carbonation using activated serpentinite based mining residues. *Appl. Geochem.* **2014**, *50*, 66–73. [CrossRef]

32. Moreno Correia, M.J. Optimisation de la Précipitation des Carbonates de Magnésium pour L'application dans un Procédé de Séquestration de CO_2 par Carbonatation Minérale de la Serpentine. Master's Thesis, Université du Québec, Québec, QC, Canada, 2018; p. 125.

33. Pasquier, L.-C.; Mercier, G.; Blais, J.-F.; Cecchi, E.; Kentish, S. Technical and economic evaluation of a mineral carbonation process using southern Québec mining wastes for CO_2 sequestration of raw flue gas with by-product recovery. *Int. J. Greenh. Gas Control* **2016**, *50*, 147–157. [CrossRef]

34. Kemache, N.; Pasquier, L.-C.; Cecchi, E.; Mouedhen, I.; Blais, J.-F.; Mercier, G. Aqueous mineral carbonation for CO_2 sequestration: From laboratory to pilot scale. *Fuel Process. Technol.* **2017**, *166*, 209–216. [CrossRef]

35. Kemache, N.; Pasquier, L.-C.; Mouedhen, I.; Cecchi, E.; Blais, J.-F.; Mercier, G. Aqueous mineral carbonation of serpentinite on a pilot scale: The effect of liquid recirculation on CO_2 sequestration and carbonate precipitation. *Appl. Geochem.* **2016**, *67*, 21–29. [CrossRef]

36. Aumento, F. *Serpentine Mineralogy of Ultrabasic Intrusion in Canada and on the Mid-Atlantic Ridge*; Department of Energy, Mines and Resources: Ottawa, ON, Canada, 1970.

37. Tertian, R.; Claisse, F. *Principles of Quantitative X-ray Fluorescence Analysis*; Wiley-Heyden: London, UK, 1982.

38. Wilson, S.A.; Raudsepp, M.; Dipple, G.M. Verifying and quantifying carbon fixation in minerals from serpentine-rich mine tailings using the Rietveld method with X-ray powder diffraction data. *Am. Mineral.* **2006**, *91*, 1331–1341. [CrossRef]

39. Bish, D.L.; Howard, S. Quantitative phase analysis using the Rietveld method. *J. Appl. Crystallogr.* **1988**, *21*, 86–91. [CrossRef]

40. Pawley, G.S. Unit-Cell Refinement from Powder Diffraction scans. *J. Appl. Crystallogr.* **1981**, *14*, 357–361. [CrossRef]

41. Raudsepp, M.; Pani, E.; Dipple, G. Measuring mineral abundance in skarn; I, The Rietveld method using X-ray powder-diffraction data. *Can. Mineral.* **1999**, *37*, 1–15.

42. Rietveld, H.M. A Profile Refinement Method for Nuclear and Magnetic Structures. *J. Appl. Crystallogr.* **1969**, *2*, 65–71. [CrossRef]

43. Turvey, C.C.; Wilson, S.A.; Hamilton, J.L.; Southam, G. Field-based accounting of CO_2 sequestration in ultramafic mine wastes using portable X-ray diffraction. *Am. Mineral.* **2017**, *102*, 1302–1310. [CrossRef]

44. Scarlett, N.V.Y.; Madsen, I.C. Quantification of phases with partial or no known crystal structures. *Powder Diffr.* **2006**, *21*, 278–284. [CrossRef]

45. Le Bail, A. Whole powder pattern decomposition methods and applications: A retrospection. *Powder Diffr.* **2005**, *20*, 316–326. [CrossRef]

46. Falini, G.; Foresti, E.; Gazzano, M.; Gualtieri, A.F.; Leoni, M.; Lesci, I.G.; Roveri, N. Tubular-Shaped Stoichiometric Chrysotile Nanocrystals. *Chem. A Eur. J.* **2004**, *10*, 3043–3049. [CrossRef]

47. Du Breuil, C.; Pasquier, L.C.; Dipple, G.; Blais, J.F.; Iliuta, M.C.; Mercier, G. Impact of particle size in serpentine thermal treatment: Implications for serpentine dissolution in aqueous-phase using CO_2 in flue gas conditions. *Appl. Clay Sci.* **2019**, *182*, 105286. [CrossRef]

48. Jolicoeur, C.; Duchenes, D. Infrared and thermogravimetric studies of the thermal degradation of chrysotile asbsestos fibers: Evidence for matrix effects. *Can. J. Chem.* **1981**, *59*, 1521–1526. [CrossRef]

49. Viti, C. Serpentine minerals discrimination by thermal analysis. *Am. Mineral.* **2010**, *95*, 631–638. [CrossRef]
50. Balucan, R.D.; Dlugogorski, B.Z. Thermal activation of antigorite for mineralization of CO_2. *Environ. Sci. Technol.* **2013**, *47*, 182–190. [CrossRef]
51. Hora, Z. Ultramafic hosted chrysotile asbestos. In *Fieldwork*; British Columbia Geological Survey: Victoria, BC, Canada, 1997; p. 4.
52. Bodénan, F.; Bourgeois, F.; Petiot, C.; Augé, T.; Bonfils, B.; Julcour-Lebigue, C.; Guyot, F.; Boukary, A.; Tremosa, J.; Lassin, A.; et al. Ex situ mineral carbonation for CO_2 mitigation: Evaluation of mining waste resources, aqueous carbonation processability and life cycle assessment (Carmex project). *Miner. Eng.* **2014**, *59*, 52–63. [CrossRef]
53. Julcour, C.; Bourgeois, F.; Bonfils, B.; Benhamed, I.; Guyot, F.; Bodénan, F.; Petiot, C.; Gaucher, É.C. Development of an attrition-leaching hybrid process for direct aqueous mineral carbonation. *Chem. Eng. J.* **2015**, *262*, 716–726. [CrossRef]
54. Birle, J.; Gibbs, G.; Moore, P.; Smith, J. Crystal structures of natural olivines. *Am. Mineral.* **1968**, *53*, 807.

Article

Proposed Methodology to Evaluate CO_2 Capture Using Construction and Demolition Waste

Domingo Martín [1], Vicente Flores-Alés [2] and Patricia Aparicio [1,*]

[1] Departamento de Cristalografía, Mineralogía y Q. Agrícola, Facultad de Química, Universidad de Sevilla, 41012 Sevilla, Spain; dmartin5@us.es

[2] Departamento de Construcciones Arquitectónicas II, Escuela Técnica Superior de Ingeniería de Edificación, Universidad de Sevilla, 41012 Sevilla, Spain; vflores@us.es

* Correspondence: paparicio@us.es

Received: 8 August 2019; Accepted: 1 October 2019; Published: 5 October 2019

Abstract: Since the Industrial Revolution, levels of CO_2 in the atmosphere have been constantly growing, producing an increase in the average global temperature. One of the options for Carbon Capture and Storage is mineral carbonation. The results of this process of fixing are the safest in the long term, but the main obstacle for mineral carbonation is the ability to do it economically in terms of both money and energy cost. The present study outlines a methodological sequence to evaluate the possibility for the carbonation of ceramic construction waste (brick, concrete, tiles) under surface conditions for a short period of time. The proposed methodology includes a pre-selection of samples using the characterization of chemical and mineralogical conditions and in situ carbonation. The second part of the methodology is the carbonation tests in samples selected at 10 and 1 bar of pressure. The relative humidity during the reaction was 20 wt %, and the reaction time ranged from 24 h to 30 days. To show the effectiveness of the proposed methodology, Ca-rich bricks were used, which are rich in silicates of calcium or magnesium. The results of this study showed that calcite formation is associated with the partial destruction of Ca silicates, and that carbonation was proportional to reaction time. The calculated capture efficiency was proportional to the reaction time, whereas carbonation did not seem to significantly depend on particle size in the studied conditions. The studies obtained at a low pressure for the total sample were very similar to those obtained for finer fractions at 10 bars. Presented results highlight the utility of the proposed methodology.

Keywords: suitable methodology for mineral carbonation; construction and demolition waste

---

1. Introduction

CO_2 emissions into the atmosphere are a growing environmental problem in different industrial sectors. The construction sector is no stranger to this problem considering gaseous emissions are derived from manufacturing processes of materials used mostly in construction, such as ceramic materials or cement [1]. Independent of the existing controversy regarding the uncertainty of the sensitivity of the climate in the international scientific community [2], the most negative forecasts consider some of the effects predicted by the uncontrolled emission of greenhouse gases as a possible increase in Earth's temperature, climatic alterations that will accelerate desertification, and a possible loss of part of the coastline due to the rise in sea levels.

In the current scenario, the challenges of reducing emissions cannot solely be met with greater energy efficiency and renewable energy resources in the generation phase. It is absolutely necessary to act on the management and treatment of emissions. For this reason, research aimed at capturing CO_2 is of vital importance in order to achieve the standards set as an objective.

Mineral-carbonation systems for CO_2 fixation are another option for Carbon Capture and Storage (CCS). Although less efficient than geological storage, they are much simpler, cheaper, and have fewer requirements, which responds to the requirement of ecological rationality raised above. The present work deals with the possibilities of using construction and demolition waste as CO_2 reservoirs. Preliminary research has shown that construction materials containing calcium and/or magnesium in their composition in the form of silicates, oxides, and hydroxides, which can react with CO_2 to give rise to carbonates, thus constituting a possible alternative for mineral carbonation from ceramic waste within options for CO_2 capture/storage [3–6]. This possibility is based on the capacity of construction and demolition waste with a high content of calcium and/or magnesium in the form of silicates, oxides, and hydroxides to fix CO_2 under optimal conditions, using a chemical reaction whose product is the formation of carbonates and silica as stable by-products. This process of carbonation occurs naturally with very slow kinetics; it is of interest to design a system and methodology to accelerate this process to make the capture of CO_2 from anthropogenic activity profitable, achieving maximal industrial and energy efficiency. There are several patents and studies that use residues with a high calcium content from different types of industrial waste for their carbonation [7–22].

The present work presents a methodological sequence for the control and validation of a viable alternative of CCS using mineral carbonation of construction and demolition waste (bricks, concrete, tiles) with a high content of silicates rich in calcium and magnesium. In this case, those with a high content of ceramic materials and cementitious conglomerate (mortar and concrete) can be considered optimal landfills for their transformation into CO_2 sinks [23,24].

A laboratory model was tested that indicated the efficiency of the system, ensuring extrapolated conditions at a landfill scale were in accordance with the difficulty involved in the reproduction of parameters that can be controlled at a laboratory scale. In this sense, it is necessary to establish a system that approximately reproduces conditions of isolation, humidity, dimensions, and others that, with generic character, can be reached in a controlled residue deposit. The physical–chemical and mineralogical mechanisms, the external conditions that have a decisive influence on the process, and the kinetics of capture and carbonation reactions that favor gas fixation and stabilization were analyzed. The control and verification procedures of the process were also analyzed, which allowed the adequate monitoring and optimization of the process.

2. Materials and Methods

2.1. Materials and Carbonation Test

The material used for the development of this methodology proposal was a brick type widely used in construction, which was specifically a clinker brick (MPC2) from Malpesa S.L. (Bailén, South of Spain) fired at 1050 °C.

The samples were subjected to a crushing process for granulometric conditioning. Fractions of less than 4 mm were selected to obtain size ratios consistent with the diameter of the reactor, obtaining homogeneous distribution so that there was a predominance of coarser fractions corresponding to conventional waste shredding.

Carbonation tests were carried out in a 0.3 L volume hermetic reactor (Parr Instruments Co., Moline, IL, USA). The fixed conditions were 10 bars of CO_2, 4:1 solid–water ratio, and room temperature. The variable conditions were reaction time (between 24 and 720 h) and particle size (<4, 2–4, and 1–2 mm). These particle-size fractions were selected because they were the most representative results of the crushing treatment. Additionally, a test was carried at low pressure (1 bar), room temperature, and a 4:1 solid–water ratio in a 5 L volume hermetic reactor of continuous flow to maintain pressure at 1 bar during the 720 h of reaction time.

Post-treatment, the samples were dried at 100 °C for 24 h, powdered, and sieved at 50 μm for subsequent analysis.

2.2. Instrumental Techniques: Methodology

Major multielemental chemical composition (in oxides) was performed with an automated Panalytical Axios model wavelength-dispersive X-ray fluorescence spectrometer (WDXRF). The samples were prepared for analysis as glass discs to reduce the "matrix effect."

The mineralogical composition of the untreated and treated samples was determined by X-ray diffraction (XRD) using a Bruker D8 Advance diffractometer (Bruker AXS, Berlin, Germany) with standard monochromatic Cu–Kα radiation at 40 kV and 30 mA with a Ni filter and Linxeye 1D detector. Routine scanning was performed with a 0.015° 2θ step size, and at 0.1 s per step from 3° to 70°. Rietveld refinement was also realized to determine the quantitative composition of the untreated bricks. In this case, scanning was performed with a 0.010° 2θ step size at 0.5 s per step in the range of 3°–120° and adding zincite (15 wt %) as an internal standard. Rietveld refinement for the present phase quantification was done with Bruker's commercial Topas v5 software (Bruker AXS, Berlin, Germany).

In addition, previous carbonation analysis was carried out by X-ray diffraction using a Bruker D8 Advance powder diffractometer (Bruker AXS, Berlin, Germany) equipped with an Anton Paar XRK 900 reactor chamber (Anton Parr GmbH, Graz, Austria) and high-sensitivity detector Bruker Vantec 1 (Bruker AXS, Berlin, Germany). This chamber was designed for X-ray diffraction experiments of up to 900 °C and 10 bar for solid state–gas reactions. Standard monochromatic Cu–Kα radiation operating at 40 kV and 40 mA was employed. Scanning was performed with a 0.022° 2θ step size at 0.2 s per step from 3° to 70°. Samples were in a CO_2-rich environment for 24 h in this reactor chamber.

Macro- and micro-observations were obtained by stereomicroscope using a Greenough Leica S8 APO (Leica Microsystems GmbH, Wetzlar, Germany) equipped with a DC300 camera (Leica Microsystems GmbH, Wetzlar, Germany) and by scanning electronic microscopy (SEM) using a JEOL 6460 LV microscope (JEOL Ltd., Akishima, Japan) equipped with energy-dispersive spectrometers (Oxford Instruments INCA, Oxford, UK).

The carbonate content of the carbonated samples was determined by two analytical methods: differential thermal and thermogravimetric analysis (DTA-TG) and an elemental analyzer. DTA-TG were performed on a TG Netzsch STA 409 PC. Samples (around 150 mg) were heated in an aluminum oxide crucible under a nitrogen atmosphere at 10 °C min^{-1} from room temperature to 1200 °C. Weight loss was measured by thermogravimetric analysis in the temperature range of 450–900 °C relative to the total carbonated decomposition. Elemental carbon content was measured using an elemental analyzer, Leco Truspec CHNS Micro (St. Joseph, MI, USA), which calculated the carbonated ratio by assuming that the whole carbon content was calcite.

Soluble Si, Ca, and Mg ions of the original and treated samples were also determined. Analysis was performed with simultaneous inductively coupled plasma optical emission spectrometry (IPC-OES) analysis using a Horiba Jobin Yvon ULTIMA 2 model instrument (Horiba Scientific, Palaiseau, France). The samples were prepared by mixing the solid-powder samples with water and stirring for 24 h, isolating the liquid phase by centrifugation, and filtering using a Nylon 0.22 μm syringe filter (MilliporeSigma, Burlington, MA, USA).

Specific surface area (BET) and microporosity were measured with a Micromeritics Gemini 2360 instrument (Micromeritics Instrument Corp, Norcross, GA, USA) using the absorption of N_2 at liquid nitrogen temperature. Before measuring, all samples were degassed using a Flow Prep 060 Micromeritics degasser (Micromeritics Instrument Corp, Norcross, GA, USA) with dry nitrogen gas at 80 °C for 12 h. Nanoporosity was measured with an ASAP 2420 instrument (Micromeritics Instrument Corp, Norcross, GA, USA) using CO_2 absorption at room temperature. Samples were degassed at 150 °C for 1.5 h and finally outgassed to 10^{-3} Torr. Macro- and mesoporosity were studied using mercury porosimeter Quantachrome Instruments Pore Master 60-GT (Quantachrome Instruments, Boynton Beach, FL, USA).

3. Results and Discussion

3.1. Proposed Methodological Sequence to Evaluate Effectiveness of Construction and Demolition Waste for CCS

A first outcome measure to take into consideration to provide the optimal construction residue is to have an important content of calcium oxide and/or magnesium as reactive compounds, since these elements are necessary for precipitation in the form of the carbonate, although the carbonation of other alkali metals is possible, as shown in the following equations [25]:

$$MO + CO_2 \rightarrow MCO_3. \tag{1}$$

This reaction is usually exothermic in nature, as per Lackner et al. (1995) [26]:

$$CaO + CO_2 \rightarrow CaCO_3 + 179 \text{ kJ/mole} \tag{2}$$

$$MgO + CO_2 \rightarrow MgCO_3 + 118 \text{ kJ/mole.} \tag{3}$$

Consequently, sample characterization requires a chemical analysis that is commonly used (WDXRF).

It has been widely described in the literature that the main possible carbonation minerals are oxides, silicates, and anhydrite; therefore, it was necessary to determine the minerals present in the sample in order to evaluate the candidate. Natural wollastonite is a widely studied mineral in several works, for example, Huijgen and coworkers [27], as a candidate for mineral carbonation. Different studies used other types of sample, such as industrial waste, that were mainly composed of wollastonite [28–30].

The proposed methodology includes a sample preselection using the characterization of chemical and mineralogical conditions, and in situ carbonation. The second part of the methodology is carbonation tests in the selected samples. The flow diagram of the proposed methodology is shown in Figure 1.

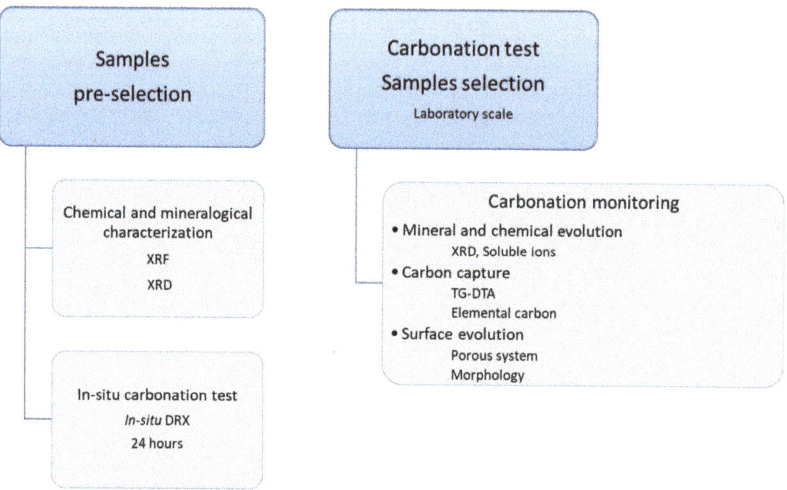

Figure 1. Flow diagram of proposed methodology.

For that reason, mineral characterization by powder X-ray diffraction is commonly used. In the same way, we propose an in situ carbonation test and process the evaluation by using a diffractometer equipped with a reaction chamber in a CO_2-rich environment. We mixed 2 g of the powdered sample and a couple of pipetted drops, and took a scan every six hours over the course of 24 h. These tests made it possible to obtain the first analysis of carbonation evolution, selecting a feasible candidate for the carbonation tests.

Once the carbonation tests were completed, the treated samples had to be analyzed by XRD, showing the presence of carbonate phases as carbonation by-products. Rietveld refinement of both the treated and the original sample allowed the quantification of the present phases, determining the percentage of new stable carbonate phases and the destruction (or partial destruction) of phases that provided the necessary calcium and/or magnesium.

Likewise, the presence of new precipitated minerals could be observed by SEM. Combined with microchemical composition from energy-dispersive spectrometers (EDS), it confirmed the possible by-product phase. If the precipitate was large enough, then the optical microscopy supply provided information on how the precipitate grew on the surface of the sample and texture, as well as estimating sizes.

As an alternative to the quantification processes, and especially when these new phases did not represent a sufficient percentage for correct quantification, to quantify the amount of CO_2 capture, two techniques were employed: elemental carbon measure and weight loss by differential thermal and thermogravimetric analysis (DTA-TG) in the temperature range of carbonate mineral decomposition. In the first case, CO_2 capture can be calculated by the direct conversion of carbon to dioxide (CO_2 (wt %) = 3.6641 × C (wt %)) for the difference of the carbon content in the original and the treated sample. In the second one, weight-loss measuring in the right temperature range corresponding to the thermal decomposition of carbonates for the difference between treated and the original sample corresponded to the percentage by weight of captured CO_2. Since it was different, if any of the minerals constituting the samples had total or partial decomposition in the same temperature range, they would not be quantified, since they would be present in both samples pre- and post-treatment, for example, carbonates that already originally existed. The evolution of soluble ion content helps what follows mineral destruction.

Another aspect to take into consideration was the evaluation of the specific surface area by Brunauer–Emmett–Teller (BET) analysis. This technique can help us understand how the specific surface of a sample evolves at the microscopic level (the texture). Complete porosity analysis from the nano- to the macroscale also allows a study of the porous system after carbonation by isotherm adsorption of CO_2 and N_2. In this sense, it was possible to analyze how the precipitate completed the sample-surface pores, reducing their size to a smaller scale from the macro- or mesoscale to the micro- or nanoscale.

Finally, it is common in the literature to use the Steinour formula [21,24,31–33] to determine the efficiency of the reaction from the theoretical maximal CO_2 sequestration value obtained from the following stoichiometric formula (Equation (4)):

$$CO_2(wt\ \%) = 0.785(\%CaO - 0.7\%SO_3) + 1.09\%MgO + 0.71\%Na_2O + 0.468\%K_2O. \qquad (4)$$

3.2. Validation of Proposed Methodology Using Ca-Silicate-Rich Brick

For each of the techniques described in the previous section, an example of the obtained results for the selected sample for this study (MPC2) and the correlations between them is presented. However, this example sample was already studied in more detail in a previous work of some of the authors in mineral carbonation processes [5].

In the MPC2 sample, CaO was around 15 wt % and 2 wt % for MgO (Table 1 and Tables S1–S3 in Supplementary Materials), which were appropriate values to be considered candidates for the mineral-carbonation process. This sample was composed of quartz, k- and alkali feldspar, plagioclases (orthoclase, anorthite, and albite), and calcium-rich silicate as wollastonite (Figure 2 and Table 2).

Table 1. Chemical composition of original brick MPC2 (wt %) by X-ray fluorescence (XRF) (modified from [5]).

Sample	SiO$_2$	Al$_2$O$_3$	Fe$_2$O$_3$	MnO	MgO	CaO	Na$_2$O	K$_2$O	TiO$_2$	P$_2$O$_5$	SO$_3$	LOI	TOTAL
MPC2	56.8	16.6	5.4	0.1	1.9	14.8	0.6	2.6	0.9	0.1	0.3	0.8	100.8
Detection Limit (DL)	0.01	0.01	0.01	0.02	0.01	0.04	0.01	0.02	0.03	0.01	0.22		
Quantification Limit (QL)	0.02	0.02	0.02	0.03	0.02	0.05	0.03	0.03	0.10	0.02	0.23		
Relative Error	0.012	0.020	0.058	0.184	0.007	0.047	0.038	0.028	0.061	0.025	0.063		

LOI: loss on ignition at 1025 °C.

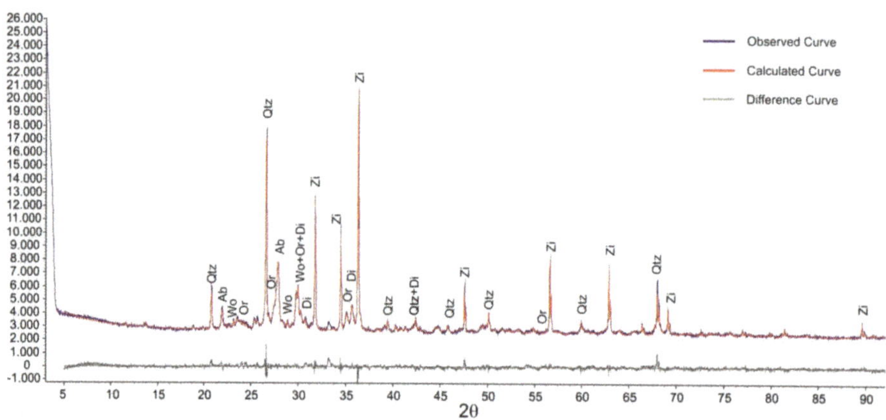

Figure 2. X-ray diffraction (XRD) pattern of MPC2 (and Rietveld refinement). Abbrv.: Qtz—quartz; Ab—Albite; Wo—wollastonite; Or—orthoclase; Di—diopside; Zi—zincite (Internal Standard).

Table 2. Mineralogical composition (wt %) of selected brick MPC2 (Qtz: quartz; Wo: wollastonite; Or: orthoclase; Ab: albite; Di: diopside; An: anorthite; Amor: amorphous phase). Rexp, Rwp, and GOF are numerical indicators of how well the Rietveld model was refined. Rwp, residual of least-squares refinement (weighted), which must be improved in refinement (with common sense); Rexp evaluates data quality; and GOF, goodness of fit parameter.

Sample	Qtz	Wo	Or	Ab	Di	An	Amor
MPC2	20.5	6.2	3.7	11.8	4.7	29.5	23.6
Rietveld Refinement	Rexp:	1.86	GOF:	1.72	Rwp:	3.21	

An in situ carbonation test (Figure 3) showed the evolution of newly grown calcite over time in a CO$_2$-rich environment. Intensity for the main calcite's peak (at d = 3.04 Å) increased directly proportional to time.

The next step was to perform carbonation tests on a laboratory scale with the selected bricks according to previous analysis, the high content of CaO, mineralogical composition rich in Ca silicate, and the presence of calcite in the in situ carbonation test. Tests were performed at room temperature and 10 bar pressure for different reaction times, and three fractions of particle sizes were representative of the total. Additionally, a test was carried out for the original sample at low pressure (1 bar), room temperature, and a 4:1 solid–water ratio in a 5 L volume hermetic reactor for 720 h of reaction time.

In the X-ray patterns of treated samples, compared with the nontreated, the most obvious differences were the presence of calcite, and the partial destruction of wollastonite and some orthoclase. The attack on wollastonite with carbonic acid allowed for the release of calcium ions and calcite precipitation [4–6,27,28,34–36].

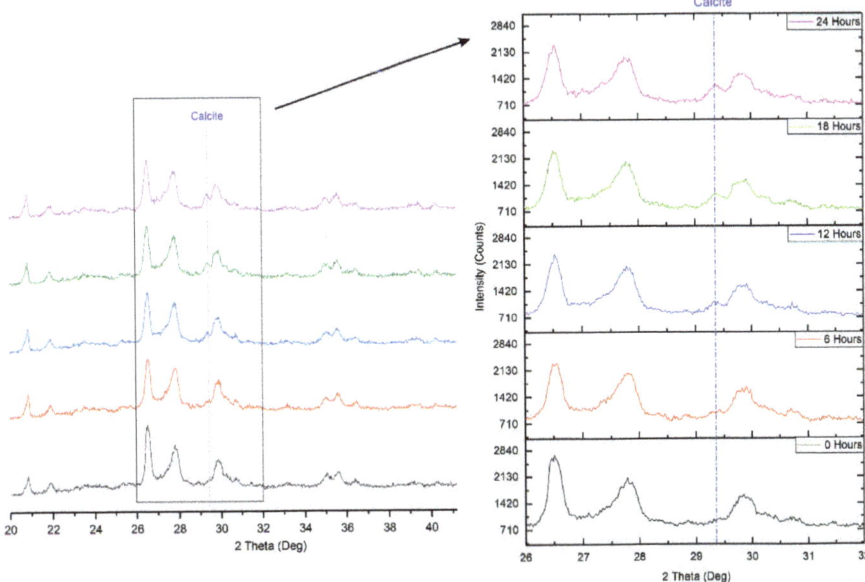

Figure 3. In situ XRD MPC2 pattern where d = 3.04 Å, reflection corresponding to precipitated newly formed calcite.

Therefore, the carbonation process occurred in the following two steps: a) silicate mineral dissolution and b) carbonate precipitation. Several studies based on the mineral carbonation of wollastonite [27,28] described the process in an aqueous carbonation route as:

1. Dissolution of CO_2 in water for the production of a (bi-)carbonate:

$$CO_2\ (g) + H_2O\ (l) \rightarrow H_2CO_3\ (aq) \rightarrow HCO_3^-\ (aq) + H^+\ (aq). \tag{5}$$

2. Leaching Ca from wollastonite by acidic attack:

$$CaSiO_3\ (s) + 2H^+\ (aq) \rightarrow Ca^{2+}\ (aq) + H_2O\ (l) + SiO_2\ (s). \tag{6}$$

3. Nucleation and growth of calcium carbonate:

$$Ca^{2+}\ (aq) + HCO_3^-\ (aq) \rightarrow CaCO_3\ (s) + H^+\ (aq). \tag{7}$$

These new carbonate crystals were observed by stereomicroscope (optical microscopy) and scanning electron microscopy (Figures 4 and 5), with a composition close to theoretical calcite, as shown by the results of elemental analysis by EDS.

Figure 4. Layers of calcite on the MPC2 surface, observed by stereomicroscope after 720 h of CO_2 treatment.

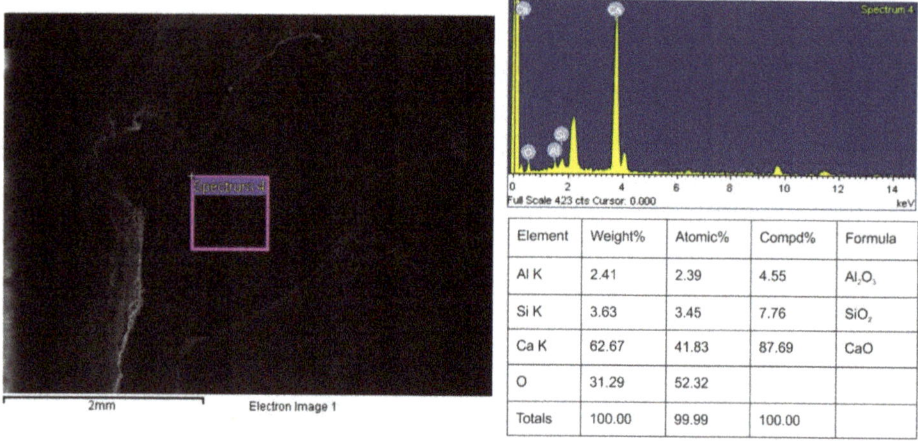

Element	Weight%	Atomic%	Compd%	Formula
Al K	2.41	2.39	4.55	Al_2O_3
Si K	3.63	3.45	7.76	SiO_2
Ca K	62.67	41.83	87.69	CaO
O	31.29	52.32		
Totals	100.00	99.99	100.00	

Figure 5. Scanning-electron-microscopy (SEM) image, and the energy-dispersive-spectrometer (EDS) spectrum and elemental quantification of MPC2.

The measurement of elemental carbon and weight loss by DTA-TG in the temperature range of the calcite decomposition (Table 3 or Figure 6a) was used to quantify the amount of captured CO_2. Both results could be expressed in terms of calcite as carbonate ore. Obviously, it was an indirect method due to it being assumed, on the one hand, that the only phase that decomposes in that temperature range was calcite and, on the other hand, that it only precipitates calcite as a carbonated material. This weight loss also corresponded to the expulsion of CO_2 and carbon content that was related to the theoretical content of CO_2 in the chemical composition of calcite (Equations (8) and (9)) [37,38].

$$\% \text{ Calcite}_{DTA-TG} = \frac{\Delta m_{450-900°C}}{CO_2\,\text{Theorical}} \times 100 = \frac{\Delta m_{450-900°C}}{43.97} \times 100 = 2.274 \times \Delta m_{450-900°C} \tag{8}$$

$$\% \text{ Calcite}_{C-Elmental} = \frac{C_{content}}{C_{Theorical}} \times 100 = \frac{C_{content}}{12} \times 100 = \frac{25}{3} C_{content} \tag{9}$$

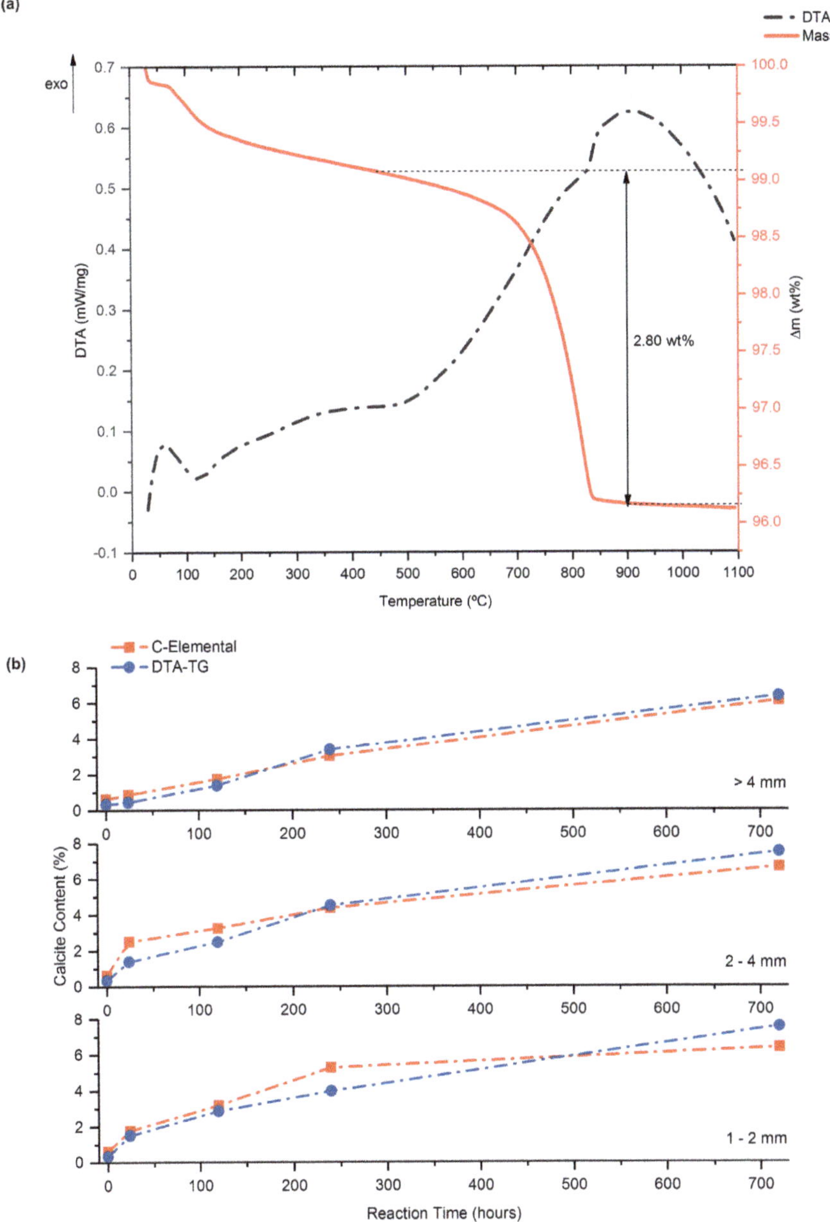

Figure 6. (**a**) MPC2 thermogravimetry (solid line) and differential thermal analyses (dashed line); >4 mm and 720 h reaction time. (**b**) Calculated calcite content by DTA-TG (circle) and by C element (square) as function of reaction time and particle size fraction (modified from [5]).

Table 3. Carbon elemental content, mass loss by differential thermal and thermogravimetric analysis (DTA-TG), and calcite content calculated by both techniques.

Particle Size	Reaction Time	C-Elemental	Mass Loss	Calcite	Calcite
(mm) MPC2	(hours)	(wt %)	(wt %)	(wt %) by C-Elemental	(wt %) by DTA-TG
Original	0	0.08	0.15	0.63	0.34
>4 mm	24	0.10	0.20	0.87	0.45
>4 mm	120	0.21	0.61	1.74	1.39
>4 mm	240	0.36	1.49	3.02	3.39
>4 mm	720	0.73	2.80	6.10	6.37
2–4 mm	24	0.30	0.61	2.51	1.39
2–4 mm	120	0.39	1.10	3.26	2.50
2–4 mm	240	0.53	1.99	4.38	4.53
2–4 mm	720	0.80	3.30	6.67	7.51
1–2 mm	24	0.21	0.67	1.79	1.52
1–2 mm	120	0.38	1.26	3.19	2.87
1–2 mm	240	0.63	1.75	5.29	3.98
1–2 mm	720	0.77	3.33	6.38	7.57

The difference compared to carbon analysis using the elemental analyzer could be attributed to the sum of experiment errors and analysis sensitivity. Yet, it was also necessary to take account of the adsorbed CO_2, which could be measured in CHNS instead of DTA. However, both followed the same tendency with respect to reaction times (Figure 6b). In both instances, calcite content was higher than the original and directly proportional to the reaction time in the studied size fractions.

The wollastonite attack with carbonic acid allowed for the release of calcium ions and calcite precipitation. Concerning the presence of soluble ions (Figure 7 and Table S4 in Supplementary Materials), the amount of Ca ions decreased over reaction time because of calcite precipitation, while Si increased as a consequence of partial silicate destruction. This partial destruction of silicates resulted in a new specific surface on the bulk (Figure 8). The increase of the specific surface had a direct relationship reaction time. The newly precipitated calcite on the sample surface also increased BET. However, both possibilities must have had a greater effect than physical CO_2 absorption that would result in a reduction of the specific surface area.

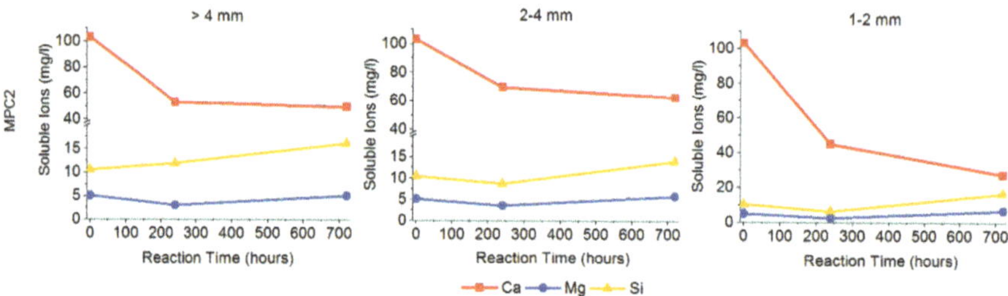

Figure 7. Soluble Ca, Mg, and Si ions measured untreated and treated after 240 and 720 h of reaction for MPC2 (modified from [5]).

After the carbonation tests, the brick samples revealed macro- and mesoporosity as determined by Hg porosity, which affects the proportion and size of the pores (Figure 9a,b), with a decrease in microporosity and increase in nanoporosity by N_2 and CO_2 absorption (Figure 9c,d). All of them were a result of two differentiated processes: (a) the action of carbonic acid destroying calcium silicates that produced an irregular surface and an increment of macro- and mesoporosity, and (b) the precipitation of carbonates that filled the micropores and probably reduced them to nanopores.

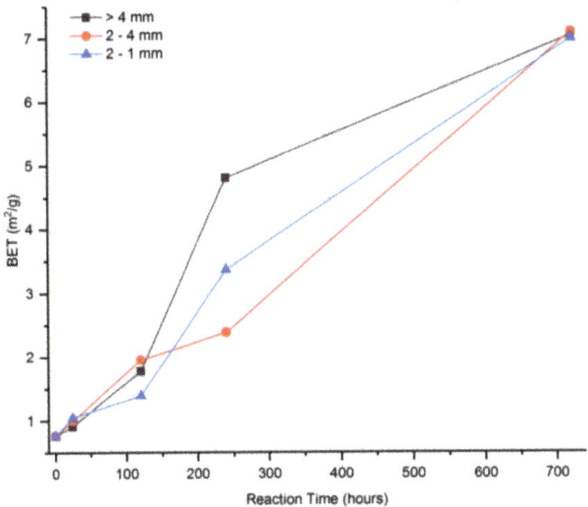

Figure 8. Specific surface evolution by specific surface area (BET) as a function of reaction time for MPC2 (modified from [5]).

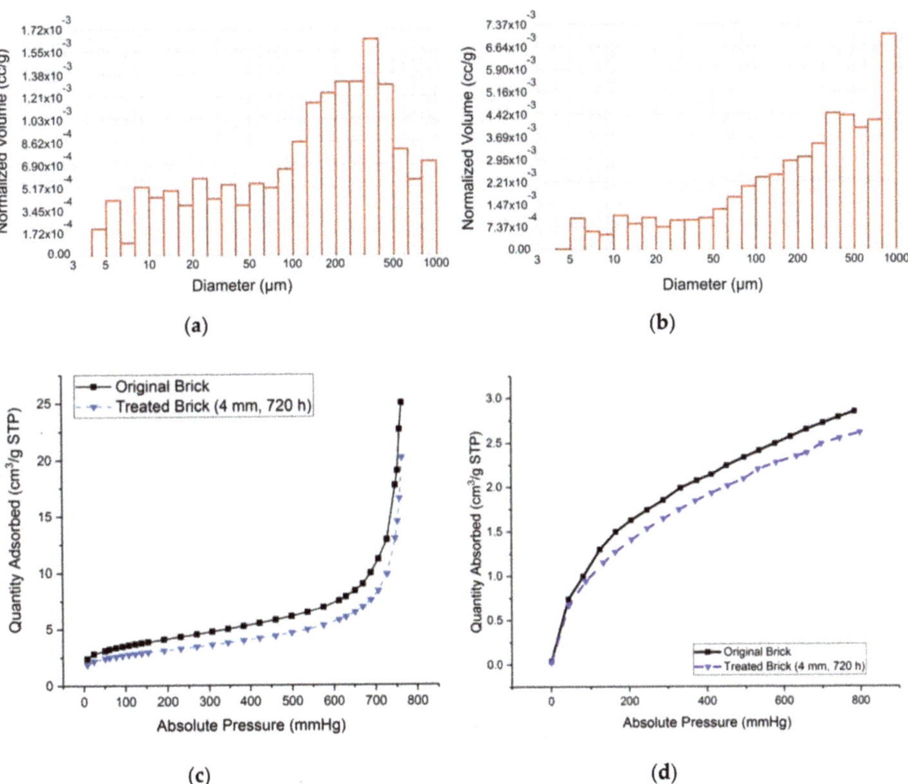

(a)

(b)

(c)

(d)

Figure 9. (**a**,**b**) Histograms of Hg porosity, (**c**) N_2 adsorption, and (**d**) CO_2 adsorption isotherms for original brick and treated brick layer.

Calculating efficiency was as the quotient between the percentage of captured CO_2 (experimental) and theoretical CO_2 by Steinour's formulae (Equation (4); Table 4). The calculated capture efficiency was proportional to the reaction time (the longer the time was, the higher the carbonation amount), whereas carbonation did not seem to significantly depend on particle size in the studied conditions. Results obtained at low pressure (1 bar) after 720 h of reactions for the total sample were very similar to those obtained for the finer fractions (2–4, 1–2 mm) at 10 bars.

Table 4. Carbonation efficiency (CE) according to Steinour's equation.

Particle Size	Pressure	Reaction Time	CO_2 Exp	Efficiency
(mm) MPC2	(bar)	(hours)	(wt %)	(%)
Original	10	0	0.15	1.00
>4 mm	10	24	0.20	1.32
>4 mm	10	120	0.61	4.08
>4 mm	10	240	1.49	9.94
>4 mm	10	720	2.80	18.67
2–4 mm	10	24	0.61	4.08
2–4 mm	10	120	1.10	7.33
2–4 mm	10	240	1.99	13.28
2–4 mm	10	720	3.30	22.02
1–2 mm	10	24	0.67	4.46
1–2 mm	10	120	1.26	8.41
1–2 mm	10	240	1.75	11.67
1–2 mm	10	720	3.33	22.19
Total Fraction	1	720	3.20	21.33

4. Conclusions

The present research showed a suitable methodology to evaluate the possibility of using ceramic construction waste (or other types of construction and demolition waste) for mineral carbonation under surface conditions in short periods of time. Even though the experiment was conducted in comparatively large scale, each test method yielded only a very small fraction. Therefore, to increase reliability each test could be conducted several times.

Specifically, Ca-rich bricks were successfully used as raw material for direct mineral carbonation by the destruction of Ca silicates. The amount of carbonation was proportional to the reaction time, whereas it did not seem to significantly depend on particle size or pressure in the studied conditions.

Acceptable carbonation efficiency was achieved under the favorable conditions of low pressure and temperature.

These results open the possibility for future studies using the proposed methodology for other types of construction and demolition waste rich in calcium silicates or other calcium compounds, which could be directly carbonated by in situ injections of CO_2.

Supplementary Materials: The following are available online at http://www.mdpi.com/2075-163X/9/10/612/s1, Table S1: List of certificated standars used for XRF calibration.; Table S2: Detection limit (L.D.), quantification limit (L.C.) and relative error for the measuremento of the mayor elements by XRF in Panalytical Axios spectrometer of the SGI Laboratorio de Rayos X (University of Seville) [September 2014]; Table S3: Mayor elements comparative between monitor standards (calibration validation) in white box and the certificate value in grey box [September 2014]; Table S4: Soluble Ca, Mg, and Si ions measured untreated and treated after 240 and 720 h of reaction for MPC2 and their standard deviations.

Author Contributions: D.M. collected and analyzed the data and wrote the paper; V.F.-A. conceived and designed the ideas; and revised the paper; P.A. made the formal Investigation, discuss the data and revised/edited the manuscript.

Funding: This research was funded by the Junta de Andalucía (Consejería de Economía y Conocimiento) (P12-RNM-568 MO project) and the contract of Domingo Martín granted by the V Plan Propio de Investigación de la Universidad de Sevilla.

Acknowledgments: Authors are grateful to editor and reviewers for their comments which improved the manuscript. XRD, XRF, ICP-OES, C-elemental and SEM analysis were performed using the facilities of the General Research Center at the University of Seville (CITIUS).

Conflicts of Interest: The authors declare no conflict of interest.

References

1. Alberola, E.; Chevallier, J.; Chèze, B. The EU emissions trading scheme: The effects of industrial production and CO_2 emissions on carbon prices. *Econ. Int.* **2008**, *4*, 93–125. [CrossRef]

2. Caldeira, K.; Jain, A.K.; Hoffert, M.I. Climate sensitivity uncertainty and the need for energy without CO_2 emission. *Science* **2003**, *299*, 2052–2054. [CrossRef] [PubMed]

3. Pan, S.-Y.Y.; Chang, E.E.; Chiang, P.C. CO_2 capture by accelerated carbonation of alkaline wastes: A review on its principles and applications. *Aerosol Air Qual. Res.* **2012**, *12*, 770–791. [CrossRef]

4. Martín, D.; Aparicio, P.; Galán, E. Mineral carbonation of ceramic brick at low pressure and room temperature. A simulation study for a superficial CO_2 store using a common clay as sealing material. *Appl. Clay Sci.* **2018**, *161*, 119–126. [CrossRef]

5. Martín, D.; Aparicio, P.; Galán, E. Accelerated Carbonation of Ceramic Materials. Application to Bricks from Andalusian Factories (Spain). *Constr. Build. Mater.* **2018**, *181*, 598–608. [CrossRef]

6. Martín, D.; Aparicio, P.; Galán, E. Time evolution of the mineral carbonation of ceramic bricks in a simulated pilot plant using a common clay as sealing material at superficial conditions. *Appl. Clay Sci.* **2019**, *180*, 105191. [CrossRef]

7. Eighmy, T.; Gardner, K.; Seager, T. Method for Sequestering Carbon Dioxide. U.S. Patent US20050238563A1, 27 October 2005. Available online: https://patentimages.storage.googleapis.com/bb/49/8c/56d81ea0f8406a/US20050238563A1.pdf (accessed on 3 October 2019).

8. Montes-Hernandez, G.; Pérez-López, R.; Renard, F.; Charlet, L.; Nieto, J.-M. Process for Sequestration of CO2 by Reaction with Alkaline Solid Waste. European Patent EP 07 1233005.5, 14 December 2007. Available online: https://patentimages.storage.googleapis.com/47/a0/49/787d3e0e245078/WO2007071633A1.pdf (accessed on 3 October 2019).

9. Mayoral, M.C.; Andrés, J.M.; Gimeno, M.P. Optimization of mineral carbonation process for CO_2 sequestration by lime-rich coal ashes. *Fuel* **2013**, *106*, 448–454. [CrossRef]

10. Santos, R.M.; Van Bouwel, J.; Vandevelde, E.; Mertens, G.; Elsen, J.; Van Gerven, T. Accelerated mineral carbonation of stainless steel slags for CO_2 storage and waste valorization: Effect of process parameters on geochemical properties. *Int. J. Greenh. Gas Control* **2013**, *17*, 32–45. [CrossRef]

11. Sipilä, J.; Teir, S.; Zevenhoven, R. Carbon Dioxide Sequestration by Mineral Carbonation Literature Review Update 2005–2007. Report VT 2008-1. p. 52. Available online: http://innovationconcepts.nl/res/literatuurGPV/mineralcarbonationliteraturereview0507.pdf (accessed on 3 October 2019).

12. Soong, Y.; Fauth, D.L.; Howard, B.H.; Jones, J.R.; Harrison, D.K.; Goodman, A.L.; Gray, M.L.; Frommell, E.A. CO_2 sequestration with brine solution and fly ashes. *Energy Convers. Manag.* **2006**, *47*, 1676–1685. [CrossRef]

13. Uibua, M.; Kuusik, R.; Andreas, L.; Kirsimäe, K. The CO_2-binding by Ca-Mg-silicates in direct aqueous carbonation of oil shale ash and steel slag. *Energy Procedia* **2011**, *4*, 925–932. [CrossRef]

14. Gunning, P.J.; Hills, C.D.; Carey, P.J. Accelerated carbonation treatment of industrial wastes. *Waste Manag.* **2010**, *30*, 1081–1090. [CrossRef] [PubMed]

15. Geerlings, J.J.C.; Van Mossel, G.A.F.; Veen, B.C.M.I. Process for Sequestration of Carbon Dioxide by Mineral Carbonation. U.S. Patent US20100196235A1, 25 May 2010. Available online: https://patentimages.storage.googleapis.com/47/b3/78/16189676247184/US20100196235A1.pdf (accessed on 3 October 2019).

16. Kawatra, S.K.; Eisele, T.C.; Simmons, J.J. Capture and Sequestration of Carbon Dioxide in Flue Gases. U.S. Patent 7.919.064B2, 5 April 2011. Available online: https://patentimages.storage.googleapis.com/3d/6c/86/beea6a2fc7f1ba/US7919064.pdf (accessed on 3 October 2019).

17. Riman, R.E.; Atakan, V. Systems and Methods for Carbon Capture and Sequestration and Compositions Derived Therefrom. U.S. Patent US8114367B2, 14 February 2012. Available online: https://patentimages. storage.googleapis.com/72/67/38/c293969a42ecee/US8114367.pdf (accessed on 3 October 2019).
18. Galán, E.; Aparicio, P. Captura y Secuestro de CO2 Mediante la Carbonatición de Residuos Cerámicos. WITO Patent WO2011089292A1, 28 July 2011.
19. Costa, G.; Baciocchi, R.; Polettini, A.; Pomi, R.; Hills, C.D.; Carey, P.J. Current status and perspectives of accelerated carbonation processes on municipal waste combustion residues. *Environ. Monit. Assess.* **2007**, *135*, 55–75. [CrossRef] [PubMed]
20. Dri, M.; Sanna, A.; Maroto-Valer, M.M. Mineral carbonation from metal wastes: Effect of solid to liquid ratio on the efficiency and characterization of carbonated products. *Appl. Energy* **2014**. [CrossRef]
21. Huntzinger, D.N.; Gierke, J.S.; Sutter, L.L.; Kawatra, S.K.; Eisele, T.C. Mineral carbonation for carbon sequestration in cement kiln dust from waste piles. *J. Hazard. Mater.* **2009**, *168*, 31–37. [CrossRef]
22. Li, X.; Bertos, M.F.; Hills, C.D.; Carey, P.J.; Simon, S. Accelerated carbonation of municipal solid waste incineration fly ashes. *Waste Manag.* **2007**, *27*, 1200–1206. [CrossRef]
23. Haselbach, L. Potential for carbon dioxide absorption in concrete. *J. Environ. Eng.* **2009**, *135*, 465–472. [CrossRef]
24. Fernández Bertos, M.; Simons, S.J.R.; Hills, C.D.; Carey, P.J. A review of accelerated carbonation technology in the treatment of cement-based materials and sequestration of CO_2. *J. Hazard. Mater.* **2004**, *112*, 193–205. [CrossRef]
25. Seifritz, W. CO_2 disposal by means of silicates. *Nature* **1990**, *345*, 486. [CrossRef]
26. Lackner, K.S.; Wendt, C.H.; Butt, D.P.; Joyce, E.L.; Sharp, D.H. Carbon dioxide disposal in carbonate minerals. *Energy* **1995**, *20*, 1153–1170. [CrossRef]
27. Huijgen, W.J.J.; Witkamp, G.J.; Comans, R.N.J. Mechanisms of aqueous wollastonite carbonation as a possible CO_2 sequestration process. *Chem. Eng. Sci.* **2006**, *61*, 4242–4251. [CrossRef]
28. Daval, D.; Martinez, I.; Corvisier, J.; Findling, N.; Goffé, B.; Guyot, F. Carbonation of Ca-bearing silicates, the case of wollastonite: Experimental investigations and kinetic modeling. *Chem. Geol.* **2009**, *262*, 262–277. [CrossRef]
29. Huijgen, W.J.J.; Ruijg, G.J.; Comans, R.N.J.; Witkamp, G.J. Energy consumption and net CO_2 sequestration of aqueous mineral carbonation. *Ind. Eng. Chem. Res.* **2006**, *45*, 9184–9194. [CrossRef]
30. Tai, C.Y.; Chen, W.; Shih, S. Factors affecting wollastonite carbonation under CO_2 supercritical conditions. *AIChE J.* **2006**, *52*, 292–299. [CrossRef]
31. Huntzinger, D.N.; Gierke, J.S.; Kawatra, S.K.; Eisele, T.C.; Sutter, L.L. Carbon Dioxide Sequestration in Cement Kiln Dust through Mineral Carbonation. *Environ. Sci. Technol.* **2009**, *43*, 1986–1992. [CrossRef] [PubMed]
32. Steinour, H.H. Some effects of carbon dioxide on mortars and concrete-discussion. *J. Am. Concr. Inst* **1959**, *30*, 905–907.
33. Yixin, S.; Xudong, Z.; Monkman, S. A new CO_2 sequestration process via concrete products production. In Proceedings of the 2006 IEEE EIC Climate Change Conference, Ottawa, ON, Canada, 10–12 May 2006.
34. Wang, W.; Xiao, J.; Wei, X.; Ding, J.; Wang, X.; Song, C. Development of a new clay supported polyethylenimine composite for CO_2 capture. *Appl. Energy* **2014**, *113*, 334–341. [CrossRef]
35. Mun, M.; Cho, H. Mineral carbonation for carbon sequestration with industrial waste. *Energy Procedia* **2013**, *37*, 6999–7005. [CrossRef]
36. Prigiobbe, V.; Hänchen, M.; Werner, M.; Baciocchi, R.; Mazzotti, M. Mineral carbonation process for CO_2 sequestration. *Energy Procedia* **2009**, *1*, 4885–4890. [CrossRef]
37. Lim, M.; Han, G.C.; Ahn, J.W.; You, K.S. Environmental remediation and conversion of carbon dioxide (CO_2) into useful green products by accelerated carbonation technology. *Int. J. Environ. Res. Public Health* **2010**, *7*, 203–228. [CrossRef]
38. Villagrán-Zaccardi, Y.A.; Egüez-Alava, H.; De Buysser, K.; Gruyaert, E.; De Belie, N. Calibrated quantitative thermogravimetric analysis for the determination of portlandite and calcite content in hydrated cementitious systems. *Mater. Struct.* **2017**, *50*, 179. [CrossRef]

Article

Experimental Simulation of the Self-Trapping Mechanism for CO$_2$ Sequestration into Marine Sediments

Hak-Sung Kim [1] and Gye-Chun Cho [2,*]

[1] Advanced Plant Laboratory, Central Research Institute, Korea Hydro and Nuclear Power Co., Ltd. (KHNP), Daejeon 34101, Korea; shield5200@kaist.ac.kr

[2] Department of Civil and Environmental Engineering, Korea Advanced Institute of Science and Technology (KAIST), Daejeon 34141, Korea

* Correspondence: gyechun@kaist.edu; Tel.: +82-42-350-3622; Fax: +82-42-350-3610

Received: 31 July 2019; Accepted: 16 September 2019; Published: 24 September 2019

Abstract: CO$_2$ hydrates are ice-like solid lattice compounds composed of hydrogen-bonded cages of water molecules that encapsulate guest CO$_2$ molecules. The formation of CO$_2$ hydrates in unconsolidated sediments significantly decreases their permeability and increases their stiffness. CO$_2$ hydrate-bearing sediments can, therefore, act as cap-rocks and prevent CO$_2$ leakage from a CO$_2$-stored layer. In this study, we conducted an experimental simulation of CO$_2$ geological storage into marine unconsolidated sediments. CO$_2$ hydrates formed during the CO$_2$ liquid injection process and prevented any upward flow of CO$_2$. Temperature, pressure, P-wave velocity, and electrical resistance were measured during the experiment, and their measurement results verified the occurrence of the self-trapping effect induced by CO$_2$ hydrate formation. Several analyses using the experimental results revealed that CO$_2$ hydrate bearing-sediments have a considerable sealing capacity. Minimum breakthrough pressure and maximum absolute permeability are estimated to be 0.71 MPa and 5.55×10^{-4} darcys, respectively.

Keywords: CO$_2$ geological sequestration; unconsolidated sediments; gas hydrates

---

1. Introduction

Carbon capture and storage (CCS) technology is essential for rapid CO$_2$ mitigation. The geological storage of carbon dioxide (CO$_2$) is a highly effective, long-term mitigation solution for the large quantities of CO$_2$ emissions [1,2]. For these reasons, to date, CO$_2$ geological sequestration (CGS) technology has been developed by several leading countries. However, most of existing CGS methods worldwide require particular geological structures to work, such as a highly pervious rock formation (e.g., sandstone layer) imbedded in impermeable layers (i.e., cap-rocks). This requirement leads to CGS application difficulties such as a shortage of proper sites, challenges in the long-range transport of CO$_2$, deep drilling and injection, and restricted storage capacity, which substantially increases the cost of using CGS. To overcome these limitations, several CGS methods that do not need cap-rock, such as carbonated water injection (CWI), have been suggested [3–6].

CO$_2$ can be stored in unconsolidated sediments under CO$_2$ hydrate-bearing sediments. CO$_2$ hydrates are ice-like solid lattice compounds composed of hydrogen-bonded water cages that encapsulate guest molecules of CO$_2$. CO$_2$ hydrates are formed in the seabed under low temperatures and high pressures [7,8]. Previous studies on natural gas hydrate-bearing sediments [9] and preliminary studies on CO$_2$ hydrate-bearing sediments [10,11] have shown that CO$_2$ migration is significantly hampered by the formation of gas hydrates, resulting in a self-trapping mechanism. Furthermore, the self-preservation response of CO$_2$ hydrates slows the CO$_2$ hydrate dissociation process [12], which serves to mend unintended fractures of CO_2 hydrate-bearing sediments, thereby severely diminishing the transport of CO_2 fluids [13,14]. Thus, it has been suggested that CO_2 hydrates can be used as primary or secondary safety factors for CO_2 geological storage in marine unconsolidated sediments [15–17]. Furthermore, unconsolidated sand sediments have advantages over consolidated rocks (e.g., sandstones) in that the CO_2 storage capacity of the former is higher than that of the latter due to the high porosity of unconsolidated sandy sediments (40–60%). In addition, the CO_2 injectability of unconsolidated sand sediments is superior because of their high permeability (0.1–10 darcys) resulting from wide and well-connected pore spaces.

Tohidi et al. (2010) performed experimental CO_2 leakage simulations through each type of unconsolidated sediment (glass-bead, sand, and sand–clay mixture), and using electrical resistance measurements and a CO_2 concentration analysis they confirmed the existence of the self-trapping effect of CO_2 hydrates [11]. Massah et al. (2018) demonstrated the sequestration of CO_2 through horizontal injection into a laboratory scale reservoir and revealed the large storage density of CO_2 hydrate formations [18]. Gauteplass et al. (2018) described CO_2 hydrate formation caused by liquid CO_2 injection into cold, water-saturated sandstone and reported that hydrate formation in the pore space resulted in blockage of CO_2 flow under most conditions [19]. However, more direct and comprehensive experimental data including temperature–pressure relations, elastic wave velocity, and dissociation tests are required for a better understanding of the behavior of CO_2 and CO_2 hydrate formation in unconsolidated sediments.

The objective of this study was to simulate CO_2 geological storage into marine unconsolidated sediments using CO_2 hydrates as a cap-rock. A large reaction cell was used to experimentally verify the CO_2 self-trapping mechanism in marine sediments and to evaluate the behavior of CO_2-stored unconsolidated sediment during CO_2 hydrate formation and dissociation.

2. Experimental Program

2.1. Soil Used

The strata of unconsolidated marine sediments typically consist of multiple layers of a different sediment types such as sand-rich sediment layers and fine-grained sediment layers. The permeability of fine-grained sediments is very low (i.e., 10^{-3}–10^{-7} darcys; [20]), therefore, fine-grained sediments can be practically considered as impermeable layers, which obstruct the upward flow of CO_2. Meanwhile, sand-rich sediments are suitable as CO_2 storage host sediments because of their relatively high permeability (0.1–10 darcys), while they are ready for permeation of CO_2. The effect of the self-trapping mechanism on a sand-rich layer is, therefore, important to CO_2 geological storage into unconsolidated sediments. In this study, fine sand (Ottawa F110; mean particle size = 120 μm, specific gravity = 2.65, permeability = 5–6 darcys, quartz 99%) was used as the host sediment sample.

2.2. Experiment Setup

The experimental design simulated CO_2 injection into a shallow marine sediment (i.e., high water pressure, low temperature) and CO_2 hydrate formation. The experimental design used in this study is shown in Figure 1. A cylindrical and rigid-wall reaction cell was made of an aluminum alloy (duralumin, AA2024). The inner diameter of the cell was 20 cm, the height of the interior was 100 cm, and the internal volume was 31.4 L. The reaction cell was originally developed for an experimental simulation of thermal stimulation on gas hydrate-bearing sediments [21]. Water and liquid CO_2 were injected from the bottom of the reaction cell using a water pump and gas booster. Pressure inside the reaction cell was controlled using a back-pressure regulator at the top of the cell. The quantities of CO_2 gas and water that flowed out of the reaction cell were measured using a water substitution system.

Various types of sensors were installed at predetermined layers (every 10 cm) within the reaction cell as shown in Figure 2. The cell contained five T-type thermocouples for temperature measurements of the cell interior, five pressure transducers for fluid pressure measurements, five pairs of piezoelectric ceramic disks (diameter: 20 mm) for compressional wave (P-wave) measurements at layers A1–A5, and four pairs of electrodes for electrical resistance measurements at layers B1–B4. For P-wave velocity measurements, square-shaped pulses with amplitude of 10 V (peak-to-peak) were used for excitation, and the input frequency ranged from 1 to 10 kHz. The electrodes were connected to an LCR meter in order to measure the electrical resistance (frequency = 50 kHz).

Cool water was circulated through copper tubes that coiled around the reaction cell. The temperature inside the reaction cell was controlled by two water coolers, which had different temperatures (i.e., Water cooler 1 at 3 °C, and Water cooler 2 at 15 °C). Figure 3 shows the temperature gradient of the inside of the reaction cell, which was formed by two separated cooling systems. The CO_2 hydrate stability zone was developed in the middle part of the reaction cell (i.e., height of 0.35–0.75 m in the reaction cell).

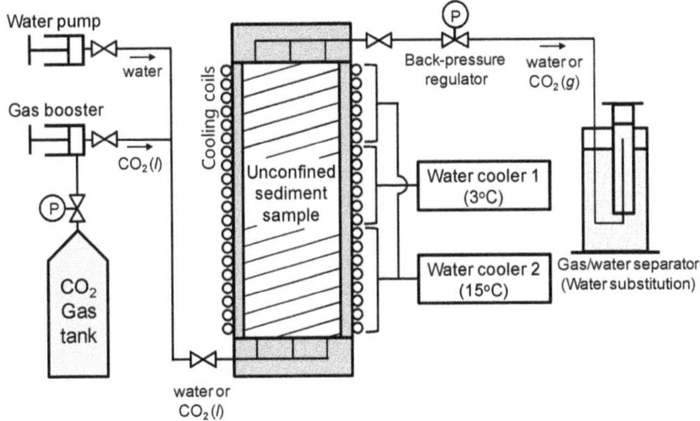

Figure 1. Conceptual drawing of the experimental design.

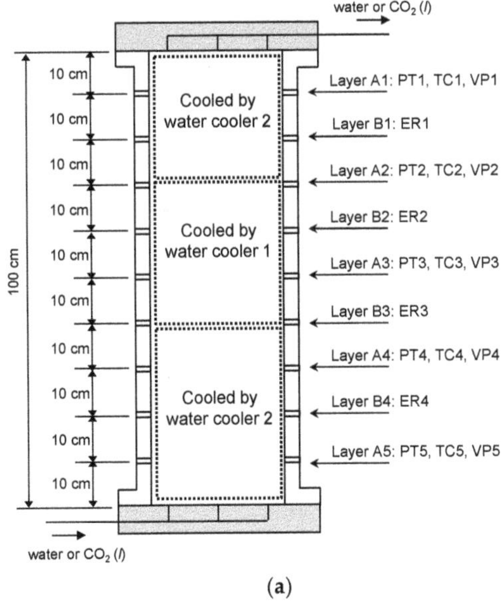

(**a**)

Figure 2. *Cont.*

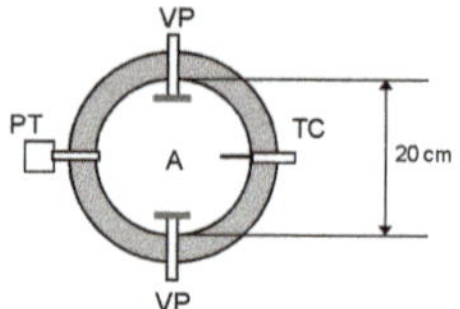

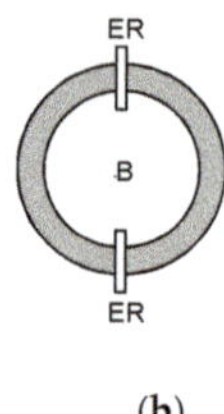

(b)

Figure 2. Conceptual drawing of the high-pressure cell used in this study; PT: Pressure transducer, TC: Thermocouple, VP: Piezoelectric plates for P-wave velocity measurements, ER: Electrode for electrical resistance measurements. (**a**) Vertical cross-sectional drawing of the cell. (**b**) Horizontal cross-sectional drawing of the layers A1–A5 and B1–B4, respectively.

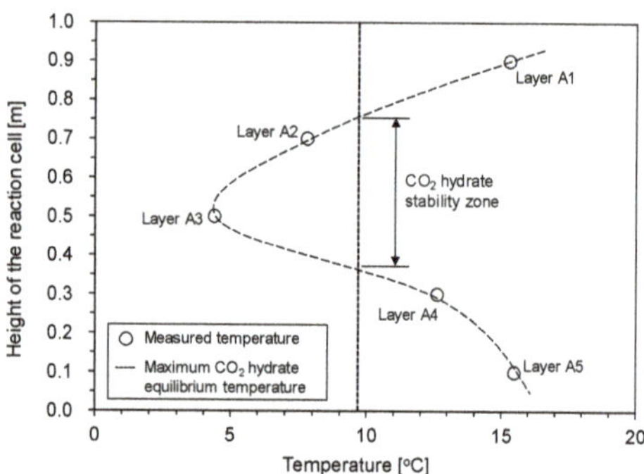

Figure 3. Temperature distribution of the high-pressure cell obtained from the preliminary test result with distilled water. The maximum CO_2 hydrate equilibrium temperature is approximated using the second quadruple point of CO_2–water mixture (the intersection of the water–CO_2 vapor–CO_2 liquid, and water–CO_2 hydrate–CO_2 vapor equilibrium line) because the water–CO_2 hydrate–liquid CO_2 equilibrium line is essentially vertical in the pressure–temperature diagram [22].

2.3. Experimental Procedure

The experiment involved three procedures: (1) The preparation of a water-saturated sample; (2) the injection of the CO_2 liquid; and (3) the depressurization of the cell. When the CO_2 liquid was injected from the bottom of the cell, the injected CO_2 moved upward in the water-saturated sediment sample due to its buoyancy. Then, CO_2 hydrates formed within the CO_2 hydrates stable zone, which was located at the middle part of the cell (refer to Figure 3). The depressurization test was performed after CO_2 hydrate formation to evaluate the behavior of CO_2 during hydrate dissociation. These procedures are detailed within this subsection.

2.3.1. Water-Saturated Sample Preparation

To simulate deep-marine sediments, a water-saturated fine sand (i.e., Ottawa F110 sand) sample was prepared. First, the sample was mixed with distilled water and packed into the cell by hand tamping. The cell was fully filled with moist sand and the porosity of the sample was 0.40. Then, the cell was slowly flushed with distilled water at pressure of ~0.5 MPa for several hours to remove remaining air bubbles. During the water flushing, a cooling process was initiated. After the completion of water flushing, the cell was pressurized to 5.5 MPa with distilled water and left for about 16 h to stabilize the temperature throughout the sample.

Our primary goal was to study the self-trapping effect induced by CO_2 hydrates, and because of that, distilled water was used as pore water instead of saline water (it allows easier hydrate formation). However, this decision means that the electrical and geochemical behavior in this experiment was different from the real behavior in the marine sediments.

2.3.2. Injection of CO_2 Liquid

Liquid CO_2 was introduced into the cell from the bottom using the gas booster. The injection pressure was 5.6 MPa, and the backpressure at the top of the cell was 5.0 MPa. Thirty-four hours later, the injection pressure and backpressure increased to 6.2 and 5.6 MPa, respectively. The CO_2 liquid was injected for more than 17 days while the temperature, pressure, P-wave velocity, and electrical resistance were measured. The formation of CO_2 hydrates during the CO_2 injection process was expected (refer to Section 3).

2.3.3. Depressurization

The cell was depressurized stepwise using a back-pressure regulator while the inlet valve was closed. Each depressurization step was 0.5 MPa for more than 16 hours. Temperature, pressure, P-wave velocity, and electrical resistance during the depressurization process were measured in the same manner as during CO_2 injection process.

3. Experimental Results

3.1. Liquid CO_2 Injection Process

When the CO_2 liquid was injected from the bottom of the cell, a CO_2 liquid plume moved upward because of a buoyancy force. Eventually, the CO_2 liquid front reached the CO_2 hydrate stability zone, and then CO_2 hydrates formed. The CO_2 hydrate-bearing sediment layer then obstructed the upward flow of the CO_2 liquid. While water and CO_2 were consumed in the CO_2 hydrate formation process in the hydrate stability zone, the CO_2 hydrate-bearing sediment layer prevented CO_2 supply. Thus, a pressure difference between the upper and lower part of the cell appeared. The P-wave velocity and electrical resistance monitoring results indicated the formation of CO_2 hydrates and the blockage of the CO_2 flow. Detailed experimental results are shown in the following sections.

3.1.1. Temperature and Pressure

Figure 4 shows the pressure of each layer in the cell over time. The CO_2 liquid was injected at 1300 min after data logging started. The pressure in each layer was scattered until 5000 min because of a difference in pressure between the injection pressure and backpressure (i.e., ~0.6 MPa, refer to Section 2.3.2), and volume change of pore fluids due to CO_2 dissolution into pore water. The pressure in each layer was very similar to one another because the pore space was well connected throughout the sample. CO_2 hydrates started to form when the injected CO_2 reached the CO_2 hydrate stability zone (herein, between layers A3 and A4; Figure 3). Then, the pressure in layers A1, A2, and A3 rapidly dropped to 3.3 MPa after 5000 min while the pressure of layers A4 and A5 were nearly constant and identical to the injection pressure. The difference in pressure between the upper and lower part of the cell was induced by the sealing (i.e., pore clogging) effect of the CO_2 hydrate bearing-sediments layer.

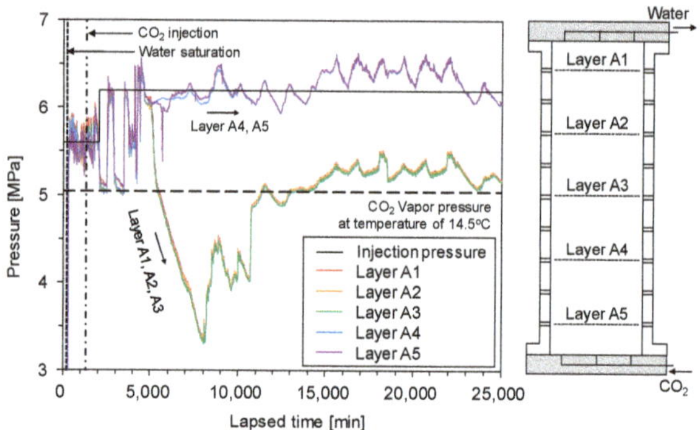

Figure 4. Pressure of the cell with lapsed time during CO_2 injection.

The sealing capacity of the CO_2 hydrate-bearing sediment layer gradually increased during the growth of CO_2 hydrates in the pore space of the sample. Meanwhile, water and CO_2 were consumed during CO_2 hydrate formation. For a constant-volume process, the consumption of CO_2 and water during CO_2 hydrate formation leads to a pressure decrease because the molar volume of the CO_2 hydrates is smaller than the original molar volume of the consumed fluids. For the lower part of the cell (represented by layers A4 and A5) the pressure was preserved because CO_2 was supplied continuously from the bottom of the cell during the experiment. For the upper part of the cell (represented by layers A1, A2, and A3), however, the pressure decreased because the CO_2 hydrate-bearing sediment layer prevented the CO_2 to be supplied from the lower part of the cell.

Figure 5 shows the pressure–temperature evolution during the CO_2 injection test. Note that layers A2 and A3 were in the CO_2 hydrate stable condition while the others were not. The pressure of the upper part of the cell (i.e., layers A1, A2, and A3) decreased when the sealing capacity of the CO_2 hydrate-bearing sediment layer increased to a level that prevented flow. Note that there were no CO_2 hydrates in layer A2 because the injected CO_2 did not reach it, even though this layer is in the CO_2 hydrate stability zone. Meanwhile, the lower part of the cell (i.e., layers A4 and A5) maintained a constant pressure level. Then, the pressure of the upper part of the cell gradually increased. There are two mechanisms for the pressure recovery of the upper part of the cell: (1) The uppermost CO_2 hydrates dissociated with the pressure decrease. Therefore, pressure was recovered restrictively via emitted CO_2 and water from the CO_2 hydrates, and (2) CO_2 hydrate saturation is limited by capillary pressure, which is determined by the pore size [23].

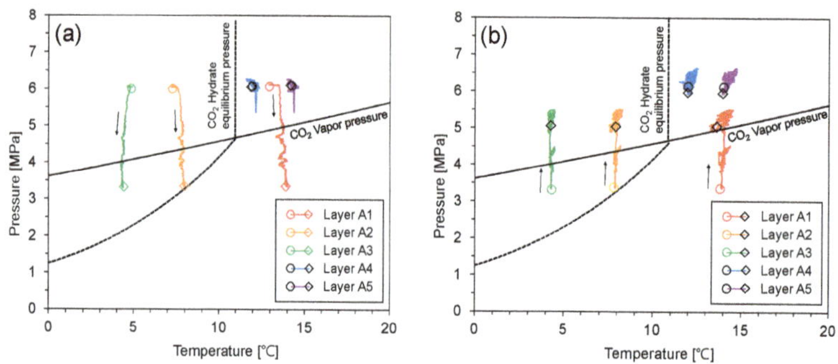

Figure 5. The pressure–temperature relationship during CO_2 injection. (**a**) 4680–8100 min, (**b**) 8100–25,300 min.

In the central part of the CO_2 hydrate-bearing sediment layer, CO_2 hydrates grew until the CO_2 hydrate saturation reached maximum CO_2 hydrate saturation. Thus, the consumption of CO_2 and water diminished because any additional formation of CO_2 hydrates was restricted. Finally, the pressure of the upper part of the cell increased to 5.2–5.5 MPa. However, the pressure of the upper part of the cell was still lower than that of the lower part of the cell. The repetitive ascending and descending pressure could be due to the continuous repetition of the CO_2 formation and dissociation process at the CO_2 hydrates front.

3.1.2. P-Wave Velocity

Figure 6 shows the results of the P-wave velocity measurements during the CO_2 injection process. Before the water injection process, the cell was partially saturated, and the P-wave velocity was about 900 m/s. When the sediment sample was saturated by distilled water, the P-wave velocity of all the layers was about 1600 m/s. Then, when the CO_2 liquid was injected, the P-wave velocities of the lower part of the cell (i.e., layers A4 and A5) decreased because the bulk modulus of the CO_2 liquid was much lower than that of the water [24,25]. Meanwhile, the P-wave velocity of layer A3 gradually increased because of the stiffening effect induced by CO_2 hydrate formation. This P-wave velocity increase in layer A3 indicates that the CO_2 hydrate bearing-sediment layer is between layers A3 and A4. On the other hand, the P-wave velocity of layer A1 and A2 did not change during CO_2 injection. This is evidence that the CO_2 hydrate-bearing sediment layer prevents any upward flow of the CO_2 liquid.

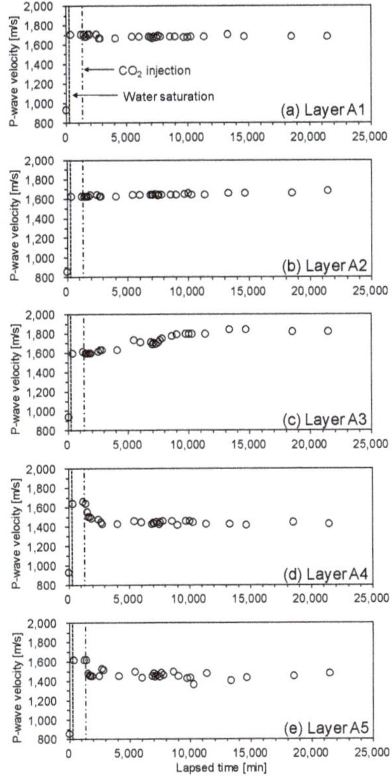

Figure 6. P-wave velocity of the unconsolidated sediment sample during CO_2 injection. (**a**) layer A1, (**b**) layer A2, (**c**) layer A3, (**d**) layer A4, (**e**) layer A5. P-wave velocity of layers A1 and A2 did not change during CO_2 injection because the CO_2 hydrate-bearing sediment layer prevented the upward flow of CO_2 liquid.

3.1.3. Electrical Resistance

Figure 7 shows the normalized electrical resistance (R/R_0) during the CO_2 injection process, where R is the measured electrical resistance and R_0 is the initial electrical resistance of the distilled water-saturated sample at each layer (i.e., layers B1–B4). For the lower part of the cell (i.e., layers B3 and B4), the electrical resistance decreased with CO_2 injection because of the dissolution of CO_2.

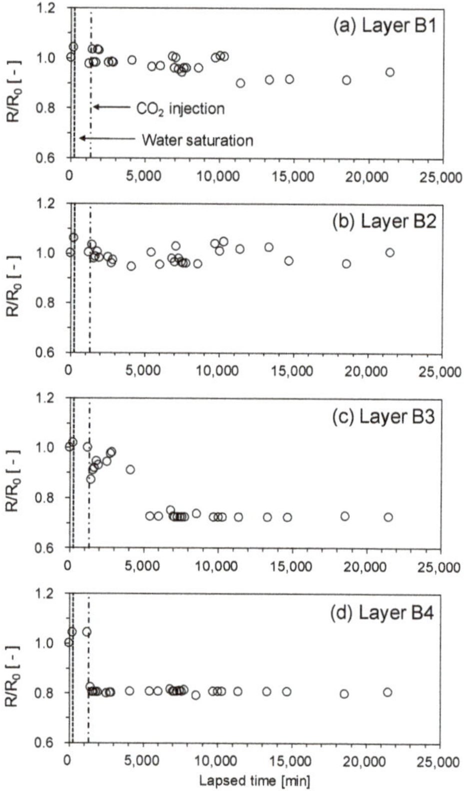

Figure 7. Normalized electrical resistance R/R_0 of the unconsolidated sediment sample during CO_2 injection. (**a**) layer B1, (**b**) layer B2, (**c**) layer B3, (**d**) layer B4.

Electrical resistance increased in-situ in the marine sediments because the conductive pore water (i.e., brine) was replaced by CO_2, which is a nonpolar molecule. This electrical resistance increase induced by the CO_2 replacement was weakened by the dissolution of CO_2 and the surface effect of the mineral grains [26]. For typical brine, the effect of CO_2 dissolution on electrical resistance is negligible because the concentration of salt (i.e., NaCl) is much larger than the ionic concentration increased by CO_2 dissolution [27]. However, if the salt concentration is low (e.g., onshore sediments), electrical resistance of in-situ sediments can decrease during the CO_2 permeation [28]. In this experiment, the effect of dissolved CO_2 was dominant on electrical resistance because the pore water was distilled.

Meanwhile, the electrical resistance of the upper part of the cell (i.e., layers B1 and B2) showed minor changes during CO_2 injection. This is additional evidence demonstrating that CO_2 liquid did not reach the upper part of the cell. Based on the change of the pressure, P-wave velocity and electrical resistance, we can presume that the CO_2 hydrate formation front is located between layers A3 and B3. Meanwhile, CO_2 hydrates formation was not observed in the electrical resistance data. The electrical resistance of in-situ water-saturated sediments increased when CO_2 hydrates formed because the electrical resistance of CO_2 hydrates is higher than pore water [29,30]. In this study, however, the change in electrical resistance was insignificant in spite of the presence of CO_2 hydrates, because distilled water was used as pore water in this experiment. This is the one of the limitations of this experiment.

3.2. Depressurization Process

3.2.1. Temperature and Pressure

Figure 8 shows the pressure of each layer in the cell over time. The pressure of the cell was reduced step-wise using a back-pressure regulator. Pressure differences between the upper and lower part of the cell remained, even though CO_2 was vaporized. This pressure discrepancy indicated that the sealing capacity of the CO_2 hydrate-bearing sediment layer was preserved. When the CO_2 hydrate dissociated completely, the pressure of each layer became equal.

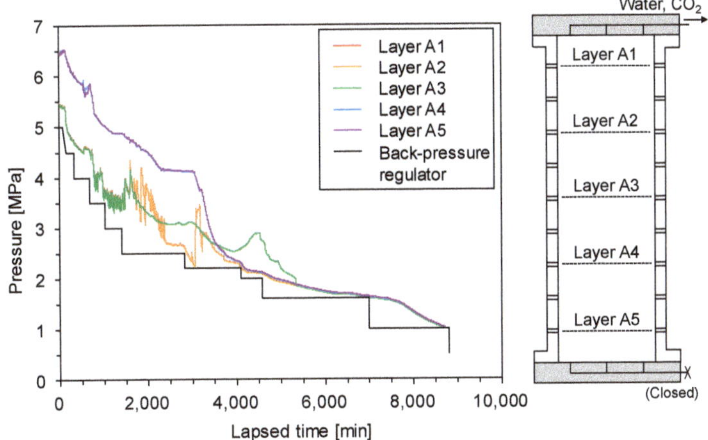

Figure 8. Pressure of the cell with lapsed time during depressurization.

Figure 9 shows the pressure–temperature relationship during the depressurization test. The pressure of the cell dropped with the pressure release using the back-pressure regulator. When the pressure of layers A4 and A5 reached the CO_2 vapor pressure, their pressure and temperature relation moved along the CO_2 vapor pressure (Figure 9a). The path of the pressure and temperature relationship of layers A4 and A5 was similar to that of the isometric process because the flow of fluids was obstructed by the remaining CO_2 hydrate-bearing sediment layer (Figure 9b). Then, the pressure and temperature relationship of layer A2 moved along the CO_2 hydrate equilibrium line (Figure 9c). This is evidence that the CO_2 hydrates re-formed and dissociated in layer A2. In the previous CO_2 injection process, CO_2 liquid did not reach layer A2 because of the sealing effect of the CO_2 hydrate-bearing sediment layer, which was located between layers A3 and B3. However, CO_2 was supplied to layer A2 when the existing CO_2 hydrates partially dissociated by depressurization. CO_2 hydrates then re-formed because layer A3 is in the CO_2 hydrate stability zone. Then, CO_2 hydrates in layer A3 dissociated by additional depressurization. During the dissociation of CO_2 hydrates in layer A3, a self-preservation effect was observed in the pressure and temperature relationship as seen in previous experimental studies on CO_2 hydrate dissociation [12,31]. Finally, CO_2 hydrates completely dissociated with step-wise depressurization (Figure 9d).

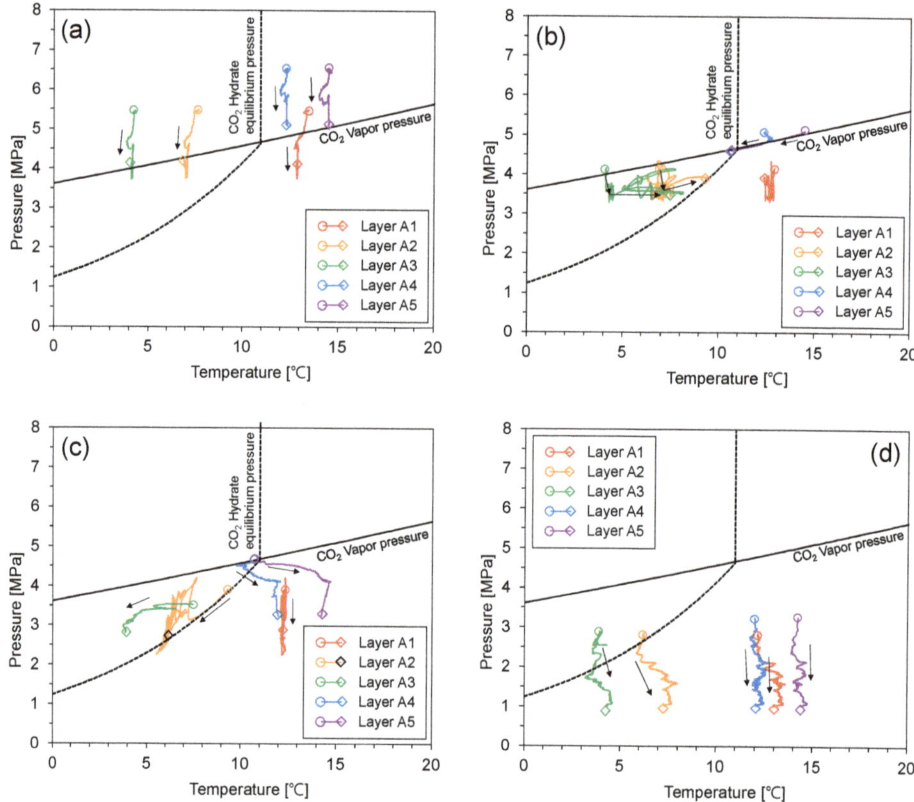

Figure 9. The pressure–temperature relationship during depressurization. (**a**) 0–980 min, (**b**) 980–1770 min, (**c**) 1770–3300 min, (**d**) 3300–8700 min.

3.2.2. P-Wave Velocity

Figure 10 shows the results of the P-wave velocity measurements taken during the depressurization process. The sealing effect of the original CO_2 hydrate-bearing sediment layer reduced because some portion of the original CO_2 hydrates dissociated. Thus, the P-wave velocity of layer A1 decreased because CO_2 intruded the upper part of the cell. On the other hand, the P-wave velocity of layers A2 and A3 suddenly increased because CO_2 hydrates formed using the CO_2 supply from the lower part of the cell. Note that layers A2 and A3 were in the CO_2 hydrate stability zone until 3300 minutes (refer to Figure 9). Meanwhile, the P-wave velocity of layers A4 and A5 decreased because CO_2 vaporized during depressurization. When the pressure was lower than the equilibrium pressure of the CO_2 hydrates (i.e., 4000–6000 min, refer to Figure 9d), the P-wave velocity of layers A2 and A3 suddenly decreased because reformed CO_2 hydrates in these layers dissociated.

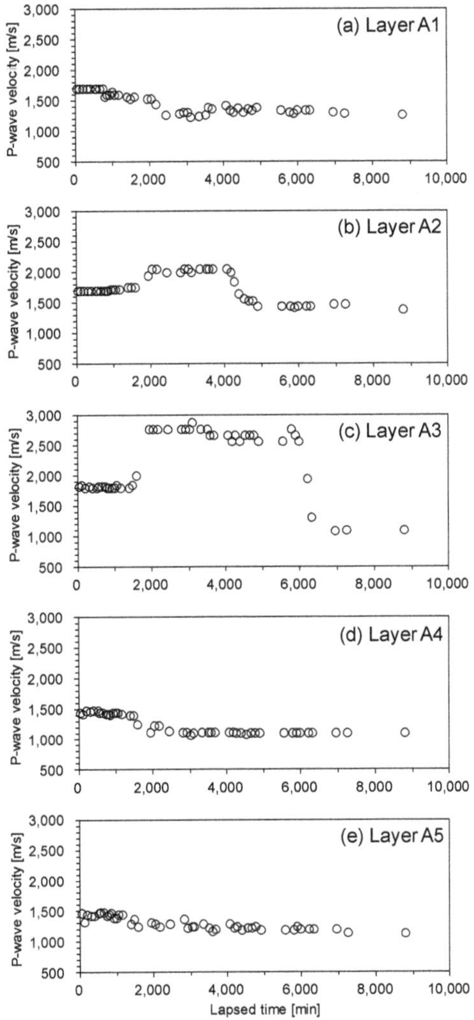

Figure 10. P-wave velocity of the unconsolidated sediment sample during depressurization. (**a**) layer A1, (**b**) layer A2, (**c**) layer A3, (**d**) layer A4, (**e**) layer A5.

3.2.3. Electrical Resistance

Figure 11 shows the normalized electrical resistance (R/R_0) during the depressurization process. The results of the electrical resistance measurements indicated that CO_2 intruded the upper part of the cell. For layers B1 and B2, electrical resistance exhibited complex behavior due to the formation of CO_2 hydrates and the movement of CO_2 gas bubbles. However, the electrical resistance decreased generally because of CO_2 dissolution. As stated before, electrical resistance of distilled water-saturated sediment decreased with CO_2 intrusion because of dissolved CO_2. Meanwhile, the electrical resistance of the lower part of the cell (i.e., layers B3 and B4) was barely affected by depressurization because the pore water of the lower part of the cell was already saturated by dissolved CO_2.

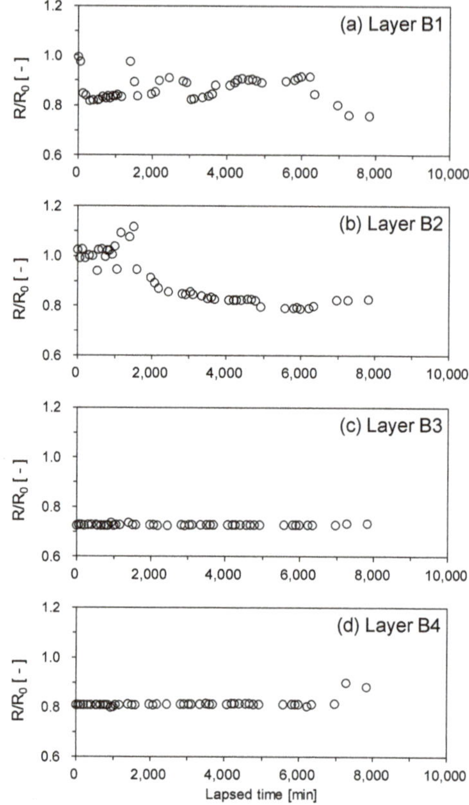

Figure 11. Normalized electrical resistance R/R_0 of the unconsolidated sediment sample during depressurization. (**a**) layer B1, (**b**) layer B2, (**c**) layer B3, (**d**) layer B4.

4. Discussion: Simple Analysis on the Sealing Capacity of CO_2 Hydrate-Bearing Sediments

There are two major sealing mechanisms for CO_2 structural trapping. One is the capillary seal, which occurs by capillary pressure between CO_2 and water in pores. The other sealing mechanism is the permeability seal, which is related to the laminar flow velocity of CO_2 in pores due to a pressure gradient. In view of the two sealing mechanisms, a simple analysis on the sealing capacity of the CO_2 hydrate-bearing sediment was performed using the experimental results.

4.1. Capillary Sealing Capacity

Capillary pressure is the difference in pressure across the interface between two fluids. In petroleum reservoirs, capillary pressure between oil and water in rock pores is responsible for trapping oil [32,33]. In the same manner, capillary pressure between water and CO_2 can trap CO_2. For a given pore structure, the CO_2 breakthrough pressure ($P_C{}^*$) induced by capillarity can be described using the Young–Laplace equation:

$$P_C{}^* = \frac{4\gamma cos\theta}{d^*},$$

(1)

where γ is interfacial tension between water and CO_2, θ is wetting angle, and d^* is the critical pore throat diameter. Several researchers have measured various temperatures and pressures for the interfacial tension between water and CO_2 [34–36].

In this study, the CO_2 hydrate-bearing sediment layer could maintain about 0.71 MPa of pressure difference between the upper and lower part of the cell (Figure 12). Thus, the minimum breakthrough pressure ($P_C{}^*_{min}$) can be assumed as the average pressure difference between the upper and lower parts of the cell ($\Delta P_{average}$), which is described in Figure 12 (i.e., $P_C{}^* \geq P_C{}^*_{min} = \Delta P_{average}$). Then, the maximum critical pore throat diameter of CO_2-hydrate bearing sediments (d^*_{max}) was calculated as 132 nm using Equation (1). The values used in this calculation are summarized in Table 1.

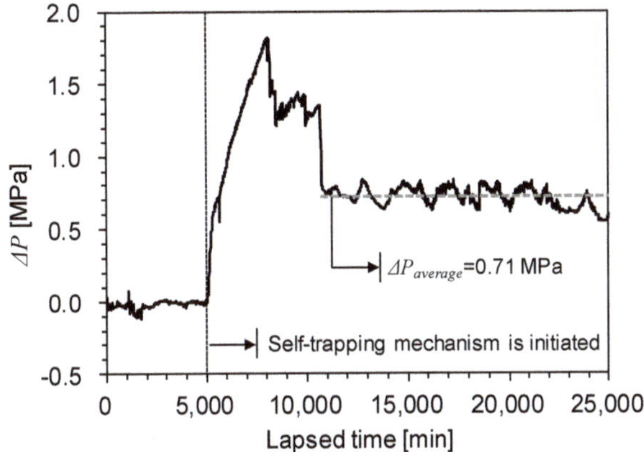

Figure 12. Pressure differences between upper (layer A1, A2, and A3) and lower (layer A4 and A5) parts of the cell (ΔP). When the CO_2 hydrates formed and stabilized, the average ΔP ($\Delta P_{average}$) was about 0.71 MPa.

Table 1. Assumed values for the calculation of d_{max}.

Coefficient	Value	Remark
γ	25×10^{-3} N/m	At a pressure of 6 MPa and temperature of 15 °C [35]
θ	20°	At a pressure of 6 MPa and temperature of 15 °C, on quartz [35]

If the pressure difference between the fluid interface exceeds $P_C{}^*$, then CO_2 breaks through the interface, and laminar flow occurs [35,37,38]. Thus, in order for the capillary sealing mechanism to work, the breakthrough pressure ($P_C{}^*$) must be larger than the buoyancy pressure of the CO_2 plume. The buoyancy pressure (P_B) that is induced by the density difference between water and CO_2 can be described as

$$P_B = gh\left(\rho_{water} - \rho_{CO_2}\right), \tag{2}$$

where g is the acceleration of gravity, h is the thickness of the CO_2-stored layer, and ρ_{water} and ρ_{CO2} are the density of water and CO_2, respectively. In a similar manner to the calculation of d^*_{max}, the minimum buoyancy pressure (P_{Bmin}) which could be maintained by the CO_2 hydrate-bearing sediment can be assumed as $\Delta P_{average}$ (i.e., $P_B \geq P_{Bmin} = \Delta P_{average}$). At a similar thermodynamic condition of the experiment in this study (i.e., pressure of 6 MPa and temperature of 15 °C), ρ_{CO2} was 784 kg/m^3 [39]. Then, the minimum thickness of the CO_2-stored layer (h_{min}) was calculated as 335 m according to Equation (2). Thus, we can presume that the capillary trapping capacity of the CO_2 hydrate-bearing sediment is high enough.

The wettability (i.e., wetting angle) could be altered by the increase of the gas hydrate saturation because the solid materials contacting pore fluids are changed from sand particles to CO_2 hydrates. For simplicity, this wettability alteration was not considered in this study. Further studies are required to evaluate the effect of wettability iteration on CO_2 capillary sealing capacity.

4.2. Permeability Sealing Capacity

When the buoyancy pressure (P_B) is higher than breakthrough pressure (P_C^*), CO_2 flow occurs. Fluid flow through soils finer than coarse gravel is laminar [40]. For laminar flow in CO_2-saturated sediments the flow velocity, v, can be expressed by Darcy's law as follows:

$$v = K\frac{\rho_{CO_2}g}{\mu_{CO_2}}i,$$ (3)

where K is absolute or intrinsic permeability of the sediments, ρ_{CO2} is the density of CO_2 fluids, g is the gravity constant, μ_{CO2} is the viscosity of CO_2 fluids, and i is the hydraulic gradient which is expressed by the difference between two hydraulic heads over the flow length. Note that the hydraulic gradient (i) is 1 for a vertical flow. Meanwhile, the average flow velocity for flow through a round capillary tube (v_0) can be described by Poiseuille's law as follows:

$$v_0 = \frac{\rho_{CO_2}gd^2}{32\mu_{CO_2}}i,$$ (4)

where d is the diameter of the capillary tube. The flow velocity determined by Poiseuille's law (v_0) is the upper limit of the flow velocity (i.e., $v \leq v_0$) in porous media because flow velocity in sediments decreases by the tortuosity of the flow channel. Therefore, the upper limit of the absolute permeability of sediments can be defined using Equations (3) and (4) as

$$K \leq \frac{d^2}{32}.$$ (5)

The maximum absolute permeability of CO_2 hydrate-bearing sediment (K_{max}) can, therefore, be calculated using the d^*_{max}, which was obtained before. K_{max} is about 5.55×10^{-4} darcy. This value is similar to the permeability of fine-grained sediments (i.e., 10^{-3}–10^{-7} darcys [20]), and can be considered as "very low" permeability [41].

4.3. Comparison with Other Materials

Estimated maximum absolute permeability (K_{max}) and minimum breakthrough pressure ($P_C^*{}_{min}$) are compared with measured absolute permeability (K) and breakthrough pressure (P_C^*) of various sediment samples, as shown in Figure 13. We presumed that the breakthrough pressure of F110 sand increases by more than 10^2 times with CO_2 hydrate formation. The minimum breakthrough pressure ($P_C^*{}_{min}$) of the CO_2 hydrate-bearing sediments estimated in this study is comparable with that of unconsolidated clays and the shale sample. Meanwhile, the actual P_C^* of CO_2 hydrate-bearing sediments in this experimental simulation may be higher than the estimated $P_C^*{}_{min}$ because the latter was estimated conservatively using the pressure difference between upper and lower parts of the cell, instead of being measured directly. In the same manner, actual K of CO_2 hydrate-bearing sediments in this experimental simulation may be lower than estimated K_{max}. This might be attributed to the K_{max} being calculated conservatively using the assumption of fluid flow in a round capillary tube without any tortuosity. To determine the range of absolute permeability and breakthrough pressure of CO_2 hydrate-bearing sediments, further experimental studies are required.

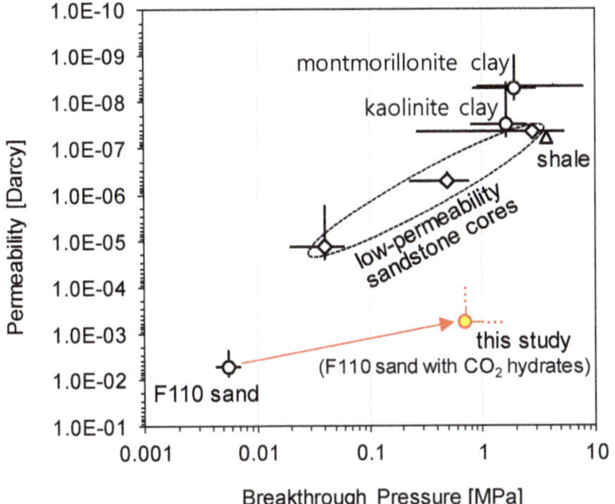

Figure 13. Absolute permeability and breakthrough pressure of various consolidated and unconsolidated sediments and estimated values in this study. Filled circle represents estimated values of minimum breakthrough pressure and maximum permeability of CO_2 hydrate-bearing sediments. Hollow circles represent F110 sand, kaolinite clay, and montmorillonite clay [42], hollow diamonds represent low-permeability sandstone cores [43], and hollow triangle represents shale [44]. The vertical and horizontal bars indicate the range of the measured values.

5. Conclusions

We performed an experimental simulation of CO_2 geological storage in marine unconsolidated sediments in this study. CO_2 hydrates were formed during the CO_2 liquid injection process, and we observed the self-trapping effect of CO_2 hydrates. In addition, simple analyses were conducted using the experimental results. The feasibility of CO_2 geological storage in marine unconsolidated sediments was experimentally verified using 1-m-height high-pressure cell. CO_2 hydrates instantly formed in the unconsolidated sediments with CO_2 introduction, and prevented any upward leakage of CO_2. The main findings are summarized as follows:

- CO_2 hydrates formed in the CO_2 hydrate stability zone of the cell during the CO_2 liquid injection process. The CO_2 hydrate-bearing sediment layer prevented any upward flow of CO_2. This self-trapping effect was confirmed by monitoring pressure, P-wave velocity, and electrical resistance.
- The original CO_2 hydrates partially dissociated during the depressurization process, and additional CO_2 hydrates instantly formed in the upper layer, which was in the CO_2 hydrate stability zone. When CO_2 hydrates dissociated, CO_2 hydrates could re-form in the upper layer (i.e., cooler layer for marine sediments) instantly. This behavior is a positive characteristic of CO_2 hydrates for use as cap-rock in CGS applications.
- The CO_2 hydrate-bearing sediment layer maintained a pressure of 0.71 MPa during the experiment. Simple analyses revealed that the capillary and permeability sealing capacity of CO_2 hydrate-bearing sediments are considerably high.

Permeability and breakthrough pressure of CO_2 hydrate-bearing sediments depend on the saturation of CO_2 hydrates in the pore system. However, CO_2 hydrate saturation was not analyzed in detail in the present study, as sufficient information regarding maximum CO_2 hydrate saturation via experimental simulation was lacking. We expect that a geophysical analysis using experimental data from a denser sensor array could overcome this limitation. Meanwhile, the electrical and geochemical behavior of the CO_2-containing sediments in this study was different from that in real marine sediments because distilled water was used as pore water instead of saline water. To overcome this limitation, an experiment using saline water will be performed for further study.

Author Contributions: Conceptualization, G.-C.C. and H.-S.K.; Methodology, G.-C.C. and H.-S.K.; Writing—Original Draft Preparation, H.-S.K.; Writing—Review & Editing, G.-C.C.; Visualization, H.-S.K.; Supervision, G.-C.C.

Funding: This research was supported by the Korea government (Ministry of Trade, Industry, and Energy) through the Project "Gas Hydrate Exploration and Production Study (19-1143)" under the management of the Gas Hydrate Research and Development Organization (GHDO) of Korea, and the National Research Foundation of Korea (NRF) grant funded by the Korea government (Ministry of Science and ICT) (No. 2017R1A5A1014883).

Conflicts of Interest: The authors declare no conflict of interest.

References

1. Hepple, R.; Benson, S. Implications of Surface Seepage on the Effectiveness of Geologic Storage of Carbon Dioxide as a Climate Change Mitigation Strategy. In Proceedings of the Greenhouse Gas Control Technologies—6th International Conference, Kyoto, Japan, 1–4 October 2002.
2. Shaffer, G. Long-term effectiveness and consequences of carbon dioxide sequestration. *Nat. Geosci.* **2010**, *3*, 464. [CrossRef]
3. Sohrabi, M.; Riazi, M.; Jamiolahmady, M.; Kechut, N.I.; Ireland, S.; Robertson, G. Carbonated water injection (CWI)—A productive way of using CO_2 for oil recovery and CO_2 storage. *Energy Procedia* **2011**, *4*, 2192–2199. [CrossRef]
4. Sohrabi, M.; Kechut, N.I.; Riazi, M.; Jamiolahmady, M.; Ireland, S.; Robertson, G. Coreflooding Studies to Investigate the Potential of Carbonated Water Injection as an Injection Strategy for Improved Oil Recovery and CO_2 Storage. *Transp. Porous Media* **2012**, *91*, 101–121. [CrossRef]
5. Bagalkot, N.; Hamouda, A.A. Diffusion coefficient of CO_2 into light hydrocarbons and interfacial tension of carbonated water–hydrocarbon system. *J. Geophys. Eng.* **2018**, *15*, 2516–2529. [CrossRef]
6. Hamouda, A.A.; Bagalkot, N. Effect of salts on interfacial tension and CO_2 mass transfer in carbonated water injection. *Energies* **2019**, *12*, 748. [CrossRef]
7. Brewer, P.G. Direct Experiments on the Ocean Disposal of Fossil Fuel CO_2. *Science* **1999**, *284*, 943–945. [CrossRef] [PubMed]
8. Inagaki, F.; Kuypers, M.M.M.; Tsunogai, U.; Ishibashi, J.; Nakamura, K.; Treude, T.; Ohkubo, S.; Nakaseama, M.; Gena, K.; Chiba, H.; et al. Microbial community in a sediment-hosted CO_2 lake of the southern Okinawa Trough hydrothermal system. *Proc. Natl. Acad. Sci. USA* **2006**, *103*, 14164–14169. [CrossRef] [PubMed]
9. Nimblett, J.; Ruppel, C. Permeability evolution during the formation of gas hydrates in marine sediments: Gas hydrate and permeability changes. *J. Geophys. Res. Solid Earth* **2003**, *108*, 2420. [CrossRef]
10. Koide, H.; Takahashi, M.; Tsukamoto, H.; Shindo, Y. Self-trapping mechanisms of carbon dioxide in the aquifer disposal. *Energy Convers. Manag.* **1995**, *36*, 505–508. [CrossRef]
11. Tohidi, B.; Yang, J.; Salehabadi, M.; Anderson, R.; Chapoy, A. CO_2 Hydrates Could Provide Secondary Safety Factor in Subsurface Sequestration of CO_2. *Environ. Sci. Technol.* **2010**, *44*, 1509–1514. [CrossRef] [PubMed]
12. Kwon, T.H.; Cho, G.C.; Santamarina, J.C. Gas hydrate dissociation in sediments: Pressure-temperature evolution: Gas hydrate dissociation in sediments. *Geochem. Geophys. Geosyst.* **2008**, *9*. [CrossRef]
13. Stern, L.A.; Circone, S.; Kirby, S.H.; Durham, W.B. Anomalous Preservation of Pure Methane Hydrate at 1 atm. *J. Phys. Chem. B* **2001**, *105*, 1756–1762. [CrossRef]
14. Kuhs, W.F.; Genov, G.; Staykova, D.K.; Hansen, T. Ice perfection and onset of anomalous preservation of gas hydrates. *Phys. Chem. Chem. Phys.* **2004**, *6*, 4917–4920. [CrossRef]
15. Koide, H.; Takahashi, M.; Shindo, Y.; Tazaki, Y.; Iijima, M.; Ito, K.; Kimura, N.; Omata, K. Hydrate formation in sediments in the sub-seabed disposal of CO_2. *Energy* **1997**, *22*, 279–283. [CrossRef]
16. House, K.Z.; Schrag, D.P.; Harvey, C.F.; Lackner, K.S. Permanent carbon dioxide storage in deep-sea sediments. *Proc. Natl. Acad. Sci. USA* **2006**, *103*, 12291–12295. [CrossRef] [PubMed]
17. Rochelle, C.A.; Camps, A.P.; Long, D.; Milodowski, A.; Bateman, K.; Gunn, D.; Jackson, P.; Lovell, M.A.; Rees, J. Can CO_2 hydrate assist in the underground storage of carbon dioxide? *Geol. Soc. Lond. Spec. Publ.* **2009**, *319*, 171–183. [CrossRef]
18. Massah, M.; Sun, D.; Sharifi, H.; Englezos, P. Demonstration of gas-hydrate assisted carbon dioxide storage through horizontal injection in lab-scale reservoir. *J. Chem. Thermodyn.* **2018**, *117*, 106–112. [CrossRef]

19. Gauteplass, J.; Almenningen, S.; Ersland, G.; Barth, T. Hydrate seal formation during laboratory CO_2 injection in a cold aquifer. *Int. J. Greenh. Gas Control* **2018**, *78*, 21–26. [CrossRef]

20. Mitchell, J.K.; Soga, K. *Fundamentals of Soil Behavior*, 3rd ed.; John Wiley & Sons: Hoboken, NJ, USA, 2005.

21. Kwon, T.H.; Oh, T.M.; Choo, Y.W.; Lee, C.H.; Lee, K.R.; Cho, G.C. Geomechanical and thermal responses of hydrate-bearing sediments subjected to thermal stimulation: Physical modeling using a geotechnical centrifuge. *Energy Fuels* **2013**, *27*, 4507–4522. [CrossRef]

22. Sloan, E.D. Gas Hydrates: Review of Physical/Chemical Properties. *Energy Fuels* **1998**, *12*, 191–196. [CrossRef]

23. Dai, S.; Santamarina, J.C.; Waite, W.F.; Kneafsey, T.J. Hydrate morphology: Physical properties of sands with patchy hydrate saturation: Patchy hydrate saturation. *J. Geophys. Res. Solid Earth* **2012**, *117*. [CrossRef]

24. Kim, H.S.; Oh, T.M.; Cho, G.C. P-wave velocity estimation of unconsolidated sediments containing CO_2. *Int. J. Greenh. Gas Control* **2015**, *33*, 18–26. [CrossRef]

25. Span, R.; Wagner, W. A New Equation of State for Carbon Dioxide Covering the Fluid Region from the Triple-Point Temperature to 1100 K at Pressures up to 800 MPa. *J. Phys. Chem. Ref. Data* **1996**, *25*, 1509–1596. [CrossRef]

26. Börner, J.H.; Herdegen, V.; Repke, J.U.; Spitzer, K. The impact of CO_2 on the electrical properties of water bearing porous media—Laboratory experiments with respect to carbon capture and storage. *Geophys. Prospect.* **2013**, *61*, 446–460. [CrossRef]

27. Fleury, M.; Deschamps, H. Electrical conductivity and viscosity of aqueous NaCl solutions with dissolved CO_2. *J. Chem. Eng. Data* **2008**, *53*, 2505–2509. [CrossRef]

28. Zhou, X.; Lakkaraju, V.R.; Apple, M.; Dobeck, L.M.; Gullickson, K.; Shaw, J.A.; Cunningham, A.B.; Wielopolski, L.; Spangler, L.H. Experimental observation of signature changes in bulk soil electrical conductivity in response to engineered surface CO_2 leakage. *Int. J. Greenh. Gas Control* **2012**, *7*, 20–29. [CrossRef]

29. Buffett, B.A.; Zatsepina, O.Y. Formation of gas hydrate from dissolved gas in natural porous media. *Mar. Geol.* **2000**, *164*, 69–77. [CrossRef]

30. Santamarina, J.C.; Ruppel, C.D. The impact of hydrate saturation on the mechanical, electrical, and thermal properties of hydrate-bearing sand, silts, and clay. In Proceedings of the 6th International Conference on Gas Hydrates (ICGH 2008), Vancouver, BC, Canada, 6–10 July 2008.

31. Falenty, A.; Kuhs, W.F. "Self-Preservation" of CO_2 gas hydrates-surface microstructure and ice perfection. *J. Phys. Chem. B* **2009**, *113*, 15975–15988. [CrossRef]

32. Berg, R.R. Capillary Pressures in Stratigraphic Traps. *AAPG Bull.* **1975**, *59*, 939–956.

33. Schowalter, T.T. Mechanics of secondary hydrocarbon migration and entrapment. *AAPG Bull.* **1979**, *63*, 720–760.

34. Chun, B.S.; Wilkinson, G.T. Interfacial tension in high-pressure carbon dioxide mixtures. *Ind. Eng. Chem. Res.* **1995**, *34*, 4371–4377. [CrossRef]

35. Espinoza, D.N.; Santamarina, J.C. Water-CO_2-mineral systems: Interfacial tension, contact angle, and diffusion- Implications to CO_2 geological storage. *Water Resour. Res.* **2010**, *46*, W07537. [CrossRef]

36. Nielsen, L.C.; Bourg, I.C.; Sposito, G. Predicting CO_2–water interfacial tension under pressure and temperature conditions of geologic CO_2 storage. *Geochim. Cosmochim. Acta* **2012**, *81*, 28–38. [CrossRef]

37. Washburn, E.W. Note on a Method of Determining the Distribution of Pore Sizes in a Porous Material. *Proc. Natl. Acad. Sci. USA* **1921**, *7*, 115–116. [CrossRef]

38. Li, S.; Dong, M.; Li, Z.; Huang, S.; Qing, H.; Nickel, E. Gas breakthrough pressure for hydrocarbon reservoir seal rocks: Implications for the security of long-term CO_2 storage in the Weyburn field: Gas breakthrough pressure for different gas/liquid systems. *Geofluids* **2005**, *5*, 326–334. [CrossRef]

39. Lemmon, E.W.; McLinden, M.O.; Friend, D.G. *Thermophysical Properties of Fluid Systems*; Linstrom, P.J., Mallard, W.G., Eds.; National Institute of Standards and Technology (NIST Chemistry WebBook): Gaithersburg, MD, USA, 1998.

40. Taylor, D.W. *Fundamentals of Soil Mechanics*; Wiley: New York, NY, USA, 1948.

41. Lambe, T.W.; Whitman, R.V. *Soil Mechanics*; Wiley: New York, NY, USA, 1969.

42. Espinoza, D.N.; Santamarina, J.C. CO_2 breakthrough—Caprock sealing efficiency and integrity for carbon geological storage. *Int. J. Greenh. Gas Control* **2017**, *66*, 218–229. [CrossRef]

43. Zhao, Y.; Yu, Q. CO_2 breakthrough pressure and permeability for unsaturated low-permeability sandstone of the Ordos Basin. *J. Hydrol.* **2017**, *550*, 331–342. [CrossRef]

44. Angeli, M.; Soldal, M.; Skurtveit, E.; Aker, E. Experimental percolation of supercritical CO_2 through a caprock. *Energy Procedia* **2009**, *1*, 3351–3358. [CrossRef]

Article

Estimates of scCO$_2$ Storage and Sealing Capacity of the Janggi Basin in Korea Based on Laboratory Scale Experiments

Jinyoung Park [1,2], Minjune Yang [1], Seyoon Kim [1], Minhee Lee [1,*] and Sookyun Wang [3]

1   Department of Earth Environmental Sciences, Pukyong National University, Busan 48513, Korea
2   BK21 Plus Project of the Graduate School of Earth Environmental Hazard System,
    Pukyong National University, Busan 48513, Korea
3   Department of Energy Resources Engineering, Pukyong National University, Busan 48513, Korea
*   Correspondence: heelee@pknu.ac.kr; Tel.: +82-51-629-6630

Received: 5 June 2019; Accepted: 19 August 2019; Published: 26 August 2019

Abstract: Laboratory experiments were performed to measure the supercritical CO$_2$ (scCO$_2$) storage ratio (%) of conglomerate and sandstone in the Janggi Basin, which are classified as rock in Korea that are available for CO$_2$ storage. The scCO$_2$ storage capacity was evaluated by direct measurement of the amount of scCO$_2$ replacing the pore water in each reservoir rock core. The scCO$_2$ sealing capacity of the cap rock (i.e., tuff and mudstone) was also compared by measuring the scCO$_2$ capillary entry pressure (Δp) into the rock core. The measured average scCO$_2$ storage ratio of the conglomerate and the sandstone were 30.7% and 13.1%, respectively, suggesting that the scCO$_2$ storage capacity was greater than 360,000 metric tons. The scCO$_2$ capillary entry pressure for the tuff ranged from 15 to 20 bar and for the mudstone it was higher than 150 bar, suggesting that the mudstone layers had enough sealing capacity from the aspect of hydromechanics. From XRF analyses, before and after 90 d of the scCO$_2$-water-cap rock reaction, the mudstone and the tuff were investigated to assure their geochemical stability as the cap rock. From the study, the Janggi Basin was considered an optimal CO$_2$ storage site based on both its high scCO$_2$ storage ratio and high capillary entry pressure.

Keywords: CO$_2$ reservoir rock; CO$_2$ sealing capacity; CO$_2$ sequestration; CO$_2$ storage capacity; CO$_2$ storage ratio; supercritical CO$_2$

1. Introduction

Eco-friendly plans and policies to reduce CO$_2$ emission are being driven forward around the world. In developed countries, CO$_2$ capture and sequestration (CCS) technology is partially commercialized and the total amount of subsurface CO$_2$ storage has been on the rise [1–4]. Since the early 2000s, the government of South Korea has been working on several projects to determine the optimal CO$_2$ storage sites on the Korean peninsula and the Janggi Basin. This basin is located in the southeastern part of the East Sea and is currently being evaluated for its potential to be one of the best onshore or offshore storage sites in Korea [5,6]. From geophysical and geological surveys, it is has become clear that the Miocene Janggi Basin consists of four small blocks (Guryongpo, Ocheon, Noeseongsan, and Yeongamri basins) [7]. The Noeseongsan block contains rudaceous sandstone and conglomerate layers that are considered promising for CO$_2$ storage sites more than 800 m deep; these also have mudstone and dacitic tuff layers above them that are able to serve as stable shield layers [8,9]. The Korean government has a plan to inject a hundred thousand metric tons of CO$_2$ in a pilot-scale onshore CO$_2$ storage test site in 2030 and the Janggi Basin is considered a suitable place for the scCO$_2$ storage test site. During stratigraphic analysis in 2015 and 2016, four sites in the Janggi Basin were drilled by the Korea Institute of Geoscience and Mineral Resources (KIGAM), and continuous drill cores to 1200 m in depth were collected at each site. From previously collected well logging data and the geo-structural interpretation results, the Janggi conglomerate formation in the Janggi Basin can be divided into four lithofacies. These consist of conglomerate and rudaceous sandstone from gravelly braided stream deposits, coarse sandstone deposited in mouth-bar or delta, muddy sandstone and shale deposited in floodplain environments, and mudstone as lacustrine deposits [8]. The western part of the basin is mainly composed of thick conglomerate and rudaceous sandstone lithofacies, which are available for use as a CO_2 storage reservoir, whereas the mudstone and muddy sandstone lithofacies constitute the eastern part of the basin. The western part of the Janggi Basin has no large faults and is considered to be an optimal CO_2 storage area. This area is now estimated to be at least about 10 km^2, assuming that the practical volume of the CO_2 storage volume in the Janggi Basin deeper than 800 m is about 0.025 km^3 [8,9].

The estimation of the CO_2 storage capacity of geological storage reservoirs is essential to determine reasonable CO_2 storage site candidates and is directly dependent on the practical amount of $scCO_2$ in non-aqueous phase, which can be stored in the pore spaces of the reservoir rock after $scCO_2$ injection. The more water displaced by $scCO_2$ in the void spaces of the reservoir rock, the more CO_2 could be stored there. There are no exact definitions for $scCO_2$ storage capacity yet, although there are several definitions of storage capacity from previous studies [10–18]. In recent research, the $scCO_2$ storage capacity was generally defined as the proportion of the volume of $scCO_2$ stored after injection, in relation to the pore volume of the CO_2 reservoir rock [19,20]. Geological exploration and numerical simulation to determine the representative amount of $scCO_2$ that could be stored within the pore spaces of a specific reservoir formation have been major subjects in CO_2 sequestration (CS) studies.

However, the pore volume saturated with water was not fully replaced by $scCO_2$ while injecting $scCO_2$ into the reservoir rock and the practical $scCO_2$ storage capacity for the specific reservoir rock can often be overestimated. The $scCO_2$ displacement of water from pore spaces of the rock during $scCO_2$ injection can be influenced by various parameters—Not just physical properties such as pore size, heterogeneity of the pore network, and injection pressure, but also mineralogical and geochemical reactivity. Thus, the best way to determine the amount of $scCO_2$ storage possible in a specific rock formation is direct measurement of $scCO_2$ displacement of water under $scCO_2$ injection conditions. This occurs using rock cores at a laboratory scale, from which the results are extended to macro scale, including the entire reservoir formation [21,22].

The $scCO_2$ storage ratio (%) (or "$scCO_2$ displacement efficiency (%)") of a reservoir formation is one of the most general parameters used in laboratory work for evaluating the CO_2 storage amount of a formation [3,10,19,20]. The $scCO_2$ storage ratio is defined as the fraction of the amount of $scCO_2$ occupying pore spaces after $scCO_2$ injection into a reservoir formation. It can be directly measured under $scCO_2$ injection conditions simulated on a laboratory scale. In 2016, Wang et al. and Kim et al. [19,20] carried out direct laboratory measurement of the $scCO_2$ storage ratio for possible reservoir rock cores, presenting the possibility that it might be used to estimate the $scCO_2$ capacity for the specific reservoir. However, this process was still in the experimental stages and the number of studies on $scCO_2$ storage capacity based on direct measurement in the laboratory are very limited. The CO_2 storage capacity for a specific reservoir formation can be calculated by multiplying the $scCO_2$ storage ratio by the total void volume of the formation, ignoring the amount of dissolved CO_2. There are many benefits of direct measurement of the $scCO_2$ storage ratio because it represents the substantive amount of CO_2 retained in a specific storage formation, under the $scCO_2$ injection condition. The estimation of the practical amount of CO_2 storage for a specific reservoir rock could be made possible by using both the $scCO_2$ storage ratio and the reservoir volume acquired from geological survey, which has almost never been tried before. In this study, laboratory experiments were performed to measure the amount of $scCO_2$ displacing water from the pore spaces of the sandstones and conglomerate cores sampled from 800–1000 m depth in the Janggi Basin, which is classified as an available CO_2 storage reservoir in Korea. The $scCO_2$ storage capacity of the Janggi Basin was calculated quantitatively, according to the measured $scCO_2$ storage ratio and with additional geophysical data on the spatial domain of the Janggi Basin.

The $scCO_2$ sealing capacity of the cap rock is another major parameter used to select successful CO_2 storage sites because it correlates to the leakage of $scCO_2$ during the anticipated duration of CO_2 sequestration. Even if a CO_2 storage site has enough $scCO_2$ storage capacity, it has to be excluded from suitable CO_2 storage sites if the $scCO_2$ leakage safety of the cap rock in the site is not assured. Layers of mudstone and dacitic tuff are repeated above the rudaceous sandstone and conglomerate layers in the Janggi Basin and it is assumed that they can play the role of cap rock to prevent the upward movement of $scCO_2$ from deeper reservoir rock [9]. In 2017, the initial capillary entry pressure of $scCO_2$ into the cap rock core surface, determined when the $scCO_2$ began to infiltrate the rock, was successfully measured in the laboratory [23]. In this study, several experimental conditions for the capillary entry pressure measurement such as the boundary condition of the high-pressure tank and the $scCO_2$ injection time on the core surface were modified, to more realistically simulate the $scCO_2$ leakage at the boundary between the reservoir and the cap rock. More information for the comparison of the experimental conditions can be drawn from [23]. The $scCO_2$ sealing capacity of the mudstone and dacitic tuff in the Janggi Basin was evaluated based on their initial $scCO_2$ capillary entry pressures under simulated $scCO_2$ injection P-T conditions. The change of mineralogical composition of the reservoir and cap rock after 90 d of $scCO_2$–water–rock reaction at 50 °C and 100 bar was also investigated by XRF analysis. This was done to observe the effect of $scCO_2$-related geochemical reactions on the sealing capacity. From the experimental results on the $scCO_2$ storage ratio for the reservoir rock, on the initial $scCO_2$ capillary entry pressure for the cap rock, and from mineralogical analyses for rock cores, the feasibility of the Janggi Basin as an available pilot-scale CO_2 storage test site where a hundred thousand metric tons of CO_2 could be injected was evaluated.

This study presents a novel and reliable method by which to select a successful CO_2 storage site based on both quantitative $scCO_2$ storage ratio and capillary entry pressure under CO_2 sequestration conditions, as well as on geochemical analyses. The results of this study will also provide ideas for further quantitative research about the CO_2 storage capacity and CO_2 leakage safety based on practical measurements of the $scCO_2$ storage ratio and initial $scCO_2$ capillary entry pressure.

2. Materials and Methods

2.1. Preparation of the scCO₂ Reservoir and Capping Rock Cores

From the well logging data for four drilling sites in the Janggi Basin, rudaceous sandstone and conglomerate layers were considered available CO_2 storage sites and the dacitic tuff and mudstone layers overlaying them as suitable cap rock [8,9]. Continuous drill cores (4.2 cm average diameter) from a drilling site 1200 m deep were acquired from KIGAM. From property analysis of these cores, three rudaceous sandstone cores (JG1-S1, JG1-S2, and JG1-S3; from 930–950 m) and three conglomerate cores (JG1-C1, JG1-C2, and JG1-C3; from 950–980 m) with average porosity of 14–18% were found and they were used for measurement of the $scCO_2$ storage ratio. For the sealing cap rock, three mudstone cores (JG1-M1, JG1-M2, and JG1-M3; from 700–760 m) and three tuff cores (JG1-T1, JG1-T2, and JG1-T3; from 800–810 m) were used for the measurement of the initial $scCO_2$ capillary entry pressure. Each rock core used in the experiment was cylindrical without cracks or fractures (4.2 cm diameter; length 5–7 cm). The geological map showing the area around the drilling site in the Janggi Basin and the rock cores used for the experiments are shown in Figure 1.

The CO_2 storage and sealing capacity of rocks depend on physico-chemical properties such as porosity, permeability, reaction rate, and mineralogical stability. The porosity of the rock cores was measured using the vacuum saturation method suggested by the International Society for Rock Mechanics (ISRM) with vacuum pressure of 1 torr and vacuum time of 80 min. For each sandstone and conglomerate core, several thin slabs (1 cm × 1 cm × 0.2 cm each) were also prepared to identify the mineral composition of each core by modal analysis. To quantify the average mineral portion of each reservoir rock, 500 locations on each thin section surface of each rock slab were observed using a point-counter installed in a polarizing microscope. For each mudstone and tuff core, mineralogical and geochemical analyses were performed using XRD (X-Ray Diffractometer; X'Pert-MPD, Philips, Almelo, The Netherlands) and XRF (X-Ray Fluorescence Spectrometer; XRF-1800, Shimadzu, Kyoto, Japan), to determine their mineralogical properties.

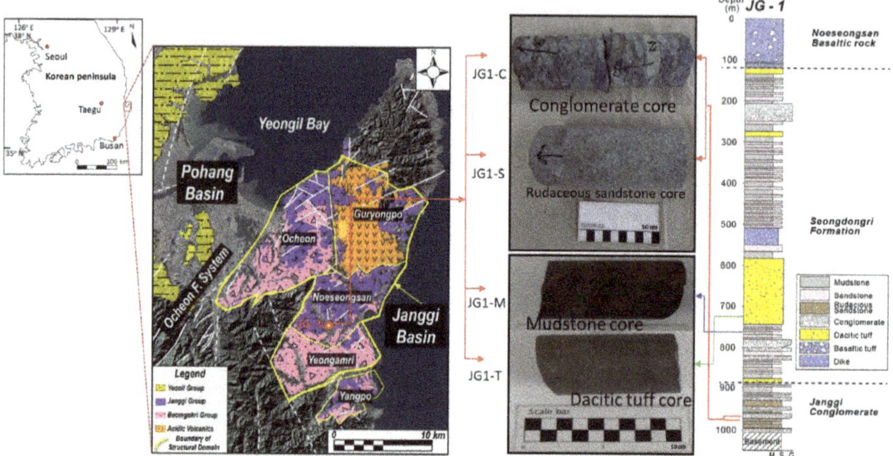

Figure 1. Geological map of the area around the drilling site (JG-1: ○) in the Janggi Basin, Korea, photographs of rock cores used for the experiment and the stratigraphic columnar section (from left to right) used for the experiment (modified from [7,8]).

2.2. Measurement of the scCO$_2$ Storage Ratio for the Conglomerate and Sandstone Cores

When scCO$_2$ was injected, water filling the void spaces of the reservoir rock was not fully replaced by the scCO$_2$ because of the difference in interfacial tension between water and the scCO$_2$ (or difference in wettability) [24]. Thus, the amount of scCO$_2$ that could be stored in the subsurface reservoir rock was less than the total void space of the rock, and was affected by various physico-chemical parameters. For selection of an optimal CO$_2$ storage site, the scCO$_2$ storage capacity of rock from that specific reservoir should be estimated, based on the real scCO$_2$ storage ratio. Moreover, the potential amount of storage for each kind of reservoir rock should also be determined before subsurface injection of the scCO$_2$. Laboratory scale measurement of the scCO$_2$ storage ratio, displacing water from the pore spaces of Janggi sandstone and conglomerate cores under the simulated scCO$_2$ injection conditions was performed. The experimental conditions were maintained (50 °C and 100 bar) to simulate the CO$_2$ storage conditions underground. The sandstone and conglomerate cores were cut (4.2 cm in diameter and 5–7 cm in length) and their cut surfaces were polished using powdered diamond paper to maintain a uniform scCO$_2$ or water injection pressure on the cut surface. A high-pressure stainless-steel cell was developed to measure the amount of scCO$_2$ stored in the pore spaces of each core after scCO$_2$ injection. It is difficult to measure the scCO$_2$ remaining in the pore space of the rock core after scCO$_2$ injection because of the leakage of the injected scCO$_2$ at the cylinder surface boundary between the cell inner wall and the rock-core wall surface. The high-pressure cell was designed with two different walls; the inner wall was composed of a thick rubber layer (1 cm thick) and the outer wall of stainless steel. The space between the inner and the outer wall of the cell was sealed with pressurized water, which was injected from outside the cell. The surface of the inner rubber wall was in tight contact with the rock core cylinder surface, when the water pressure in the space was much higher than the scCO$_2$ injection pressure (Δp > 100 bar). The rock core top and bottom head surfaces were held using a screw-type steel holder with a hole in the middle for scCO$_2$ or water injection/drainage in the rock core. It was possible to shut off the bypass of injected $scCO_2$ or pore water through the boundary between the core cylinder surface and the cell inner wall, allowing the $scCO_2$ (or water) to flow only through pore spaces within the rock core. Figure 2 shows photographs of the high-pressure cell and the schematic diagram for the cross-section of the cell used in the experiment.

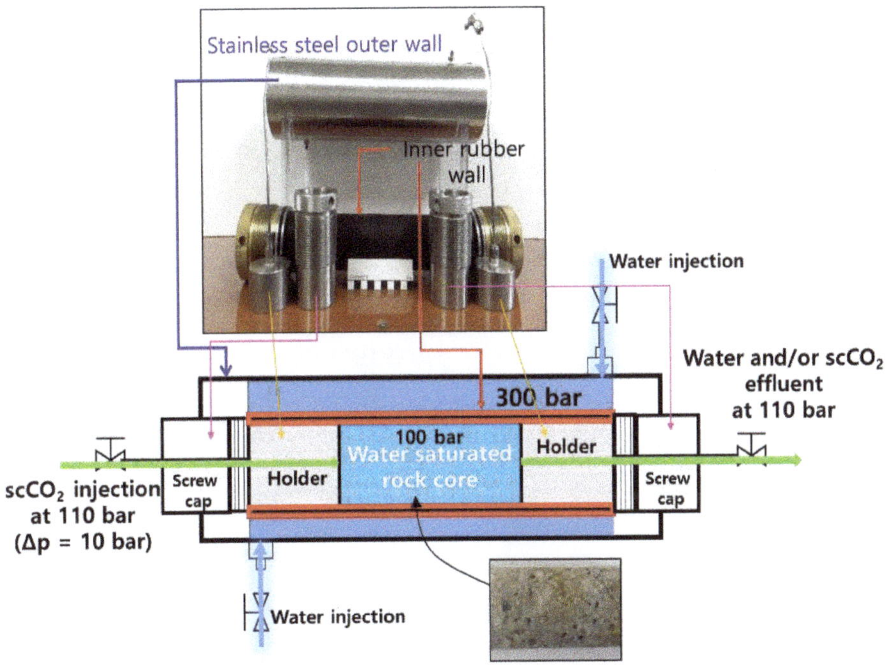

Figure 2. Photograph of the high-pressure cell used for the experiment to measure the supercritical CO_2 ($scCO_2$) storage ratio.

For the experiment, each core was fully dried at 50 °C in an oven and then weighed. The dried core was fixed by two core holders inside the high-pressure stainless-steel cell. The outer wall of the high-pressure cell was covered by a heating jacket to maintain the cell wall at the cell temperature of 50 °C. Distilled water was injected into the sealed space between the inner wall and the outer wall of the cell by a syringe pump (Isco-D260; Teledyne Isco, Inc., Nebraska, NE, USA), which was maintained at 300–350 bar. Then, the distilled water was flushed through the core at 100 bar (the injection pressure) for three pore volumes of the core to fully saturate the core with water. Next, $scCO_2$ was injected through the influent opening into the cell to displace water from the pore spaces of the core at 110 bar (Δp = 10 bar between the injection pressure and the pore water pressure in the core), while more than two pore volumes of $scCO_2$ were flushed from the core at 110 bar (assuming that displacement of water by the $scCO_2$ was successful within a few days). All of the effluent water was stored in a small stainless storage cell and its mass was weighed to measure the amount of water displaced by the $scCO_2$ in the rock core. A high-pressure stainless-steel chamber (5 L capacity) was connected to the effluent of the cell to consider the boundary condition of the reservoir rock when the $scCO_2$ was flushed from the rock core in the experiment. The water in the pores was compressed as the pore pressure increased due to $scCO_2$ injection and enough water or $scCO_2$ volume had to be provided in the chamber for the replacement of all the $scCO_2$ during the experiment. All of the high-pressure cells were maintained at 50 °C and 110 bar, after the CO_2 injection, to simulate the subsurface CO_2 storage conditions. Figure 3 shows the procedure of the experiment for $scCO_2$ exchange in the rock core.

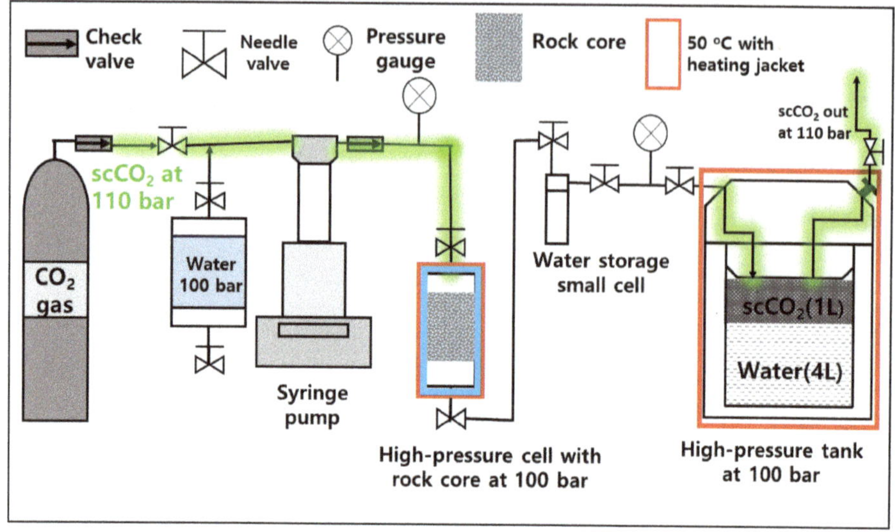

Figure 3. Schematic of the experiment to measure the $scCO_2$ storage ratio.

When the amount of water drained from the core was measured, the $scCO_2$ storage ratio for the specific rock core under the $scCO_2$ injection condition (in this case, 100 bar and 50 °C) could be calculated using Equation (1).

$$\text{The } scCO_2 \text{ storage ratio (\%) for the rock } (\varepsilon) = \left(1 - \frac{W_s - W_{out}}{W_s}\right) \times 100 \tag{1}$$

where W_s is the volume of water saturating the core and W_{out} is the water volume displaced by $scCO_2$ during the $scCO_2$ injection.

From Equation (1), the $scCO_2$ storage capacity of the conglomerate and the sandstone formations in the Janggi Basin were estimated via Equation (2) with the volume of the stratum, the average porosity, the specific gravity of the $scCO_2$, and the storage ratio of $scCO_2$ [3,14].

$$\text{The storage capacity (ton)} = \sum (V \times \varphi \times \rho \times \varepsilon), \tag{2}$$

where V is the volume of the conglomerate or sandstone layer (estimated from the previous geological survey data), φ is the average porosity, ρ is the specific gravity of the $scCO_2$, and ε is the $scCO_2$ storage ratio. The feasibility of the Janggi Basin as a CO_2 reservoir formation was evaluated based on the $scCO_2$ storage capacity of rudaceous sandstone and conglomerate drill core samples from the Janggi Basin (total of six cores).

2.3. Measurement of the Initial scCO₂ Capillary Entry Pressure for Mudstone and Dacitic Tuff

When $scCO_2$ is injected into reservoir rock, it is distributed as a separate phase from water in pore spaces and begins to move slowly upward from the lower part of the reservoir rock, due to buoyant force, because of its lower density as compared to water. Some of the injected $scCO_2$ might reach the boundary between the cap rock and the reservoir rock during continuous injection. At the early stage of $scCO_2$ injection, the amount of $scCO_2$ reaching the boundary is not much and the cap rock can prevent the intrusion of $scCO_2$ because of its low permeability. As the amount of $scCO_2$ at the boundary and its buoyant pressure increase because of continuous $scCO_2$ injection, advective and the diffusive intrusion of $scCO_2$ into the cap rock occurs. This pressure initiates seepage of the $scCO_2$ from the reservoir rock, threatening the leakage safety of the CO_2 storage site. The sealing capacity of the cap rock is directly dependent on the initial $scCO_2$ seepage (or intrusion) pressure on the cap rock surface [22]. The direct measurement of the initial $scCO_2$ capillary entry pressure of the cap rock was performed at a laboratory scale, to evaluate the $scCO_2$ shielding capacity of the cap rocks in the study area. The mudstone and dacitic tuff rock cores sampled during the deep drilling expedition (from 700–800 m) at the onshore site in the Janggi Basin were used for the experiment (Figure 1).

Rock cores without cracks or fractures were cut (4.2 cm diameter, 5–6 cm length) and were then fully dried at 50 °C in an oven for 7 d. Each rock core was fixed in the high-pressure stainless-steel cell (which was used in the same way as in the previous experiment, Section 2.2). Each core was saturated with distilled water at 100 bar of pore water pressure. The effluent of the cell was connected to a large tank filled with 3 L of water and 2 L of $scCO_2$ at 100 bar and 50 °C, simulating the subsurface $scCO_2$ injection conditions. The initial $scCO_2$ capillary entry pressure into the rock core head (higher than 100 bar: Δp = injection pressure − 100 bar), was controlled at the influent port with the regulator of the core holder in the cell bottom, until the $scCO_2$ began to penetrate the rock. The $scCO_2$ injection pressure was set at 110 bar and the injection pressure was increased by 10 bar until the $scCO_2$ began to penetrate the rock core head. At the outset, the $scCO_2$ injection pressure on the core head surface was set at 110 bar (Δp = 10 bar) and any $scCO_2$ intrusion into the core was observed for 10 days. If no $scCO_2$ intrusion occurred, the injection pressure was increased by 10 bar for 10 more days, to monitor any $scCO_2$ intrusion into the core. This process was repeated until the $scCO_2$ began to penetrate the rock core head. When the $scCO_2$ begin to intrude and the $scCO_2$ injection pressure started to decrease, the $scCO_2$ injection pressure was maintained until $scCO_2$ was flushed from the end of the rock core. This pressure was regarded as the initial $scCO_2$ capillary entry pressure (Δp) of the rock core. The $scCO_2$ shield capacity of each kind of cap rock core (dacitic tuffs and mudstones here) was evaluated by comparing their initial $scCO_2$ capillary entry pressures (Δp).

The mineralogical changes of mudstone and tuff were also measured to evaluate their geochemical stability for 90 d of the $scCO_2$–water–rock reaction under CO_2 storage conditions (100 bar and 50 °C). The rock core was pulverized using a mortar, and ten grams of powdered rock materials were mixed with 100 mL of distilled water in a high-pressure stainless-steel cell (capacity of 150 mL), in which the inner wall was coated with a Teflon layer. The void spaces in the cell were filled with $scCO_2$ using a syringe pump. Then $scCO_2$–water–rock reactions were allowed to occur in the cell at 100 bar and 50 °C, simulating the subsurface storage conditions. The total reaction time was 90 d and XRD analysis was conducted before and after the reaction to identify any mineralogical changes of the cap rock due to the $scCO_2$–water–rock reaction.

3. Results and Discussion

3.1. Measurement of the scCO2 Storage Ratio for the Conglomerate and Sandstone Cores

Results of the modal analyses for the rudaceous sandstone and conglomerate cores are shown in Table 1 and the photomicrographs of their thin sections are shown in Figure 4. The conglomerate was mostly composed of rhyolitic and andesitic rock fragments (average 74.5%), followed by quartz, clay/accessory minerals, feldspar, micas, and calcite (in descending order). The sandstone mainly consisted of quartz, rock fragments, clay/accessory minerals, feldspar, micas, and calcite. Their average proportions were 32.5%, 23.8%, 18.3%, 19.4%, 2.5%, and 2.5%, respectively (Table 1). In previous studies it was observed that calcite, feldspars, chlorite, micas, and clay minerals bearing Ca and Mg might control the geochemical reactions with CO_2 in the storage site, thereby regulating the physical properties of the reservoir rock [25–28]. The results from the XRD analyses (not shown in this paper) showed a mineral composition similar to that indicated by the modal analysis.

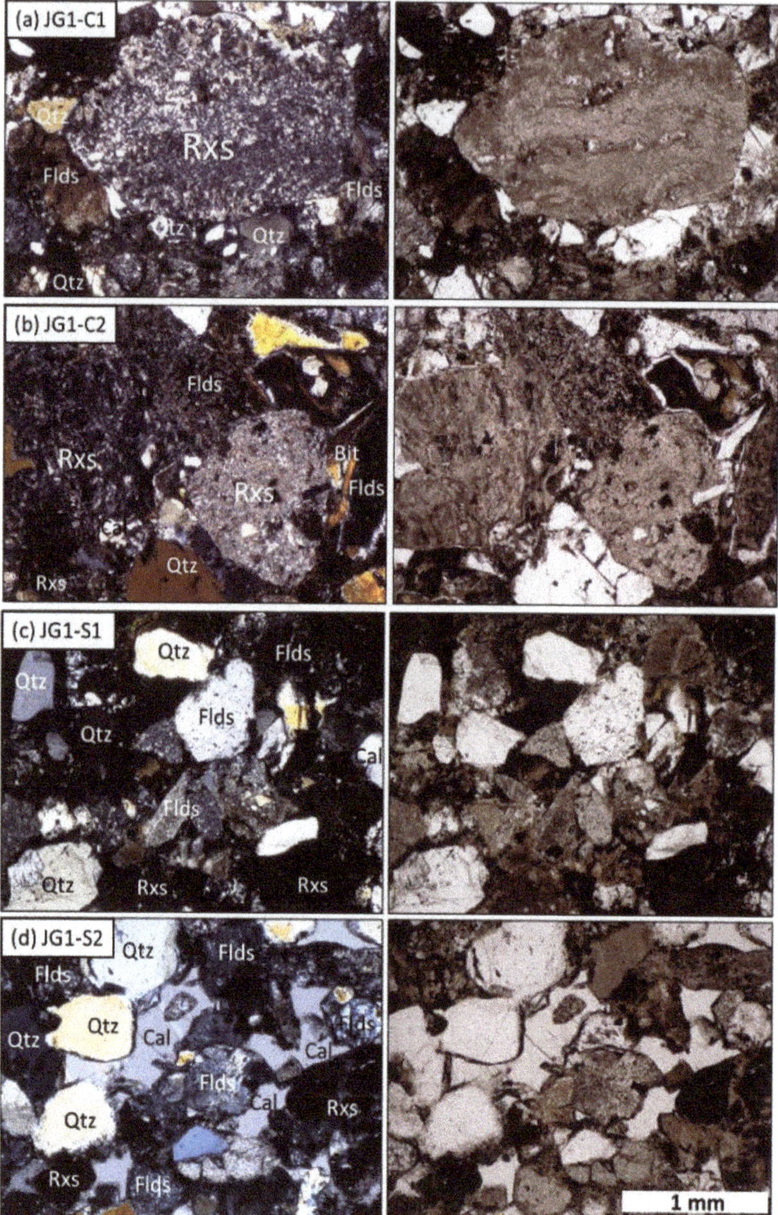

Figure 4. Photomicrographs of thin sections of the conglomerate (**a**,**b**) and the rudaceous sandstone (**c**,**d**); left column—closed mode; right column—open mode; Qtz—quartz; Flds—feldspars; Cal—calcite; Bit—biotite; and Rxs—rock fragments.

The porosity of the conglomerate and the sandstone cores were measured and are shown in Table 2. The average porosity of the conglomerate and the rudaceous sandstone was 17.8% and 14.5%, respectively, suggesting that they fall within the porosity range of typical CO_2 storage formations in other places, where large amounts of $scCO_2$ have been injected [22,29–34]. The $scCO_2$ storage ratio of each conglomerate and sandstone core was measured and their averages are shown in Table 2. The calculated average scCO$_2$ storage ratio of the Janggi conglomerates was 30.7%, suggesting that 30.7% of the void space in the conglomerate was filled with scCO$_2$, while the pressure difference between the scCO$_2$ injection and the pore water was maintained at 10 bar. The average scCO$_2$ storage ratio of rudaceous sandstones was 13.0%, which was about two-fifths that of the conglomerate. From these results, both kinds of rock had great capability for storing CO$_2$ in their pore spaces, but the conglomerate was considered to be a better option for a CO$_2$ storage site than the rudaceous sandstone.

Table 1. Petrographic detrital modal analysis of the rock cores.

Rock Type	Mineral Portion (%)							
	Quartz	K-Feldspar	Plagioclase	Rock Fragment	Fine Matrix and Clay	Calcite	Micas	Others
JG1-C1	7.8	1.8	1.2	80.6	8.0	0.2	0.0	0.4
JG1-C2	17.8	6.6	4.8	57.4	10.8	0.4	1.2	1.0
JG1-C3	6.4	1.8	1.0	85.4	3.8	0.4	0.4	0.8
Average ± standard deviation	10.7 ± 6.2	3.4 ± 2.8	2.3 ± 2.1	74.5 ± 15.0	7.5 ± 3.5	0.3 ± 0.1	0.5 ± 0.6	0.7 ± 0.3
JG1-S1	33.6	8.8	12.0	24.8	16.0	2.2	1.2	1.4
JG1-S2	31.6	11.6	8.2	24.2	18.4	0.6	4.4	1.0
JG1-S3	32.2	10.4	7.0	22.4	20.4	4.6	1.8	1.2
Average ± standard deviation	32.5 ± 1.0	10.3 ± 1.4	9.1 ± 2.6	23.8 ± 1.2	18.3 ± 2.2	2.5 ± 2.0	2.5 ± 1.7	1.2 ± 0.2
JG1-M1	40.8	3.0	1.4	0.4	47.4	1.2	2.6	3.2
JG1-M2	43.8	2.8	0.6	0.0	47.4	0.4	2.4	2.6
JG1-M3	33.6	2.4	0.4	0.2	58.6	0.2	2.2	2.4
Average ± standard deviation	39.4 ± 5.2	2.7 ± 0.3	0.8 ± 0.5	0.2 ± 0.2	51.1 ± 6.5	0.6 ± 0.5	2.4 ± 0.2	2.7 ± 0.4
JG1-T1	26.0	12.2	11.6	18.6	19.6	1.4	3.8	6.8
JG1-T2	19.4	10.4	14.8	11.6	32.4	1.2	4.2	6.0
JG1-T3	24.5	9.7	13.2	16.7	22.0	2.4	3.1	8.4
Average ± standard deviation	23.3 ± 3.5	10.8 ± 1.3	13.2 ± 1.6	15.6 ± 3.6	24.7 ± 6.8	1.7 ± 0.6	3.7 ± 0.6	7.1 ± 1.2

* JG1-C—conglomerate cores; JG1-S—rudaceous sandstone cores, JG1-M—mudstone cores and JG1-T—dacitic tuff cores.

Table 2. The scCO$_2$ storage ratios of conglomerate and rudaceous sandstone cores at the Janggi Basin.

Core Number	Rock Type	Porosity (%)	The scCO$_2$ Storage Ratio (%)
JG1-C1		19.23	23.56
JG1-C2	Conglomerate	18.92	35.41
JG1-C3		15.12	31.20
Average		17.76	30.72
JG1-S1		10.54	13.12
JG1-S2	Rudaceous sandstone	16.44	16.28
JG1-S3		17.26	9.64
Average		14.75	13.01

The scCO$_2$ storage ratio values were used for the estimation of the CO$_2$ storage capacity for the Janggi Basin (Table 3). From previous studies [7–9] it was known that the Janggi Basin extends horizontally over about 0.25 km^2 below a depth of 800 m and that the thickness of both the conglomerate and sandstone layers available for the CO$_2$ reservoir is 100 m (about 1:1 ratio), assuming that the maximum volume of the CO$_2$ storage formation in the Janggi Basin under 800 m in depth is about

0.025 km^3. The parameter values used to calculate the CO_2 storage capacity and the calculated $scCO_2$ capacity for the two formations are shown in Table 3. The $scCO_2$ storage capacity of the reservoir rocks around the probable $scCO_2$ injection site in Janggi Basin was calculated to be 368,742 metric tons, demonstrating that the conglomerate and sandstone formations in Janggi Basin have great potential for use as a pilot test site for CO_2 storage in Korea (to receive more than 100,000 metric tons of CO_2 injection at the test storage site). As mentioned earlier, the amount of dissolved CO_2 in the pore water of each core was ignored while calculating the amount of $scCO_2$ storage in Table 3. When the amount of dissolved CO_2 in the pore water was considered, the already substantial CO_2 storage capacity of the Janggi Basin was increased. Thus, if enough $scCO_2$ was secured in the conglomerate and the sandstones, the Janggi Basin could be evaluated as a positive storage site, regardless of the amount of CO_2 dissolved in the pore water.

Table 3. The $scCO_2$ storage capacity of conglomerate and rudaceous formations at the Janggi Basin (unit—metric ton).

Rock Type	The Calculated Amount of $scCO_2$ Storage for A Conglomerate and Rudaceous Formation (metric ton)
Conglomerate	V(50 m × 250 m × 1000 m) × φ(0.1776) × ρ*(400 kg/m^3) × ε(0.3072) = 272,793.6
Rudaceous sandstone	V(50 m × 250 m × 1000 m) × φ(0.1475) × ρ*(400 kg/m^3) × ε(0.1301) = 95,948.8

* Density of $scCO_2$ at 100 bar and 50 °C (from [35,36]).

3.2. Measurement of the Initial scCO$_2$ Capillary Entry Pressure for Mudstone and Dacitic Tuff

The mineralogical changes of the mudstone and the dacitic tuff after 90 days of $scCO_2$–water–rock reaction were investigated by XRF analyses and their results are shown in Table 4. The results indicated that even after 90 d of reaction, the proportions of the major constituents of the two kinds of rock were not significantly changed, except for a minor decrease of SiO_2 and CaO. These originated from increased dissolution of calcite, Ca–feldspar, and Ca-bearing silicates, which were similar to results of previous studies [37–39]. It is suggested that the mudstone and the dacitic tuff in the Janggi Basin is likely to maintain significant stability against $scCO_2$-involved geochemical reactions during CO_2 storage.

Table 4. XRF analysis of mudstone and dacitic tuff before and after 90 days of $scCO_2$–water–rock reaction.

Composition	Ratio (wt. %)					
	JG1-M1		JG1-M2		JG1-T1	
	Before	After 90-day Reaction	Before	After 90-day Reaction	Before	After 90-day Reaction
SiO_2	57.20	56.51	56.36	55.65	57.01	56.61
Al_2O_3	20.84	20.89	17.88	17.94	18.21	18.33
$TiO2$	0.78	0.78	0.62	0.62	0.76	0.76
Fe_2O_3	5.32	5.29	4.99	5.06	7.04	7.10
MnO	0.09	0.08	0.12	0.12	0.15	0.15
MgO	0.79	0.92	0.71	0.81	1.95	2.04
CaO	1.14	1.08	1.16	1.09	5.03	4.64
Na_2O	1.63	1.73	1.32	1.36	2.55	2.60
K_2O	2.30	2.40	2.04	2.09	0.95	0.94
P_2O_5	0.09	0.09	0.15	0.15	0.05	0.05
LOI	9.61	10.07	14.44	14.96	6.12	6.58
Total	99.80	99.85	99.79	99.85	99.82	99.80

Results from the measurement of the initial $scCO_2$ capillary entry pressure (Δp) for the mudstone and the tuff core are shown in Figure 5. For all of tuff cores, the $scCO_2$ began to intrude into the rock core at 115 bar (Δp = 15 bar) and continuous $scCO_2$ injection into the core occurred at a Δp higher than 20 bar. This suggests that the initial $scCO_2$ capillary entry pressure (Δp) of the dacitic tuff ranged from 15 to 20 bar, under the conditions of 100 bar and 50 °C. In the tuff cores, 8–10% of the void space (0.7–0.9 mL) was filled by $scCO_2$ at a Δp higher than 20 bar. In a previous study (Kim et al. 2019), the average initial $scCO_2$ capillary entry pressure of the sandstone and the conglomerate in the Janggi Basin was lower than 10 bar (mostly < 5 bar), which was less than one-third that for the dacitic tuff. For the mudstone cores in the Janggi Basin, the $scCO_2$ did not penetrate the core surface and the stored $scCO_2$ was less than 0.005 mL, even when the injection pressure was 250 bar (Δp = 150 bar) for 30 d. This suggests that the initial $scCO_2$ capillary entry pressure for the mudstone core was much higher than 150 bar (10 times higher than for the tuff). Based on the initial $scCO_2$ capillary entry pressure, the mudstone formation in the Janggi Basin was much more suitable than the tuff formation as a shield against $scCO_2$ leakage from the reservoir rock.

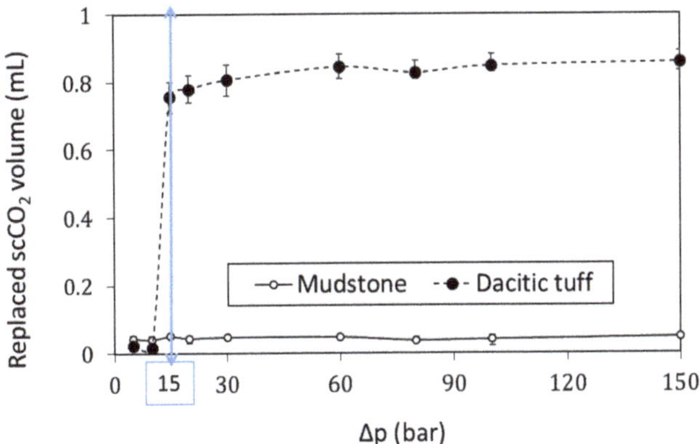

Figure 5. Initial $scCO_2$ capillary entry pressure (Δp) and the volume of $scCO_2$ stored in mudstone and dacitic tuff cores.

4. Conclusions

Determining an optimal storage site and choosing the main parameters to be considered has been one of the main issues for the geological sequestration of CO_2. During the last two decades, evaluation of the storage capacity and the leakage safety has been considered essential for optimal storage site selection. However, previously, only rough estimations were made of the reservoir and the capping rock using conventional parameters such as porosity and permeability, or by using large-scale information acquired from geophysical exploration and geological field observation. The quantitative evaluation of CO_2 storage capacity and of the risk of CO_2 leakage has been very limited, even at a laboratory scale. Recently, a direct technique for the measurement of the $scCO_2$ storage ratio and the initial $scCO_2$ capillary pressure of the rock was developed, but its application to CO_2 storage site selection has almost never been attempted in previous studies. As mentioned earlier, the Janggi Basin was selected as a feasibility testing site to store at least 100,000 metric tons of CO_2 before starting large-scale injection (more than 1,000,000 metric tons) on the Korean peninsula. For the testing site, it was important that the CO_2 storage capacity of the Janggi Basin was, at the least, larger than 100,000 metric tons. This study presents an easy and effective technique by which to evaluate the CO_2 capacity of reservoir rock and the leakage safety of the cap rock, and finally, to successfully select the Janggi basin as an optimal CO_2 storage site in Korea. The scale-up estimation of the CO_2 storage capacity for the conglomerate and the rudaceous sandstone in the Janggi Basin of Korea was performed on the basis of direct measurement of the $scCO_2$ storage ratio. The safety risk of $scCO_2$ leakage for the cap rock of dacitic tuff and mudstone in the Janggi Basin was also quantitatively evaluated by measuring the initial $scCO_2$ capillary entry pressure.

The experimental results successfully demonstrated that the conglomerate and sandstone formations of the Janggi Basin are suitable as a geological storage test site, for injection of a hundred thousand tons of CO_2 from the viewpoint of storage capacity. If the amount of dissolved CO_2 in the pore water could be considered, a more precise estimate of the CO_2 storage capacity for the specific reservoir formation could be estimated; related research in this area is already in progress.

It was also verified that the mudstone formation in the Janggi Basin is adequate to prevent the seepage of buoyant $scCO_2$ from the reservoir site because its initial capillary entry pressure (Δp) was higher than 150 bar. From the XRF analysis before and after the experiment, reliable evidence for the geochemical stability of the tuff and the mudstone was also provided. These quantitative measurements of the $scCO_2$ storage ratio and the initial $scCO_2$ capillary entry pressure applied in this study could be used to determine practical CO_2 storage sites and could also provide meaningful information for future decisions regarding $scCO_2$ injection conditions.

This study focused on the hydromechanical measurement of the $scCO_2$ in the pore spaces of rock core, over just a few months and, thus, did not deal with the effect of geochemical reactions (CO_2–water–rock) on the $scCO_2$ capacity and the capillary entry pressure with a long-term view. Recent research indicates that physical or chemical changes in the properties of rock involved in CO_2 sequestration could arise from mineralogical or geochemical reactions within a shorter time after CO_2 injection, than previously expected. For successful selections of optimal CO_2 storage subsurface sites, the geochemical stability of reservoir or cap rocks should be considered an important parameter, along with the storage capacity and the capillary entry pressure.

Author Contributions: M.L. and J.P. conceived and designed the methodology and the experiments; J.P., M.L., and S.K. performed the experiments and the data analyses; S.W. and M.Y. contributed materials and data interpretation; M.L. wrote the paper to prepare the submission.

Funding: This research was supported by the Korea Agency for Infrastructure Technology Advancement (KAIA) grant funded by the Ministry of Land, Infrastructure and Transport (Grant 19CTAP-C151965-01) and by the grant (2019002470002) from Korea Ministry of Environment as "The SEM (Subsurface Environmental Management) projects".

Acknowledgments: The authors would like to express their gratitude to the anonymous reviewers for their critical comments and advice.

Conflicts of Interest: The authors declare no conflict of interest.

References

1. Michael, K.; Golab, A.; Shulakova, V.; Ennis-King, J.; Allinson, G.; Sharma, S.; Aiken, T. Geological storage of CO_2 in saline aquifers—A review of the experience from existing storage operations. *Int. J. Greenh. Gas Control* **2010**, *4*, 659–667. [CrossRef]

2. Leung, D.Y.C.; Caramanna, G.; Maroto-Valer, M.M. An overview of current status of carbon dioxide capture and storage technologies. *Renew. Sustain. Energy Rev.* **2014**, *39*, 426–443. [CrossRef]

3. Bachu, S. Review of CO_2 storage efficiency in deep saline aquifers. *Int. J. Greenh. Gas Control* **2015**, *40*, 188–202. [CrossRef]

4. *IPCC Special Report on the Impacts of Global Warming of 1.5 °C above Pre-Industrial Levels and Related Global Greenhouse Gas Emission Pathways, in the Context of Strengthening the Global Response to the Threat of Climate Change, Sustainable Development, and Efforts to Eradicate Poverty;* IPCC: Incheon, Korea, 2018.

5. Kim, M.-C.; Kim, J.-S.; Jung, S.; Son, M.; Sohn, Y.K. Bimodal volcanism and classification of the Miocene basin fill in the northern area of the Janggi-myeon, Pohang, Southeast Korea. *J. Geol. Soc. Korea* **2011**, *47*, 585–612.

6. Song, C.W. A study on potential geologic facility sites for carbon dioxide storage in the Miocene Pohang Basin, SE Korea. *J. Geol. Soc. Korea* **2015**, *51*, 53–66. [CrossRef]

7. Kim, M.-C.; Gihm, Y.S.; Son, E.-Y.; Son, M.; Hwang, I.G.; Shinn, Y.J.; Choi, H. Assessment of the potential for geological storage of CO_2 based on its structural and sedimentologic characteristics in the Miocene Janggi Basin, SE Korea. *J. Geol. Soc. Korea* **2015**, *51*, 253–271. [CrossRef]

8. Gu, H.-C.; Hwang, I.G. Depositional history of the Janggi Conglomerate controlled by tectonic subsidence, during the early stage of Janggi Basin evolution. *J. Geol. Soc. Korea* **2017**, *53*, 221–240. [CrossRef]

9. Gu, H.-C.; Gim, J.-H.; Hwang, I.G. Variation in depositional environments controlled by tectonics and volcanic activities in the lower part of the Seongdongri Formation, Janggi Basin. *J. Geol. Soc. Korea* **2018**, *54*, 21–46. [CrossRef]

10. Bachu, S.; Bonijoly, D.; Bradshaw, J.; Burruss, R.; Holloway, S.; Christensen, N.P.; Mathiassen, O.M. CO_2 storage capacity estimation: Methodology and gaps. *Int. J. Greenh. Gas Control* **2007**, *1*, 430–443. [CrossRef]

11. Lindeberg, E.; Vuillaume, J.F.; Ghaderi, A. Determination of the CO_2 storage capacity of the Utsira formation. *Energy Procedia* **2009**, *1*, 2777–2784. [CrossRef]

12. Kopp, A.; Class, H.; Helmig, R. Investigations on CO_2 storage capacity in saline aquifers. Part 1. Dimensional analysis of flow processes and reservoir characteristics. *Int. J. Greenh. Gas Control* **2009**, *3*, 263–276. [CrossRef]

13. Pingping, S.; Xinwei, L.; Qiujie, L. Methodology for estimation of CO_2 storage capacity in reservoirs. *Pet. Explor. Dev.* **2009**, *36*, 216–220. [CrossRef]

14. Goodman, A.; Hakala, A.; Bromhal, G.; Deel, D.; Rodosta, T.; Frailey, S.; Small, M.; Allen, D.; Romanov, V.; Fazio, J.; et al. U.S. DOE methodology for the development of geologic storage potential for carbon dioxide at the national and regional scale. *Int. J. Greenh. Gas Control* **2011**, *5*, 952–965. [CrossRef]

15. Knopf, S.; May, F. Comparing methods for the estimation of CO_2 storage capacity in saline aquifers in Germany: Regional aquifer based vs. structural trap based assessments. *Energy Procedia* **2017**, *114*, 4710–4721. [CrossRef]

16. Elenius, M.; Skurtveit, E.; Yarushina, V.; Baig, I.; Sundal, A.; Wangen, M.; Landschulze, K.; Kaufmann, R.; Choi, J.C.; Hellevang, H.; et al. Assessment of CO_2 storage capacity based on sparse data: Skade Formation. *Int. J. Greenh. Gas Control* **2018**, *79*, 252–271. [CrossRef]

17. Doughty, C.; Pruess, K.; Benson, S.; Hovorka, S.D.; Knox, P.R.; Green, C.P. Capacity investigation of brine-bearing sands of the Frio formation for geologic sequestration of CO_2. *Lawrence Berkeley Natl. Lab.* **2001**. Available online: http://hdl.handle.net/2152/64418 (accessed on 26 August 2019).

18. Zhou, Q.; Birkholzer, J.T.; Tsang, C.F.; Rutqvist, J. A method for quick assessment of CO_2 storage capacity in closed and semi-closed saline formations. *Int. J. Greenh. Gas Control* **2008**, *2*, 626–639. [CrossRef]

19. Wang, S.; Kim, J.; Lee, M. Measurement of the $scCO_2$ storage ratio for the CO_2 reservoir rocks in Korea. *Energy Procedia* **2016**, *97*, 342–347. [CrossRef]

20. Kim, S.; Kim, J.; Lee, M.; Wang, S. Evaluation of the CO_2 storage capacity by the measurement of the $scCO_2$ displacement efficiency for the sandstone and the conglomerate in Janggi Basin. *Econ. Environ. Geol.* **2016**, *49*, 469–477. [CrossRef]

21. *Annual Report 2008: NETL (National Energy Technology Laboratory) Carbon Sequestration ATLAS of the United States and Canada*; DOE: Washington, DC, USA, 2008.

22. Tasianas, A.; Koukouzas, N. CO_2 storage capacity estimate in the lithology of the mesohellenic trough, Greece. *Energy Procedia* **2016**, *86*, 334–341. [CrossRef]

23. An, J.; Lee, M.; Wang, S. Evaluation of the sealing capacity of the supercritical CO_2 by the measurement of its injection pressure into the tuff and the mudstone in the Janggi Basin. *Econ. Environ. Geol.* **2017**, *50*, 303–311.

24. Iglauer, S. Optimum storage depths for strucural CO_2 trapping. *Int. J. Greenh. Gas Control* **2018**, *77*, 82–87. [CrossRef]

25. Gaus, I. Role and impact of CO_2-rock interactions during CO_2 storage in sedimentary rocks. *Int. J. Greenh. Gas Control* **2010**, *4*, 73–89. [CrossRef]

26. Park, J.; Baek, K.; Lee, M.; Wang, S. Physical property changes of sandstones in Korea derived from the supercritical CO_2–sandstone–groundwater geochemical reaction under CO_2 sequestration condition. *Geosci. J.* **2015**, *19*, 313–324. [CrossRef]

27. Jun, Y.S.; Giammar, D.E.; Werth, C.J. Impacts of geochemical reactions on geologic carbon sequestration. *Environ. Sci. Technol.* **2013**, *47*, 3–8. [CrossRef]

28. Park, J.; Baek, K.; Lee, M.; Chung, C.-W.; Wang, S. The use of the surface roughness value to quantify the extent of supercritical CO_2 involved geochemical reaction at a CO_2 sequestration site. *Appl. Sci.* **2017**, *7*, 572. [CrossRef]

29. IPCC Special Report on Carbon Dioxide Capture and Storage. *Prepared by Working Group III of the Intergovernmental Panel on Climate Change*; Cambridge University Press: Lodon, UK, 2005.

30. Bachu, S.; Gunter, W.D.; Perkins, E.H. Aquifer disposal of CO_2: Hydrodynamic and mineral trapping. *Energy Convers. Manag.* **1994**, *35*, 269–279. [CrossRef]

31. Wigand, M.; Carey, J.W.; Schütt, H.; Spangenberg, E.; Erzinger, J. Geochemical effects of CO_2 sequestration in sandstones under simulated in situ conditions of deep saline aquifers. *Appl. Geochem.* **2008**, *23*, 2735–2745. [CrossRef]

32. Zhao, J.; Lu, W.; Zhang, F.; Lu, C.; Du, J.; Zhu, R.; Sun, L. Evaluation of CO_2 solubility-trapping and mineral-trapping in microbial-mediated CO_2–brine–sandstone interaction. *Mar. Pollut. Bull.* **2014**, *85*, 78–85. [CrossRef]

33. Dawson, G.K.W.; Pearce, J.K.; Biddle, D.; Golding, S.D. Experimental mineral dissolution in Berea Sandstone reacted with CO_2 or SO_2–CO_2 in NaCl brine under CO_2 sequestration conditions. *Chem. Geol.* **2015**, *399*, 87–97. [CrossRef]

34. Mediato, J.F.; García-Crespo, J.; Izquierdo, E.; García-Lobón, J.L.; Ayala, C.; Pueyo, E.L.; Molinero, R. Three-dimensional reconstruction of the caspe geological structure (Spain) for evaluation as a potential CO_2 storage site. *Energy Procedia* **2017**, *114*, 4486–4493. [CrossRef]

35. Span, R.; Wagner, W. A new equation of state for carbon dioxide covering the fluid region from the triple-point temperature to 1100 K at pressures up to 800 MPa. *J. Phys. Chem. Ref. Data* **1996**, *25*, 1509–1596. [CrossRef]

36. Spycher, N.; Pruess, K. CO_2–H_2O mixtures in the geological sequestration of CO_2. II. Partitioning in chloride brines at 12–100 °C and up to 600 bar. *Geochim. Cosmochim. Acta* **2005**, *69*, 3309–3320. [CrossRef]

37. Sendula, E.; Páles, M.; Szabó, B.P.; Udvardi, B.; Kovács, I.; Kónya, P.; Freiler, Á.; Besnyi, A.; Király, C.; Székely, E.; et al. Experimental study of CO_2-saturated water-illite/kaolinite/montmorillonite system at 70–80 °C, 100–105 bar. *Energy Procedia* **2017**, *114*, 4934–4947. [CrossRef]

38. Rezaee, R.; Saeedi, A.; Iglauer, S.; Evans, B. Shale alteration after exposure to supercritical CO_2. *Int. J. Greenh. Gas Control* **2017**, *62*, 91–99. [CrossRef]

39. Luo, X.; Ren, X.; Wang, S. Supercritical CO_2–water–shale interactions under supercritical CO_2 simulation conditions. *Energy Procedia* **2018**, *144*, 182–185. [CrossRef]

MDPI
St. Alban-Anlage 66
4052 Basel
Switzerland
Tel. +41 61 683 77 34
Fax +41 61 302 89 18
www.mdpi.com

Minerals Editorial Office
E-mail: minerals@mdpi.com
www.mdpi.com/journal/minerals